Coping, Health and Organizations

Coping, Health and Organizations

EDITED BY

PHILIP DEWE, MICHAEL LEITER AND TOM COX

London and New York

First published 2000 by Taylor & Francis
11 New Fetter Lane, London EC4P 4EE

Simultaneously published in the USA and Canada
by Taylor & Francis Inc
29 West 35th Street, New York, NY 10001

Taylor & Francis is an imprint of the Taylor & Francis Group

Typeset in Times by
J&L Composition Ltd, Filey, North Yorkshire
Printed and bound in Great Britain by
TJ International Ltd, Padstow, Cornwall

British Library Cataloguing in Publication Data
A catalogue record for this book is available from the British Library

Library of Congress Cataloging in Publication Data
A catalogue record for this book has been requested

ISBN 0–748–40824–X (hbk)
ISBN 0–748–40823–1 (pbk)

Contents

Contributors

A. J. Baglioni Jr
Social Development Research Group
University of Washington
Seattle
USA

Julian Barling
School of Business
Queens University
Kingston
Ontario
Canada

John R. Berridge
Manchester School of Management
UMIST
Manchester
UK

Ronald J. Burke
School of Business
York University
Toronto
Canada

Victor J. Callan
Graduate School of Management
University of Queensland
Australia

Cary L. Cooper
Manchester School of Management
UMIST
Manchester
UK

Robin S. Cox
Domestic Violence Unit
Vancouver Police Department
Canada

Tom Cox
Institute of Work, Health and Organizations
University of Nottingham
UK

Philip Dewe
Department of Human Resource Management
Massey University
New Zealand

Jeffrey R. Edwards
Kenan-Flager Business School
University of North Carolina
Chapel Hill
USA

Hege R. Eriksen
Department of Biological and Medical Psychology
University of Bergen
Norway

Esther R. Greenglass
Department of Psychology
York University
Toronto
Canada

Michael P. Leiter
Department of Psychology
Acadia University
Nova Scotia
Canada

Bonita C. Long
Department of Counselling Psychology
University of British Columbia
Vancouver
Canada

Catherine Loughlin
Division of Management
University of Toronto
Ontario
Canada

Miranda Olff
Department of Prevention
Trimbos Institute
Utrecht
The Netherlands

Astrid M. Richardsen
Department of Psychology
University of Tromsø
Norway

Michael Rosenbaum
Department of Psychology
Tel Aviv University
Israel

Deborah J. Terry
School of Psychology
University of Queensland
Australia

Louise Thomson
Institute of Work, Health and Organizations
University of Nottingham
UK

Holger Ursin
Department of Biological and Medical Psychology
University of Bergen
Norway

Author biographies

Tony Baglioni received his PhD in Community Psychology from the University of Virginia in 1986. He has worked at the University of Virginia, the University of Queensland, Australia, and is currently the Senior Statistician at the Social Development Research Group at the University of Washington, USA. His research interests have included responses to relocation among the ageing, and the relationship between stress, coping, and outcomes. He is currently involved in the study of the prevention of anti-social behaviours among adolescents.

Julian Barling is Professor of Organizational Behaviour and Psychology, and Chair of the PhD and MSc programmes at the School of Business, Queen's University, Kingston, Ontario, Canada. He is currently the Associate Dean of Research and Graduate Programs for the Queen's School of Business and is Editor of the *Journal of Occupational Health Psychology* 2000–2004. Professor Barling has written over 100 research articles, numerous book chapters and several books, most recently *Young Workers: Varieties of Experience*, edited with Kevin Kelloway, (1999, American Psychological Association), and is co-editor (with Kevin Kelloway) of the Sage Publication series, *Advanced Topics in Organizational Behaviour*. Dr. Barling has recently been named as Editor-Elect of the American Psychological Association's *Journal of Occupational Health Psychology*. In 1997 he received Queen's University's annual award for 'Excellence in Research'.

John Berridge is currently Director of International Studies at the Manchester School of Management, UMIST, and has researched, taught and administered in universities in Britain, Europe, North America and Australia. His main concentration has been the application of behavioural sciences to management and the human resources function, especially in cross-cultural contexts. He has published widely in journals, and is the author of six books, of which the most recent is *Employee Assistance Programmes and Workplace Counselling* (Wiley, 1997). He edited the international journal *Employee Relations* for ten years from 1990. His current research is in employee counselling, support and welfare, and their contribution to corporate strategy in current labour market contexts.

Ronald J. Burke (PhD, University of Michigan) is Professor of Organizational Behaviour, School of Business, York University, Canada. He serves on the Editorial Board of about a dozen academic journals and was the Founding

Editor of the Canadian Journal of Administrative Sciences. He is a prolific writer, and his work has appeared in a wide range of journals. His current research interests include organizational restructuring and downsizing, gender issues in organizations, workaholism in organizations, and career development processes.

Victor J. Callan is Professor and Head of the Graduate School of Management at the University of Queensland. He has a BA(Hons) degree, 1st class, from the University of New South Wales Sydney, and a PhD in social psychology from the Australian National University, Canberra. Victor has published nine books and over 150 refereed journal papers in the areas of social, health and organizational psychology. His major research interests continue to be leadership, communication and management of organizational change. He has also consulted in these areas for some of Australia's largest public and private sector organizations.

Cary L. Cooper is BUPA Professor of Organizational Psychology and Health in the Manchester School of Management, and Pro-Vice-Chancellor (External Activities) of the University of Manchester Institute of Science and Technology (UMIST). He is author of over 80 books (on occupational stress, women at work and industrial and organizational psychology), has written over 300 articles for academic journals, and is a frequent contributor to the media. He is Founding Editor of the *Journal of Organizational Behavior*, and Co-Editor of *Stress Medicine* and the *International Journal of Management Review*. He is a Fellow of the British Psychological Society, The Royal Society of Arts, the Royal Society of Medicine, the Royal Society of Health and the (American) Academy of Management. Professor Cooper is President of the British Academy of Management, and Companion of the (British) Institute of Management. He recently published a major report for the European Union's European Foundation for the Improvement of Living and Work Conditions on *Stress Prevention in the Workplace*.

Robin S. Cox received her MA in Counselling Psychology from the University of British Columbia in 1996. Her master's thesis utilised feminist critical theory to explore feminist therapy and its impact on the feminist anti-violence movement. She is currently employed as a community counsellor and advocate working in collaboration with a police-based domestic violence unit. Her research interests include policy and practice issues pertaining to violence against women, organizational learning and health paradigms and their impact on working women, and collaborative feminist research methodologies.

Tom Cox is currently Professor of Organizational Psychology and Director of the Institute of Work, Health and Organizations in the University of Nottingham Business School. The Institute is a World Health Organization (WHO) Collaborating Centre in Occupational Health, and a Lead Topic

Centre for the European Agency, Bilbao, on good practice in safety and health relating to work stress. Tom Cox is a chartered occupational psychologist specialising in occupational health, and a Fellow of the British Psychological Society, the Royal Society for Health and the Royal Society for the Arts, Commerce and Manufactures. He is a member of the Health and Safety Commission's (HSC) Occupational Health Advisory Committee and some-time expert advisor on work stress and workplace violence to the European Commission. His research focuses on psychological, social and organizational issues in occupational health and he has a particular interest in work-related stress. Tom Cox has published widely in this area for more than 25 years. He is Managing Editor of the international journal *Work and Stress*.

Philip Dewe is currently Head of the Department of Human Resource Management in the College of Business at Massey University, Palmerston North, New Zealand. Professor Dewe was educated at Victoria University in Wellington, New Zealand and at the London School of Economics. He was for a number of years a Senior Research Officer with the Work Research Unit in the United Kingdom. Professor Dewe's research has primarily been in the area of work stress and coping. He has held visiting research and teaching posts at Erasmus University (Rotterdam), Birkbeck College (London) and at the London School of Economics, and has been recently appointed to a Readership at Birkbeck College (London).

Jeffrey R. Edwards (PhD, Carnegie Mellon University) is Professor of Management at the Kenan-Flagler Business School at the University of North Carolina. He was previously on the faculty of the University of Michigan and the University of Virginia. His research examines person-environment fit, stress and coping in organizations, the interface between work and family, and research methodology. He holds editorial positions with *Organizational Behavior and Human Decision Processes, Management Science*, the *Journal of Organizational Behavior, and Organizational Research Methods*, is Past Chair of the Research Methods Division of the Academy of Management, and has been elected to the Society of Organizational Behaviour.

Hege Randi Eriksen (PhD) is a postdoctoral fellow at the Department of Biological and Medical Psychology of the University of Bergen, Norway. Her basic training is from the Norwegian University of Sport and Physical Education. (Her Masters thesis was on the effects of physical exercise on epilepsy). She has studied the psychophysiology of psychological defence mechanisms at the Department of Anatomy and Cell Biology, UCLA (Professor M.B. Sterman), and received her PhD from the Faculty of Psychology at the University of Bergen (1998). She studies the effects of interventions on low back pain and on subjective health.

Esther Greenglass has been teaching and conducting research at York University in Toronto since 1968. Professor Greenglass has published several books, journal articles and research reports in the areas of health psychology and the psychlogy of women. Some of her areas of interest are psychology of women, Type A behaviour, stress, burnout and coping. Most recently she has developed the Proactive Coping Inventory (with Ralf Schwarzer), a coping measure which focuses on the positive aspects of human potentional and takes a proactive rather than a reactive approach to the study of stress management. Her ongoing research with Ronald Burke focuses on psychological effects of hospital restructuring on nurses.

Michael P. Leiter is Professor of Psychology and Vice President (Academic) of Acadia University in Canada. He is Director of the Center for Organizational Research and Development, which applies high quality research methods to human resource issues confronting organizations. He developed the Staff Survey for assessing the way people perceive complex organizations. This approach to organizational life arose from his extensive work with organizations undergoing major organizational change in North America and in Europe. He received degrees in psychology from Duke University (BA), Vanderbilt University (MA), and the University of Oregon (PhD). He teaches courses on organizational psychology and on stress at Acadia University. The research centre provides a lively bridge between university studies and organizational consultation for himself and his students.

Bonita C. Long is Professor of Counselling Psychology at the University of British Columbia, where she earned her PhD in interdisciplinary studies. She has published numerous articles on workplace stress and coping and has co-edited *Women, Work, and Coping: A Multidisciplinary Approach to Workplace Stress* (1993), published by McGill Queen's University Press. Her current projects focus on the influence of interindividual and intraindividual processes related to coping with occupational stress on a daily basis, and examine the dialectics of resistance and accommodation within the dominant stress discourse.

Catherine Loughlin completed her PhD in Industrial/Organizational Psychology at Queen's University, Kingston in 1998, winning provincial and national academic awards as well as an Outstanding Research Award. She has spent the last year at the Queen's School of Business on a research fellowship from the Social Sciences and Humanities Research Council of Canada. Dr Loughlin has published empirical papers in the *Journal of Experimental Psychology* and the *Journal of Organisational Behaviour*, and co-authored book chapters on acute work stress, chronic work stress, workplace health and safety, and the quality of youth employment. Currently she is at the University of Toronto and teaches in the area of organizational behaviour.

Miranda Olff is a Senior Scientist in the Department of Prevention at Trimbos Institute (Netherlands Institute of Mental Health and Addiction). Until recently she was doing research in the Department of Clinical Psychology at the University of Groningen on the effect of stress, and the treatment of depression on the immune and endocrine system as well being a Visiting Professor in the Department of Psychology at the University of Lund, Sweden where she worked in the area of stress, coping and health. Among her many professional activities she took part in the European Astronaut Selection Study for the European Space Agency. Dr Olff has published widely on the subjects of stress, coping and subjective health, and is at present working on projects in areas such as prevention of work stress and subjective health problems, and coping styles and psychobiological mechanisms in several populations.

Astrid Richardsen is Associate Professor and Department Chair at the Department of Psychology, University of Tromsø in Norway. She completed her PhD in Clinical Psychology at McGill University in Montreal in 1986, and has collaborated for several years with Professor Ron Burke at York University, Canada, on studies of work and stress. Since 1992 she has been teaching organizational and work psychology at the University of Tromsø, and her research has included studies of stress and burnout among various professional groups, and organizational support and health among managerial and professional women.

Michael Rosenbaum has a PhD in clinical psychology from the University of Illinois at Champaign-Urbana (USA). He is a Professor of Psychology at the Department of Psychology at Tel Aviv University in Israel and Head of the William S. Schwartz Laboratory for Health Behaviour Research. His areas of research include self-control, learned resourcefulness, openness to experience, health psychology, stress and coping, and cognitive-behavior therapy.

Deborah Terry is an Associate Professor in the School of Psychology at the University of Queensland, Australia. She received a BA and PhD (1989) from the Australian National University, and after a one-year postdoctoral fellowship in the School of Psychology at the University of Queensland, accepted a lecturing position at this university (1991). She has research interests in applied psychology (organizational and health psychology) and social psychology (attitudes and group processes). She has published widely in these areas, and is co-editor of *The Theory of Reasoned Action: Its Application to AIDS-preventive Behaviour* (1993), *Attitudes, Behavior, and Social Context: The Role of Group Norms and Group Membership* (1998), and *Social Identity Processes in Organizational Contexts* (forthcoming).

Louise Thomson, BSc MSc, is a Research Psychologist in Occupational Health Psychology at the Institute for Work, Health and Organizations, University of

Nottingham Business School. She graduated from University College London and has post-graduate qualifications in Occupational Psychology from the University of Nottingham. She has been the principal researcher for a number of projects on the association between organizational and psychosocial factors at work and aspects of employee well-being. Her other research interests concern employee absence: its measurement and causes, and management strategies to reduce it.

Holger Ursin is Professor of Physiological Psychology, Department of Biological and Medical Psychology, University of Bergen, and adjungated professor to the Norwegian University of Sports in Oslo. He is at present a member of the Board of Regents, University of Bergen, Chairperson for the Department of Social Affairs and Health Evaluation, and Counselling Committee for Back Pain, and member for Norway of the Scientific Committee for Antarctic Research (SCAR) Working Group on Human Biology and Medicine. He has published articles and books about the psychology, physiology, endocrinology and immunology of stress and emotions, and is currently engaged in research on the prevalence, aetiology, prevention and treatment of subjective health complaints.

Preface:
Coping, health and organizations

Michael Leiter, Louise Thomson, Philip Dewe and Tom Cox

The way in which their work is designed and managed can challenge people working in organizations. This book is about those challenges, about how people cope with them, and about how coping can moderate the effects of work on health. To fulfil their obligation to help people cope, managers and researchers need a thorough understanding of the fundamentals of coping in the work context. The chapters that follow approach these fundamentals from diverse and complementary perspectives. The objective is to develop a model of coping that is of practical benefit while furthering the development of theory.

Part 1 of the book is concerned with theoretical and psychometric considerations. It begins with a review and critique by Dewe of the various meanings and measures associated with the notion of coping at work. In asking the question 'What is it that we are trying to achieve when measuring coping?', Chapter 1 reviews a number of alternative measurement approaches, the contexts in which they should be applied, and the inconsistencies that exist in research today. In answering this question, Dewe's chapter examines the progress that has been made in the measurement of coping with work stress within the context of good psychometric practice: theoretical context, item derivation, scale instructions and scoring, and scale classification.

Dewe's chapter is followed by one written by Edwards and Baglioni that argues for reliable, valid measures of coping, and that compares measurement techniques driven by empirical considerations with those driven by theory. Using statistical techniques such as Confirmatory Factor Analysis, they compare the empirically-driven Ways of Coping Questionnaire (Folkman and Lazarus (1988) Manual for the Ways of Coping Questionnaire, Palo Alto, CA: Consulting Psychologists Press) with the theoretically-driven Cybernetic Coping Scale (Edwards and Baglioni (1993) 'The measurement of coping with stress: construct validity of the Ways of Coping Checklist and the Cybernetic Coping Scale', *Work and Stress* 7:17–31) providing further information on the validity and reliability of the two scales. Edwards and Baglioni conclude by saying that the evidence suggests clear advantages of the theoretical approach compared with the empirical one. Applied in the present context, the theoretical approach to measurement development provides greater assurance that measures address the real diversity of coping in occupational settings. Among other things, it is important that the theoretical constructs shaping measurement development are responsive to the changing demands and opportunities that confront employees in dynamic workplaces.

The final chapter in the section is by Rosenbaum, and focuses on the self-regulation of experiences of coping, and, in particular, on two types of self-control responses: openness and construction. Rosenbaum approaches such coping by differentiating between openness to experience as an unfolding event and a constructive response that fits experience into re-existing frames of reference. While acknowledging the somewhat artificial quality of this dichotomy, he explores its implications for a framework of coping behaviour. Some coping responses are attempts to exert control over events, framing them within a problem solving approach based upon one's understanding of events. Other coping responses, such as relaxation procedures, focus more closely upon one's response to events. Rosenbaum argues that both perspectives are consistent with the development of greater self-control, and are required to cope effectively. That is, effective coping requires an ability to accept, openly and non-judgementally, the stresses that one encounters, and also to reframe and interpret the experience of stress constructively on the basis of one's existing theories and constructs. This chapter takes a somewhat different approach to the previous two, based on a philosophical differentiation of psychological experiences. It explores the implications of such an approach for a framework to understand coping behaviour.

The varieties of coping considered in this book encompass both the openness and the construction perspectives described by Rosenbaum. However, reflecting the balance in the literature, the book gives much more emphasis to the constructivist approach than to the open perspective. For the most part the chapters depict people as active agents who address difficulties by taking control both cognitively and behaviourally. The environments considered by the authors are often seen as too hostile to permit open responses, such as relaxation and accepting experience. The authors depict the challenges of unemployment, discrimination and large-scale disasters as requiring active intervention from people as individuals, members of working groups or entire organizations.

The second part of the book focuses on particular work-related problems and their relationship to coping.

Chapter 4 by Loughlin and Barling reviews the literature on the effects of acute workplace disasters on employee health and well-being, and discusses how individuals cope with such situations. The authors examine the effects of both natural and technological disasters on the victims and the rescue workers and volunteers involved. Evidence concerning the importance of control, support and organizational responsibility in coping with disasters is explored.

Greenglass' chapter examines the relationship between work and family, in particular focusing on coping with work-family conflict. A number of models and hypotheses are considered that concern the effects that might arise in relation to the work-family interface, such as the buffering model of work, the scarcity hypothesis, the spillover model, and the social exchange approach. Strategies to cope with the negative effects of work-family role conflict are described both at the individual level (e.g. influencing the role environment, social support) and at the organizational level (e.g. flexible work schedules, sick leave for dependent care). Furthermore, Greenglass proposes that alternative career models are needed to enhance the positive effects of multiple roles.

The chapter by Long and Cox focuses on the potential impact of occupational stressors on women's physical and mental health. Taking a social constructivist approach, it reviews the research that focuses on employed women's ways of coping with stress, and describes problems with the research and conceptualization of stress and coping. Long and Cox argue that research has tended to conceptualize women's experience of occupational stress and coping strategies as singularly different from that of men on the basis of gender rather than other factors such as cultural context, social expectations or other individual differences. They posit that women's occupational stress and coping takes place in a cultural context, reflecting an unequal distribution of power and resources. Therefore they emphasize the need for researchers to consider these contextual issues of power and status, rather than focus on the individual woman and her coping pattern.

Leiter is the first of a number of authors to consider coping in terms of the Nottingham organizational health model. According to this model, the availability and effectiveness of coping responses is closely related to the healthiness of the work setting. Each of the three primary social environments of organizations (problem-solving, task and development) presents distinct challenges and facilitates distinct coping responses. The development environment is the specific focus of Leiter's chapter considering the impact of role expansion on hospital workers. Effective coping with the demands of skill enhancement requires not only effective learning strategies, but the capacity to interact within a social context characterized by conflicting demands and expectations. The development of mentoring relationships was a key means for middle managers to cope with the demands of role expansion. Not only did these relationships provide direct advice and training relevant to new challenges, they provided a means to communicate with management. Talking with a mentor about training needs indicates to senior management shortfalls in the existing training procedures, while providing an indicator of the strains being experienced by employees.

The final chapter in Part 2 by Eriksen, Olff and Ursin describes the diverse impact of unresolved stress experiences on health of both the individual and the organization. This perspective emphasizes the significance of non-specific symptoms of illness for individual well-being and on attendance and performance at work. The chapter reviews the effectiveness of different levels of interventions to reduce this impact, including health promotion, organizational interventions such as job redesign, and individual coping strategies such as physical exercise and relaxation. Eriksen and colleagues argue that more scope and experimental rigour is needed in such research studies in order to establish the link between individual and organizational health.

The third and final section deals with organizational interventions. The first chapter, by Cox and Thomson provides a framework for those that follow. It considers the relationship between stress and the healthiness of organizations. Using the concept of organizational healthiness, which combines the socio-technical theory of organization with the individual health analogy, Chapter 9 examines the organizational context to work-related stress, employee health and coping. Cox and Thomson argue that employee health can be affected by organizational healthiness via the two interacting mechanisms of the design and management of work, and the experience of stress and the organization's impact on employee behaviour at work, including coping.

Burke and Richardsen's chapter reviews eleven organizational-level inter-ventions to reduce occupational stressors, and evaluates current knowledge of their effectiveness. Chapter 10 describes the two approaches to organizational-level interventions: the enhancement of individuals and their resources, and the reduction of work-placed sources of occupational stress. Of the eleven interventions reviewed, the majority were based on the second approach, and the authors conclude that the interventions were generally found to have positive effects. Given these indications and the limited success of individual-level interventions, Burke and Richardsen encourage researchers to design scientific studies to implement and evaluate organizational-level stress management interventions.

The chapter by Berridge and Cooper examines the role of Employee Assistance Programmes (EAPs) in coping with the stress of new organiza-tional challenges. The authors define the EAP, review its historical develop-ment and rapid growth, and its current coverage in the US, Britain and Australia. Four principal EAP models are described and evaluated and the controversy over the appropriate method for evaluating EAPs discussed. Berridge and Cooper highlight the fact that, despite the seemingly high level of acceptance of EAPs, there is a variety of critical arguments concerning aspects of their philosophical and therapeutic bases, their organizational impact, their legal position, and their role in the management of people at work. Much of the focus of these programmes is developing a supportive relationship providing personal counselling in a group or individual format. Relaxation training, self-esteem building and self acceptance play an import-ant role in these types of interventions. They tend to develop employees' capacity to adjust to organizational realities rather than develop a perspective to re-evaluate or change the organization. The organization is accepted as a necessary if not benign reality with which individuals had best learn to cope effectively. The target of organizational interventions is developing an environment in which people feel sufficiently secure to support an open perspective. Coping occurs within a social context that facilitates some responses while discouraging others. Unhealthy organizational environments impede the use of effective coping by threatening the well-being and self-esteem of an organization and by undermining their attempts to use active coping responses.

This chapter is followed by one by Burke and Leiter discussing contemporary organizational realities as newly emerging sources of stress, and their influence on personal and professional efficacy. The chapter reviews the available literature regarding four of these emerging sources of stress: mergers and acquisitions, budget cuts, job future ambiguity, and occupational lock-in. Burke and Leiter propose that these new stressors require coping on the level of both the individual (e.g. EAPs, career counselling and retraining) and the organization (such as communication, controlling the speed of change and conflict and resistance management). They suggest that the development of a problem-solving perspective that encourages action-oriented coping in the form of building a new psychological contract is particularly important.

The last chapter in this section is by Terry, Callan and Sartori, and examines the utility of a stress-coping model of employee adjustment to organizational change. They present findings from recent research testing the utility of the model that suggests that both situational appraisals and coping responses mediated the effects of organizational merger event characteristics and coping resources on employee adjustment. Their research points to the need for organizations to consult with employees prior to change about the implementation of the change, and during the change to ensure that there is effective leadership, that there are adequate sources of support, and that employees are kept informed about the change process.

To conclude the book, Dewe, Cox and Leiter describe an agenda for the future for work stress and coping researchers, focusing on three issues: definitions of stress and coping, theoretical context and measurement. They call for the affirmation of a definition of stress that best provides a structure to systematic theoretical and research enquiry that will make a significant contribution to understanding the stress process. The growing acceptance that stress should be viewed as a transaction between the individual and the environment is recognized as providing such a research opportunity. They argue that the theoretical context within which work-stress research should be carried out must be process-oriented and provide the framework for investigating the dynamics of work stress and coping. Theoretical models need to describe the mechanisms through which the stress process develops and by which the individual and environment are linked. Finally, the authors challenge researchers to examine what alternative methodologies can provide in the measurement of the stress-coping process. The issue concerns whose reality we are measuring when we measure stress and coping, where current methodologies actually take us, and what more we could understand by using other measures of the stress process.

The wealth of research and thinking about coping in organizational settings provides a solid base for articulating with increasing thoroughness theories that integrate coping behaviour within a framework of organizational functioning. It was a primary goal of this book to bring together current work that can contribute to that development. We hope that the integration of ideas presented here help to support ambitious research efforts to understand coping more thoroughly. This research effort requires qualitative research strategies to ensure the development of ideas that remain sensitive to the major changes in the nature of work in post-industrial societies. It also requires extensive, longitudinal, quantitative research efforts that investigate ideas that have been developed and continue to be. Studies that track coping behaviour over time, as well as research that investigates the impact of interventions intended to enhance coping effectiveness will make significant contributions to understanding the nature of coping itself and its development.

PART ONE

Theoretical and psychometric considerations

CHAPTER ONE

Measures of coping with stress at work: a review and critique

PHILIP DEWE

Despite three decades of investigations into the phenomena of stress, and an explosion of research into how individuals cope (Aldwin and Revenson, 1987), the coping literature remains diverse (Endler and Parker, 1990), difficult to organize (Edwards, 1988) and fraught with problems (Cohen, 1987). Coping studies that focus on work stress have been described as being in their infancy (Kuhlmann, 1990), limited somewhat in their scope (George, Brief and Webster, 1991) and few in number (Latack, 1986). On the other hand, advice as to the direction coping should take has been relatively consistent. Reviewers (Burke and Weir, 1980; Dewe, Cox and Ferguson, 1993; Edwards, 1988; Newton, 1989; Schuler, 1984) have had little difficulty in identifying the issues that need to be addressed. These include the meanings of stress and coping, the measurement of coping strategies and the role of coping in the stress process.

Defining work stress

It is almost traditional, as Vingerhoets and Marcelisson (1988) suggest, for writers on stress to begin by pointing to the lack of agreement when defining the term. The fact that stress has been discussed so often has led some authors (Beehr and Frariz, 1987) to express their frustration and suggest that there must surely be more important issues that need confronting other than the choice of a label or term. At another level others argue (Eulberg, Weekley and Bhagat, 1988) that when defining stress we should avoid the temptation to be all-inclusive, and adopt a 'mid-range' focus that allows for more clarity and precision. This is a point reinforced by Kasl (1983) who suggests that the more focused the research the less reliant we need to be on the 'troublesome concept of stress'.

Yet clarifying what we mean by stress is important because the definition influences how we research stress and explain our results. The choice of a particular definition has typically been determined by the research question addressed (Parker and DeCotiis, 1983). Thus stress continues to be defined as a stimulus, response or as a result of some imbalance or interaction between the individual and aspects of the environment (McGrath, 1970; Cox, 1978). The different uses of the term have been important in establishing a state of knowledge about the various components of the stress process, but their somewhat artificial separation has not facilitated an understanding of the transactional nature of stress or the mediating psychological processes through which the transaction takes place (Dewe, 1992: 96).

This is not to say that the transactional processes has gone unrecognized or that its explanatory potential remains untested. Rather it is more the case that stress definitions traditionally direct attention towards external events and away from those processes within the individual through which such events are evaluated (Duckworth, 1986). At the empirical level, then, work-stress research continues to be influenced by definitions that focus on and somewhat artificially separate the different components of the stress process. At the conceptual level there is a growing acceptance that work stress should be viewed as relational in nature (Lazarus and Launier, 1978), involving some sort of transaction between the individual and the environment. Stress there-fore arises from the appraisal that particular demands are about to tax individual resources for dealing with them, thus threatening well-being (Holroyd and Lazarus, 1982). Defining stress in this way draws attention to the process of appraisal and coping, links the individual and the environment and provides a common pathway for research which so far has proved elusive in work-stress research.

Finally, having to define stress should not become some 'rite of passage' that all work-stress researchers should pass through. Rather it should be viewed not so much as a serious attempt to affirm a belief in the reality of stress but more as a mechanism for confronting the all-important question of why we believe in the current representation of stress (Newton, 1995: 10). The issue, as Newton points out, is to consider whether what is being defined is a reflection of stress discourse rather than the experience of stress itself (1995: 10). This of course is not just an issue of definition: it is also one of method-ology. Our reliance on quantitative methodology has also limited theoretical developments, leaving some authors to conclude that any meaningful progress towards a better understanding of stress will come only when we change how we define stress and the methodologies we use to investigate it (Brief and Atieh, 1987). These concerns have led to the belief that current research methods have assumed such an identity and power that they are now some-what divorced from the phenomena they are purporting to investigate, leading to a need to question the structured reality imposed by such approaches.

There is a dualism in organizational research such that a distinction can be made between two quite different approaches to organizational analysis. The traditional and more generally accepted approach is to emphasize concrete reality through observable and measurable social facts, actions and outcomes (Pfeffer, 1990; Staw, 1990). The other approach draws attention to meaning. What distinguishes this approach from the more traditional one is its emphasis on the social construction of reality; shared meanings and beliefs that interpret, legitimate and justify organizational actions. This is what Staw refers to as a distinction between rationality and justification or, as Morley and Luthans (1984) suggest, the difference between an outsider's and an insider's view of organizational behaviour. Why is all this important? Because it draws attention to two important issues. The first requires researchers to contemplate what it is that is being measured when we measure stress. The second is to consider how we should go about it (Morley and Luthans, 1984). Both require asking 'where are current methodologies taking us?' and 'what can alternative methodologies provide?' (Van Maanen, 1979). If we are to consider whose reality it is when we are defining and measuring stress then the challenge is, as Newton (1995) suggests, to think beyond current representations of stress and move towards a language of stress that is less individualistic and more concerned with the wider social and power relations of the workplace that reflect and shape individual stressful experiences.

Defining coping

As with stress, coping has been defined in a number of different ways. These include coping as a psychoanalytical process, as a personal trait or style, as a description of specific strategies, as a process and as a taxonomy of strategies (Dewe, Cox and Ferguson, 1993). From the debate surrounding the various approaches, the theme that emerges, at least in a work setting, is that coping is part of the transaction between the person and the environment where that transaction is appraised as stressful (Latack and Havlovic, 1992). The use of a transactional framework as the context for defining coping is usually taken to reinforce the importance of coping as a major factor in the appraisal process. Within this context coping should he seen as: (a) relational in that it reflects the relationship between the individual and the environment (Folkman, 1982); (b) a dynamic process in contrast to the more traditional trait-oriented approaches (Cox, 1987; Edwards, 1988; Folkman et al., 1986) and (c) integrative in that it links the other components of the stress process (Cox and Ferguson, 1991; Schuler, 1984).

A broad integrative approach is usually called for when defining coping (Latack and Havlovic, 1992). This has, particularly in a work setting, a number of advantages. These include:

- (a) recognizing that the targets towards which coping is directed can be emotional (internal) as well as situational (external), including identifying those aspects that individuals find taxing (Latack and Havlovic, 1992);
- acknowledging that the meaning individuals give to events (primary appraisal) includes not just threat and harm/loss but challenge and benefit as well (Cox and Ferguson, 1991);
- focusing researchers towards not just asking 'How much stress is there?' to 'Why is stress a problem?' (Newton, 1989) or 'What is it about work that may stimulate stress and coping processes?' (Brief and George, 1991);
- directing attention towards coping actions and behaviours, allowing coping strategies to be considered independently of their effectiveness, reducing the temptation to declare one strategy more effective than another and providing a more robust set of measures through which effectiveness can be judged (Dewe and Guest, 1990; Latack and Havlovic, 1992).

Consistent with these ideas, coping can be defined as 'cognitive and behavioural efforts to master, reduce or tolerate those demands from transactions that tax or exceed a person's resources.' Because the focus in this chapter is on work stress, the transactions referred to are work or work-related encounters that tax individual resources and abilities. Because coping efforts are constantly changing it should not be assumed from this definition that coping is 'typically placed' between the encounter and some outcome (Dewe, Cox and Ferguson, 1993) in a simple linear fashion. This is not the intention here. Coping involves a complex process of thoughts and actions and in view of the transactional nature of any encounter this complexity should be recognized. It is these dynamics that the definitional process is trying to describe.

Three important ideas are reflected in transactional definitions of stress. Building on the work of others these have been identified (Dewe, Cox and Ferguson, 1993: 6) as: (a) stress is a dynamic cognitive state, (b) representing an imbalance and (c) requiring some restoration or resolution of that imbalance. Within this framework stress does not reside solely in the individual or in the environment but in the relationship between the two. The process that links the individual and the environment involves two important appraisals. The first (primary appraisal) gives meaning to any encounter. It is in essence where the individual makes a judgement that 'something is at stake' (Folkman, 1982). For the individual it is the process through which the importance of a particular encounter is evaluated and its relevance to well-being assessed (Cox, 1987). The second (secondary appraisal) refers to 'what can be done?' (Folkman, 1982). It is where the individual evaluates how much control they have over the situation and what resources are available to them.

These two processes are highly interdependent and shape an encounter (Folkman, 1982). Identifying these two processes shifts the focus of work stress to exploring and investigating what people actually think and do in a stressful situation (Holroyd and Lazarus, 1982) and away from arbitrarily separating the different components of that process. If the future direction of work-related stress research is, as a number of authors point out (Brief and Atieh, 1987; Payne, Jick and Burke, 1982; Schuler, 1985), towards a better understanding of the transaction between the individual and the environment then such processes should, at the empirical level, be explicitly recognized and investigated through systematic measurement (Dewe, 1991).

Coping with stress

The growth over the last two decades of research into coping, published in the general psychological literature, reflects the importance of coping as a moderating variable between a stressor and strain (Endler and Parker, 1990). In this field a number of researchers have attempted to identify the basic dimensions of coping (Pearlin and Schooler, 1978; Billings and Moos, 1984; Folkman and Lazarus, 1980; Endler and Parker, 1990). From these sorts of study three issues emerge as confronting researchers. They have been identified (Dewe, Cox and Ferguson, 1993: 8) as:

● establishing the appropriate methods for measuring coping strategies;
● agreeing on how coping strategies should be classified and described; and
● considering the utility of these schemata across different subject groups.

While the most common technique involves self-report measures, 'widely divergent approaches' (Endler and Parker, 1990: 844) have been used to identify and classify the different coping strategies. To some (Carver, Scheier and Weintraub, 1989) the difference in approach can be summarized in terms of theory-based development and classification versus empirically derived solutions where the strategies emerge from the statistical analysis. To others the debate extends to issues surrounding the generality that accompanies theoretically based measurement versus the specificity derived from empirical measurement (Amirkhan, 1990). Recurring psychometric difficulties (lack of theoretical support, poor replication of strategies and low reliabilities) have only added to the intensity of the debate. To some extent this has led to researchers looking for a middle ground that integrates the theoretical with the empirical approach (Amirkhan, 1990).

If there is one thing to emerge from this research then it is the use of terms like approach-avoidance (Roth and Cohen, 1986) and problem-focused/ emotion-focused (Folkman and Lazarus, 1980) to describe coping strategies. Whether or not there is consensus among researchers as to the utility of these descriptions is debatable. At times the two descriptors are viewed as mirroring each other (Amirkhan, 1990). At other times avoidance, for example, is regarded as a third category to complement problem- and emotion-focused strategies (Endler and Parker, 1990). The issue is not that simple, despite the fact that the popularity of the problem-/emotion-focused distinction rests

somewhat on its relative unambiguousness (Fleishman,1984). The identification of social support as a third category has, for example, cast doubt on the
adequacy of a two-dimensional view of coping strategies (Amirkhan, 1990), as
has the emergence of a mixed category made up of both problem- and
emotion-focused strategies (Folkman and Lazarus, 1980).

Researchers suggest that the problem-/emotion-focused distinction is no
longer well supported as it stands (Cox and Ferguson, 1991) and point to
the popularity of three-dimensional solutions (Moos and Billings, 1982;
Pearlin and Schooler, 1978). Indeed, Cox and Ferguson go so far as to
suggest that there are now, perhaps, at least two additional dimensions
worth considering, one concerned with reappraisal and the other with
avoidance (1991: 22). None of these distinctions may be exhaustive of
coping options and may to a large extent simply reflect the difficulties
mentioned above: that is the generality versus the specificity of the measures
used, the stressful situations involved and the methodologies employed.
Furthermore, to arrive at what Latack and Havlovic (1992: 401) describe
as a level of 'comprehensiveness' when classifying coping strategies a
distinction should also be made between the focus of coping (problem-
emotion) and the method of coping (e.g. cognitive-behavioural).

However, it is clear that questions like 'how best should coping strategies be
classified?' and 'what is the appropriate methodology?' will only be answered
as improvements are made in how coping is conceptualized and, more
importantly perhaps, through consistent attention to issues of measurement
and methodology. In view of the importance of stress in the workplace and the
fact that research into coping with stress at work is gaining momentum
(Edwards, 1988) we are prompted to ask about how work-stress researchers
have tackled the measurement of coping strategies and whether there is any
consistency among the different approaches. Though a number of measures
have been specifically designed for the workplace (e.g. Burke, 1971; Dewe and
Guest, 1990; Latack, 1986; Parasuraman and Cleek, 1984), choosing from
among them is difficult (Edwards and Baglioni, 1993) because of a lack of
psychometric detail or attendance to broader issues of construct validity
(Latack, 1986). The rest of this chapter explores the measurement of coping
in a work setting and examines how work-stress researchers have dealt with
issues of methodology, classification and utility.

Coping with stress at work

Reviewers have been quick to point to the growth of research into work-
related stress in general and coping in particular. What is less clear is how
much progress has been made. Coping reviews, perhaps as a result of the need
to organize the research literature more cohesively (Edwards, 1988), perhaps
in an attempt to identify common research pathways have: established
evaluative frameworks for coping with work stress (Latack and Havlovic,

1992); identified difficulties when conceptualizing and measuring coping either generally (Dewe, Cox and Ferguson, 1993) or in terms of specific approaches (Oakland and Ostell, 1996) and identified a range of measurement issues that need to be resolved (Newton, 1989; O'Driscoll and Cooper, 1994; Edwards and Cooper, 1988). The major reason that emerges from these reviews in terms of our lack of progress in understanding the role of coping in the stress process centres around the problem of measurement (O'Driscoll and Cooper, 1994).

Research into coping with work stress has been described as rather disappointing (Bar-Tal and Spitzer, 1994). Despite the literature reviews, the identification of measurement issues and the explanatory potential of the transactional approach, issues still remain about how coping strategies should be measured in a work context. More precisely the question 'what is it we are trying to achieve when coping is being measured?' has still not been fully answered. This review sets out to examine the progress that has been made in coping measurement. It does this by exploring the research findings within the context of the different steps necessary to develop a measure of coping. In this way it may be possible to identify a number of alternative measurement approaches, the context within which each should best be applied and a better understanding of the inconsistencies in findings that have emerged from this sort of research.

Reviewing coping research in a work setting is not without its difficulties. Because the literature is so diverse and voluminous (Burke and Weir, 1980) criteria have to be established for any review of progress. Here as elsewhere (Dewe, Cox and Ferguson, 1993) these criteria involved limiting examination in general to the academic literature, and identifying articles where necessary on a citational basis and selecting those that, in general, were measurement-oriented and directed towards identifying strategies used by individuals to cope with work stress. In reality this meant carrying out a search using both of the search engines ABI/Inform and Psychological Literature. In both cases the key words *work*, *stress* and *coping* were used. It was possible from a pool of 391 articles to identify forty-five that fell within the criteria.

Only a small number, however, were primarily concerned with the development of coping measures. In the main, while measurement was not the main focus of many of the studies it was an issue, and was frequently dealt with in such depth as to warrant inclusion here. This review has been limited in general to white collar employees but acknowledges the work on coping using, for example, police (Hart, Wearing and Headey, 1995), nurses (Dewe, 1987) and bus drivers (Kuhlmann, 1990). Areas such as social support, self-management (e.g., relaxation, meditation, exercise, etc.) organizational development and employee intervention programmes were also beyond the scope of this review. It focuses largely on measurement issues, and works through these by adopting the following strategy. The policy has been to focus on the measurement process, identifying those issues that have to be confronted when scale construction is the objective. In this way the review begins by looking at the theoretical context that researchers use when considering coping. It moves from there to considering issues surrounding item derivation, scoring responses, classification (including replication) and the role of coping in the appraisal process. Each step in this process will be an attempt to provide some answers to the question raised earlier of 'what is it that we are trying to achieve when we are measuring coping?'

Theoretical context

When it comes to investigating coping within a work setting the issues explored can only be described as diverse. It is possible by examining the research objectives of those studies reviewed to identify a number of themes. By far the most common approach is one where the main aim of the research is to explore the relationship between coping and other individual, contextual or outcome variables. In many of these studies the underlying and often unstated objective would seem to be to explore the coping process. A number of approaches have been used. Some workers have examined the influence of personality, gender and race on the use of different coping strategies (Greenglass, 1988; Greenglass, Burke and Ondrack, 1990; Strutton and Lumpkin, 1992; Walsh and Jackson, 1995; Stroman and Seltzer, 1991). Others have looked at the strategies used when coping with specific stressful situations (Feldman and Brett, 1983; Leana and Feldman, 1990; Burke and Belcourt, 1974), chronic work stress (Nelson and Sutton, 1990) or task complexity (Puffer and Brakefield, 1989) or common work stressors like role conflict and role ambiguity (Havlovic and Keenan, 1991). Another approach has been to explore the relationship between the coping process and adaptational outcomes (Kuhlmann, 1990) or individual responses and assessment to work stress and burnout (Thornton, 1992).

Other themes to emerge which express the context within which coping takes place include:

- the effectiveness of different coping strategies (Bhagat, Allie and Ford, 1991; Howard, Reichnitzer and Cunningham, 1975);
- the nature of the relationship between specific styles of coping and different outcomes (Koeske, Kirk and Koeske, 1993; Leiter, 1990; Long, 1993; Parasuraman and Cleek, 1984) including the role of coping styles over the longer term (Newton and Keenan, 1990);
- a consideration of coping from the perspective of the individual's employing organization (George, Brief and Webster,1991);
- identifying innovative coping responses to organizational stress (Bunce and West, 1996);
- broadening the context in which coping has been studied (Newton and Keenan, 1985).

A number of authors have also been concerned with modelling the coping process. The aim here has been to

- empirically validate alternative theoretical models (Edwards, Baglioni and Cooper, 1990);
- explore a range of issues around the transactional model (Dewe,1989);
- test directly the theoretical framework of Lazarus (Long, Kahn and Schutz, 1992);
- examine the system within which occupational stress, strain and coping interact (Osipow and Davis, 1988) and
- examine coping strategies as a personal variable to the model of burnout (Leiter, 1991).

Finally, the measurement of coping has been explored in a number of ways. These include considering the appropriate research design for examining stress and coping (Kinicki and Latack, 1990); developing an assessment of coping that could be used on a daily basis (Stone and Neale, 1984); evaluating coping behaviour scales related to job stress (Latack, 1986) or evaluating an approach to measuring coping with stress in work settings (Dewe and Guest, 1990); and considering the comparative merits of different coping measures (Edwards and Baglioni, 1993).

From the studies reviewed it is clear that how coping should be measured has received a less than complete treatment. This is not to say that coping measurement has not been an issue. In many of the studies though, the objective for the researchers has been to understand better the stress process and while coping measurement has, at times, received attention it has not always received the emphasis that would allow a coherent approach to measurement to emerge.

The theoretical context also varies from study to study. At least three different types of approach can be identified. The first is where the research is explicitly set within transactional theory (e.g. Feldman and Brett, 1983; Long, 1993; Latack, 1986). The second approach is where some sort of transactional process is inferred from the differing effects of coping strategies, the functions coping strategies perform or the conceptualization of coping effectiveness (e.g. Greenglass, 1988; Greenglass, Burke and Ondrack, 1990; Nelson and Sutton, 1990; Osipow and Davis, 1988). The third approach is where the theoretical context is expressed through the way in which coping is defined (e.g. Kinicki and Latack, 1990; Parasuraman and Cleek, 1984; Strutton and Lumpkin, 1992).

Researchers appear therefore to accept, albeit somewhat tacitly, that the context within which coping should best be explored is a transactional one where stress does not reside solely in the individual or in the environment but in the transaction between the two. The transactional theory and its application to a work setting involves an understanding of individual processes, the context in which they operate, and the manner in which they link the individual to the environment (Lazarus, 1991). There is general agreement that stress essentially occurs at the individual level and that there are important benefits from studying individual patterns of and reactions to stress. However there is some debate as to the usefulness of this approach, particularly since researchers also have an obligation to try to identify those working conditions which are likely to adversely affect most workers exposed to them (Brief and George, 1991: 16). The application of the transactional approach to a work setting is not without its problems, not the least of which is whether it supports our role and obligation as researchers.

What is not disputed though is that whatever approach we adopt we need conceptual guidance and theory that best captures that context within which such phenomena occur (Brief and George, 1991). The different approaches are not incompatible. The field of work stress can only benefit from the careful and thoughtful application of the transactional approach (Harris, 1991). Identifying those 'common pathways' that link the individual to the environment may in fact provide the mechanism through which individual processes and general adverse conditions are drawn together, allowing solutions to be implemented and programmes devised that better serve those whose work experiences we study.

In terms of defining coping, most studies let the research 'speak for itself'. In these studies coping has been implicitly defined either through the type of question asked or the methodology used to classify strategies. In this way coping was 'how individuals handled the event they described,' or 'the sorts of things they did to cope with the stress and strain of their job.' At times researchers, while not actually defining coping, did describe the functions coping activities serve in such detail that a surrogate definition could be said to emerge (Leiter, 1991; Oakland and Ostell, 1996). At other times the definition could be inferred on the basis of how effective coping strategies were in relation to outcome measures (Greenglass, Burke and Ondrack, 1990). When definitions of coping were given, a number of approaches could be identified. The first was to simply define coping using definitions from other authors. The most frequently cited definitions were those of Lazarus and Folkman (1984); Lazarus and Launier (1978); Aldwin and Revenson (1987); and Folkman et al. (1986).

The second approach was either to define coping relatively simply as 'coping with job stress' (Greenglass, 1988), or to define it in terms of 'an individual's efforts to manage the psychological demands . . .' (Kuhlman, 1990) or be more precise about the nature or the individual effort: 'cognitive or behavioural efforts . . .' (Strutton and Lumpkin, 1992); 'active or passive attempts . . .' (Dewe, 1987) or define it specifically in terms of the problem-emotion focused functions that coping performs (Greenglass and Burke, 1991). What emerges from these definitions is a number of themes that reflect the transactional notion of stress. These include cognitive or behavioural efforts (mode), problem- or emotion-directed coping (focus), and settings where perceived situational demands threaten well-being or tax or exceed a person's physical or psychological resources.

Selection of Scale Items

Coping has been measured in a number of different ways and there are no simple answers as to which approach is best (Cohen, 1987). However there is a recognition that such measurement requires the use of methodologies that capture and describe what individuals actually think and do (Holroyd and Lazarus, 1978; Folkman, 1982). Just asking people about how they cope is not without its problems either (Cohen, 1987; Edwards, 1988). In an earlier review (Dewe et al., 1993) studies were divided into those that were empirically based and those that were theory driven. This distinction, while important, does tend to mask the range of approaches that have been adopted. Bearing in mind that for many of the studies measurement was not the primary objective a number of approaches to selecting scale items can be identified.

Some researchers, for example, provide only minimal information on the origin of the items and on the literature reviewed (Kiev and Kohn, 1979; Leana and Feldman, 1990). Others simply use existing coping measures. In these cases workers have adopted one of the following approaches:

- used all the items in the scale (Decker and Bergen, 1993; Greenglass, 1988, 1993; Koeske, Kirk and Koeske, 1993; Long, 1993; Osipow and Davis, 1988; Strutton and Lumpkin, 1992; Thornton, 1992);
- selected particular scale components that best reflect the research focus or goal (Greenglass, Burke and Ondrack, 1990; Kuhlman, 1990; Leiter, 1991);
- made some refinements or adapted existing scales (Bhagat, Allie and Ford,1991; Havlovic and Keenan, 1991);
- drawn the items from existing measures and substantially revised and supplemented them in accordance with a particular theoretical perspective (Edwards and Baglioni, 1993);
- drawn on existing scales but developed a new scale as 'coping must be assessed relative to a specific stressful transaction' (Kinicki and Latack, 1990).

In cases where the selection of items was based on a review of the literature it was possible to identify a number of different approaches. These included:

- drawing items from the literature on stress and coping as well as from the popular literature (Puffer and Brakefield, 1989);
- being influenced by earlier articles on coping but selecting items a priori to generate a number of broad categories of coping that reflect the coping literature and the work of others but with a specific stressful situation in mind (Feldman and Brett, 1983);
- evaluating items from extant coping inventories and from the empirical literature on strategies of coping (Stone and Neale, 1984);
- reviewing previous research and measures of coping and supplementing this with structured interviews (George, Brief and Webster, 1991) and in

addition with information from discussions with colleagues interested in the area (Latack, 1986).

A number of researchers have used open-ended questions to derive scale items. Here again a variety of approaches can be identified based around the focus of the question used. Some have focused their approach around simply asking respondents to generate a list of possible behaviours by employees faced with stressful working conditions (Parasuraman and Cleek, 1984). Others focus on the job by asking respondents, for example, 'what do you do to cope with the stress and strain of your particular job?' (Shinn *et al.*, 1984; Stroman and Seltzer, 1991) or 'how they cope when their job occasionally demands a great deal of them' (Burke and Belcourt, 1974), while others ask individuals to describe a particular incident and then report how they coped with it. In these cases researchers asked 'how did you handle the incident you described above?' (Newton and Keenan, 1985) or 'can you think of a time at work when you felt under stress? Can you tell me what happened and how you managed to cope with it?' (Dewe and Guest, 1990) or in terms of the problem described, 'what did you do to feel better or handle the problem?' (Schwartz and Stone, 1993).

Others (e.g. Burke and Belcourt, 1974; Dewe and Guest, 1990), often in addition to other job-related coping questions, have asked individuals to focus on the emotional discomfort involved and for example, answer 'if, like most people, you occasionally get fed up with your job and feel tense and frustrated how do you cope?' (Dewe and Guest, 1990). Others, depending on their research objectives, have varied the focus of the questions used and asked individuals to describe strategies they found 'useful' or 'effective–ineffective' (Burke, 1971). Finally, based on the Lazarus and Folkman (1984) recommendation that coping is best tested indirectly by focusing on the stress that evolves, researchers (Erera-Weatherley, 1996) have assessed coping from responses to questions about stressors experienced by respondents.

In trying to identify what people actually think and do when faced with a stressful situation, researchers have used a range of approaches and methodologies. The results illustrate the tremendous diversity and complexity of coping behaviours. Despite the fact that for many of the studies the emphasis has not been on measurement as such it is still possible to identify a number of methodological issues that need addressing. If the goal is to establish 'what it is that people actually think and do' are items then best derived directly or indirectly? If directly, then should the question assess coping relative to a particular stressful transaction (Kinicki and Latack, 1990) or be more general in focus? If more general, then should some distinction be made between coping with the job and coping with the emotional discomfort? What impact does this have on the strategies that are derived (Newton and Keenan, 1985) and are we, by selecting a particular approach, restricting the utility of those strategies or somehow commenting on their efficacy? If, on the other hand, there is also theoretical support for strategies to be derived indirectly but in relation to a particular stressful incident, then is it this approach that provides a more independent confirmation of what people actually think and do free from the problems of efficacy and connotations of success. From the literature reviewed, there does seem to be an interesting theoretical split between directly or indirectly asking about coping and whether the questions used should be stressor specific or more general in focus.

Asking people about how they cope is but one approach. The literature also supports the selection of strategies on an a priori basis using the body of knowledge that has grown around the subject. These studies often begin with a set of hypothetical categories and test this taxonomy in terms of its apparent applicability to particular stressful encounters.

These studies offer a methodical and detailed analysis of the available literature and then a well-argued rationale for item selection. This approach does raise the issue of meaning when attempting to establish or classify coping items and the question of whether meaning should be imposed or allowed to emerge from the methodology and the subsequent analysis. Dealing with this issue has led some researchers (Schwartz and Stone, 1993) to develop techniques that allow individuals themselves to identify the focus of their thoughts and actions as opposed to having it imposed or specified by the researcher. Another issue here is that if items are selected on the basis that they best reflect different categories of coping strategies then does this method simply restrict the utility of the items derived in this way. If the need is to understand coping strategies, not in isolation but within some contextual encounter as coping definitions suggest (see Erera-Weatherley, 1996: 158), then should the theory that guides item selection be more directed to the context within which coping takes place and not so much in terms of theory that reflects the focus or mode of a coping strategy.

Scale instructions and scoring

A number of approaches can also be identified when item scaling is the focus. Again it is important to remember that many researchers did not have the goal of scale development per se but were attempting to better understand the nature of coping and its role in the coping process. Within this context researchers, depending on their research goals, adopted a variety of strategies. Scale instructions differed in terms of both their specificity and their focus. Items were measured across a range of metrics. These include, for example, how likely, how frequently, how true and the extent to which they used different coping strategies. At other times respondents were asked to check the strategies they used. Not surprisingly, scale intervals varied as well. Scales ranged from three to seven points with differences in scoring keys, first as to whether 0 or 1 was the preferred number with which to begin a scale and second in the descriptions used to identify scale intervals, even when scales were purporting to measure the same thing.

To elaborate: a number of studies simply report that they used existing scales (Strutton and Lumpkin, 1992) or modified versions of existing scales (McDonald and Korabik, 1991) or some but not all of the components of existing scales (Greenglass and Burke, 1991). By far the most common approach was to get respondents to comment on the frequency with which they used different coping strategies. Instructions varied. At times they were expressed in terms of:

- the frequency with which respondents took such actions (Nelson and Sutton, 1990);
- how frequently they had used the strategy (Dewe, 1989; Dewe and Guest, 1990; Koeske *et al.*, 1993);
- the frequency of choosing the listed coping strategies when confronted with stress at work (Kuhlmann, 1990);
- the frequency with which they engaged in different types of coping behaviour (Leana and Feldman, 1990);
- how often they did the following when they felt tense because of their job (Havlovic and Keenan, 1991).

At times 'use' was the key word, and respondents were asked to indicate the degree to which they utilized each coping strategy (Bhagat *et al.*, 1991) or the degree to which it was used to deal with the stressor (Long, 1993). At other times while use was not part of the instructions it was clear from the way that the scale was scored (Edwards and Baglioni, 1993) that this was what was intended. In another set of variations, respondents were asked to indicate the 'extent' to which they engaged in coping activities (Leiter, 1990) or the extent to which they had used a particular coping action (Thornton, 1992). Here as well, scoring keys (e.g. 'hardly ever do this/almost always do this') also reflected use (Kinicki and Latack, 1990; Latack, 1986) when the scale instructions were not always reported. Getting respondents to check as many coping strategies as applicable or indicate which coping strategies they used from a list of possible strategies was another way of establishing the use of coping strategies (Kiev and Kohn, 1979; Puffer and Brakefield, 1989) as were the instructions when respondents were asked to check an item if they had used it within the last two months to deal with stressful situations (Parasuraman and Cleek, 1984).

Respondents were also asked to comment on 'how true' it was of them when it came to coping with stress (Decker and Borgen, 1993). 'Likely' was another key word and respondents were asked to indicate how likely they were to do each of certain things when they experienced stress at work (Greenglass, 1988; Greenglass *et al.*, 1990). Instructions also varied in terms of how general or specific they were. Some, for example, asked respondents to focus on or identify a particular event (e.g. Edwards and Baglioni, 1993; Kinicki and Latack, 1990; Leana and Feldman, 1990; McDonald and Korabik, 1991; Thornton, 1992). These instructions sometimes specified the time period within which the event had to occur (Koeske *et al.*, 1993; Long, 1993;

Wolfgang, 1991). The more general approach instructed respondents, for example, to focus on 'when they experienced stress at work' (Greenglass, 1988; Greenglass et al., 1990) or 'were confronted with stress at work' (Kuhlmann, 1990) or have 'difficulties with their job responsibilities' (Leiter, 1990; 1991) or 'felt tense because of their job' (Havlovic and Keenan, 1991). At times both general and specific instructions have been used (Latack, 1986).

Finally as mentioned earlier a range of scoring keys was used. Again the variation can be partly explained in terms of differing research objectives. However it is interesting to note that even when measuring the same thing a range of different scoring keys can be used. For example, when frequency was the key then the lower end of the scale could begin with 1 or 0 and describe 'never', 'not at all' or 'hardly ever do this'. Similarly the top end of the scale ranged from 3 to 7 and described 'almost totally' or 'almost always do this' or 'quite often' or 'always'. Likert scales were not the only approach. At times respondents were asked to consider a list of coping strategies and check those that they had used (Kiev and Kohn, 1979; Puffer and Brakefield, 1989; Stroman and Seltzer, 1991). At other times the use of a coping strategy was determined through the content analysis of open-ended questions (Erera-Weatherly, 1996; O'Driscoll and Cooper, 1994; Shinn et al., 1984) or by respondents indicating whether or not they employed a particular category of coping and if so described what they did (Schwartz and Stone, 1993).

Attempting to explore coping and its role in the stress process has led researchers to use a number of different strategies when devising scale instructions and scoring. Despite this level of diversity most studies have however been concerned with how often a strategy was used and the relationship between the use of a strategy and individual outcomes. The strength of this type of approach lies in the fact that it is process-oriented, allowing individuals to express what they actually thought and did. It is also non-evaluative, making no prior judgements about the effectiveness of a particular strategy and thereby meeting measurement criteria first established by Folkman and Lazarus (1980). In respect of the latter criteria, Latack and Havlovic also make it quite clear that coping measures should specifically address coping, not coping effectiveness (1992: 494), allowing effectiveness to be independently assessed through a range of outcome variables.

Yet it appears that our preoccupation with measuring the frequency with which a coping strategy is used is something of a vexed issue in work-stress research. Oakland and Ostell for example, suggest that a qualitative approach to analysing coping data reveals that coping efficacy not frequency is the important factor regarding outcomes and suggest that research should focus on refining current measurement strategies to integrate ways of assessing coping effectiveness (1996: 151). Indeed these authors go further and raise a number of issues that question whether current quantitative measures of coping can in fact actually capture the complexity of the coping process. The issues they raise are as much a comment on the gains that may be derived when both qualitative and quantitative methodologies are used as they are about current measures being 'limited both in their descriptive and in their explanatory power' (1996: 151). A number of potential limitations can be identified. The first concerns the specificity of scale instructions and the question of whether the more general the instructions the more likely it is that coping styles (stable coping actions) are being measured. The need to distinguish between coping style and coping behaviour has long been identified as an issue in coping research (Dewe *et al.*, 1993) and reflected in the type of instructions that accompany coping measures (Newton, 1989).

A second and somewhat related issue is that more general instructions may not prompt respondents to make such an obvious distinction between acute and ongoing coping (Dewe *et al.*, 1993; Newton, 1989). The consequences of inadequately identifying the stimuli that necessitate or arouse coping and the need to examine coping within the context of a particular encounter (Erera-Weatherly, 1996) are measurement issues that require attention being given at the very least to the type of instructions provided and their impact on the coping strategies derived. Having to make decisions between context-specific and more general measures is not uncommon for work-stress researchers. What now needs as much deliberation when deciding on the nature of the instructions is the limitation on explanatory potential. Clearly this is a difficult issue. One way around it is to adopt what Latack and Havlovic (1992) describe as a 'middle range' position. That is, to identify categories of work stressor that are common enough to warrant a specific measurement focus while still allowing for the development of more generalizable measures.

Finally, there is the issue of scaling. As mentioned, the most common approach is to measure the frequency with which a coping strategy is used. This practice, as already discussed, has raised some discussion particularly when qualitative and quantitative results are compared (Bar-Tal and Spitzer, 1994; Oakland and Ostell, 1996). While this debate will undoubtedly continue, it does, for as long as frequency is the preferred metric, draw attention to two somewhat parallel measurement issues. First, there is a need to ensure that coping items reflect how a person copes rather than coping effectiveness. This distinction reflects the kind of conceptual clarity necessary when item specificity is being considered (Latack and Havlovic, 1992). Second, it does mean that researchers will need to consider, if coping strategies continue to be measured in terms of their frequency, the appropriate methodology for measuring their effectiveness. While coping effectiveness is almost always assessed in terms of the relationship between use and outcomes, this somewhat broad approach does not, for example, develop or contribute to our understanding of the adequacy of coping resources (Oakland and Ostell, 1996), reappraisal of (Bar-Tal and Spitzer, 1994), issues of control and power (Handy, 1986). Neither does it answer the question of effective for whom and at what cost.

Scale classification

Of all the issues confronting researchers when considering the measurement of coping, the classification of coping strategies has attracted the most attention. Typically this involves classifying strategies according to their focus–the target towards which the coping behaviour is directed (Latack and Havlovic, 1992). By far the most popular approach is to distinguish between those strategies that are problem-focused and those that are emotion-focused (Lazarus and Folkman, 1984). Because of the variety of methods that emerge within these two categories and in an attempt to establish some order, Latack and Havlovic (1992) have suggested that coping strategies should now be further distinguished in terms of their mode. The utility of these distinctions and the variety of approaches taken when describing coping strategies are outlined below.

The classification of coping strategies has been achieved in a number of different ways. At the theoretical level this involves developing categories that are: (a) broadly reflective of the coping literature, (b) guided by a literature review or (c) based around a particular theory. The empirical approach typically works from open-ended questions or existing frameworks and then uses content or factor analysis to derive a solution that can be applied back to understanding, validating or confirming a particular schema. There are, of course, variations on these themes. At times, depending on the research goals, the approach has been simply to use established coping instruments either in part or in total or to combine theory and practice, with the result that a number of classification schemes have been developed.

While it could be argued that attempts to classify coping actions all have their roots in the distinction first made by Lazarus and Folkman (1984) between problem- and emotion-focused strategies, alternative taxonomies need to be considered because in considering alternative approaches we are confronted not just by issues of reliability and validity (O'Driscoll and Cooper, 1994) but also by the equally substantive issues of context specificity and replication (Trenberth, Dewe and Walkey, 1996). Despite the wide range of research objectives, the authors of most of the studies reviewed have been confronted with classifying or naming the coping strategies they have used. What emerges is discussed below.

As mentioned, the most well known distinction is that made between problem- and emotion-focused strategies. This distinction has been used and derived in a number of different ways in work-stress research. Using different versions of the Lazarus and Folkman (1984) Ways of Coping Checklist (WCC) researchers have for example: (a) analysed their data according to the eight sub-categories of problem- and emotion-focused coping identified by Folkman *et al.* (1986) (Thornton, 1992) or the five sub-categories identified by the Vitaliano *et al.* (1985) version of the WCC (McDonald and Korabik, 1991) and (b) subjected the items in each of the problem- and emotion-focused categories to separate factor analyses to identify sub-categories

(Strutton and Lumpkin, 1992). Other authors for example, using different instruments and approaches, have done the following:

- where their objective was to assess the differing effects of problem- and emotion-focused coping simply used this distinction, identifying in the process a third category that they describe as appraisal-focused coping (Nelson and Sutton, 1990);
- used the problem-emotion focused distinction to categorize their open-ended data (Shinn *et al.*, 1984) or classified their data, distinguishing between adaptive (problem-focused) and maladaptive (emotion-focused) strategies (Parasuraman and Cleek, 1984); or
- used factor analysis to identify a range of coping strategies and described or grouped these according to whether their focus was problem- or emotion-directed (Dewe and Guest, 1990), or problem-focused or symptom-focused (Leana and Feldman, 1990).

The Bhagat *et al.* (1991) analysis of the adapted Pearlin and Schooler (1978) measure revealed two factors labelled problem-focused and emotion-focused coping.

Other researchers, using a self modified version of the WCC, identified two factors that they describe as disengagement and engagement coping (Long, 1993: Long, Kahn and Schultz, 1992) or, using a version of the WCC and following the work of Latack (1986), were able to distinguish between active or control-centred coping and avoidance coping (Walsh and Jackson, 1995). Reviewers (O'Driscoll and Cooper, 1994) have already commented on the considerable variation in the internal structure of the Ways of Coping Check-list. What, however, is just as intriguing as the range in reliability coefficients, item loadings and resulting components is the number of ways the checklist has been used and the ways in which researchers have applied the problem-focused/emotion-focused distinction to their data. It may well be time to consider this distinction as perhaps it was first intended; as a general rubric. Accepting this may then enable researchers to focus more on exploring the structure and nature of the different components that make up each category (Latack and Havlovic, 1992) and the context within which they occur, thereby providing an opportunity to consider the more fundamental issue of whether our interest is in context-specific measures or those that can be replicated, and the implications that this has for measurement.

This is not the only schema for distinguishing between coping strategies. Latack (1986), for example, has identified three coping behaviours from her work on how people cope with job stress: control (proactive strategies), escape (avoidance strategies) and symptom management. Refinements and adaptations by Kinicki and Latack (1990) and Havlovic and Keenan (1991) tend to confirm the distinction between control and avoidance behaviours and expand on the components that make up each construct. Analysis by Leiter (1990; 1991) also supports the presence of a two-factor structure reflecting control and escape behaviour. As discussed by Latack (1986), this control–

escape distinction also appears to be present in different configurations of problem-emotion focused coping.

Based around the Wong and Reker (1984) measure, coping strategies have been variously described as: problem-focused or instrument-focused coping, emotional or palliative coping, preventive coping and existential coping (Greenglass, 1988). Support for this structure and for scale reliabilities has also been reported by Greenglass (1993); Greenglass and Burke (1991) and Greenglass *et al.* (1990). A number of schemes have emerged from adaptations of the Moos *et al.* (1987) instrument. Redefinition and redevelopment of the Moos *et al.* (1987) scale by Koeske *et al.* (1993) produced two factors describing control and avoidance coping. Analysis by Puffer and Brakefield (1989) and Wolfgang (1991) using versions of the Moos scale identified active cognitive and behavioural and avoidant cognitive and behavioural strategies and active cognitive, active behavioural and avoidance strategies respectively.

Using the personal resources questionnaire from the Osipow and Spokane (1984) measure of occupational stress, Osipow and Davis (1988) described four sub-scales, measuring coping as recreation, self care, social support and rational/cognitive. The reliability of these four sub-scales has also been reported by Decker and Borgen (1993). Other researchers (e.g. Burke and Belcourt, 1974; Newton and Keenan, 1985; Stroman and Seltzer, 1991), analysing open-ended data, have described a range of coping strategies but have not gone so far as to apply any higher order analysis to describing their results. When the different classification attempts are considered, reviewers (Dewe *et al.*, 1993; Latack and Havlovic, 1992; O'Driscoll and Cooper, 1994) are right to draw attention to a range of psychometric issues that still require attention and resolution. These have been well identified and include the amount of variance explained by different coping components, their internal reliability and their construct validity. Although the debate surrounding reliability coefficients and their appropriateness when applied to coping scales has been well summarized (Wolfgang, 1991) it does raise, somewhat obliquely perhaps, the question of 'what is it we are we trying to achieve when classifying coping strategies?' In a sense this comes down to considering the appropriate methodology for describing and classifying coping strategies. If our aim is to develop coping data that can be replicated then what are needed are coping dimensions that generalize across samples. The choice, if it has to be made, may well be between coping measures that reflect the replicable characteristics of the questionnaire and those that capture the specific characteristics of the sampled group.

As pointed out by Trenberth *et al.* (1996), researchers wishing to explore the unique characteristics of the sample group require a different approach from that used when the research focus is to identify replicable characteristics of the questionnaire. When replication is the research goal the work of Trenberth *et al.* (1996) clearly identifies two replicable factors that describe problem- and emotion-focused coping. Researchers wishing to explore coping strategies within the unique context of the sample group may well, as suggested by

Latack and Havlovic (1992), wish to consider the comprehensiveness of their findings, exploring in more detail the mode of coping used. Beginning to understand better and identify the variety in coping modes in this way may provide a more coherent base for ordering the different dimensions, allowing a more complete picture to emerge when coping focus becomes the issue.

Conclusions

Previous reviewers (Dewe *et al.*, 1993; Latack and Havlovic, 1992; O'Driscoll and Cooper, 1994) have clearly, and in some detail, identified conceptual and methodological difficulties when measuring coping with work stress. When it comes to the way forward we could follow the example of Latack and Havlovic (1992) and offer not so much a 'conceptual blueprint' but some 'strategic pathways' that operate as decision tools when considering how best to approach the measurement of coping. Issues of methodology still maintain their importance, but set within a strategic context they may help more directly to identify what those issues are and how best they can be resolved.

The essential question still remains. That is 'what is it we are trying to measure when measuring coping?' If the response to this question is to develop a measure of coping that can be used to generalize findings across samples, then the usefulness of any instrument and its sub-scales may need to be judged on the extent to which its sub-scales are built around replicable characteristics of the questionnaire itself rather than the unique characteristics of the responding group. In this case the methodological focus would be on employing techniques that explore the stable characteristics of the questionnaire itself. Efforts in this direction would, as Edwards and Baglioni (1993) discuss, require a shift away from typically relying on reliability estimates to yield construct validity, to confirmatory procedures that convincingly evaluate the stability of underlying scale dimensions.

On the other hand, when the response to the question is to explore the context within which coping takes place then a different but not mutually exclusive strategy is required. In this case the focus would be on identifying those sub-scales that reflect the unique characteristics of the sampled group. Research can then focus on the structure of the coping components, exploring their meaning through the idiosyncratic nature of the sample. Issues of reliability and construct validity remain important, as do those of theoretical scope and empirical rigour. What distinguishes this approach from others is the recognition that describing and classifying coping sub-scales is bound by the characteristics of the sample group. This avoids the confusion that often accompanies coping research when assuming that replicable components will emerge when this has clearly not been the aim of the research.

One other issue remains. It is the dilemma of stressor-specific versus general coping measures. Irrespective of whether the aim is to develop a measure that can be replicated across samples or one that reflects the unique characteristics of the sampled group we are left with having to decide on the focus of the coping measure. The strategy here may well be to follow the guide of Latack and Havlovic (1992) and consider their 'middle range' proposal. If this was the preferred approach then it would be necessary for researchers to identify those situations or categories of situation that are sufficiently common to represent a 'useful specific focus' (Latack and Havlovic, 1992: 501). Examples of this approach are already well established in the literature. Again, depending on the research objective the issue of scale replication would need to be considered and form part of the decision-making process. Identifying those situations that affect the well-being of most workers would also meet the responsibility of work-stress researchers and, as Brief and George (1991) suggest, provide an opportunity first to explore the nature of those conditions, thus better serving those individuals who work in such contexts, and second to confront work-stress researchers with the need to identify those conditions by posing research questions that capture the theoretical orientation of the transactional approach.

All this assumes of course that coping is more 'efficiently measured' by self-report coping questionnaires. Where qualitative approaches have been used to measure coping, researchers have drawn a number of conclusions:

- The indirect examination of coping through the content analysis of the stresses that give rise to it provides a more independent mechanism for ensuring the link between stressors and coping, better expresses its transactional nature and exposes meaning rather than imposes it (Erera-Weatherley (1996));
- Issues such as coping efficacy, resource adequacy and the dynamic nature of the process are better captured through semi-structured interviews which, when analysed, illustrate a changing complex process that cannot be adequately expressed through more traditional coping questionnaires (Oakland and Ostell, 1996);
- The critical incident procedure offers an opportunity to integrate better the analysis of the stressful encounter, individual coping responses and the consequences of coping behaviours, thus providing more comprehensive information about the stress-coping process (O'Driscoll and Cooper, 1994: 352).

This review illustrates the diversity and complexity of approaches to the measurement of coping. Established instruments and conventional self-report measurement techniques have broadened our understanding of the nature and structure of coping and its role in the transactional process. Their use, however, often expresses what appears to be an urgency in work-stress research to 'get on' and explore the process of coping, regarding this as the more challenging and perhaps more potentially powerful route to improving our understanding and increasing our knowledge. Yet what in the long term may prove to be more potent is the need now to engage in a period of quiet reconstruction, to thoughtfully consider where traditional methodologies are taking us and to reflect on what alternative methodologies can provide. Well-established constructs will need to be rethought and alternative methodologies explored. Work-stress research, as Bhagat and Beehr (1985) point out, offers a fertile ground for attempting such investigations and for establishing the legitimacy of alternative methodologies, thus ensuring that the next decade of work-stress studies will be marked by innovative and creative methodologies that continue to advance our knowledge and meet our obligations as researchers.

References

ALDWIN, C. N. and REVENSON, T. A. (1987) 'Does coping help? A re-examination of the relation between coping and mental health', *Journal of Personality and Social Psychology* **53**: 337-348.

AMIRKHAN, J. H. (1990) 'A factor analytical derived measure of coping. The coping strategy indicator'. *Journal of Personality and Social Psychology* **59**: 1066–1074.

BAR-TAL, Y. and SPITZER, A. (1994) 'Coping use versus effectiveness as moderating the stress-strain relationship', *Journal of Community and Applied Social Psychology* **4**: 91–100.

BEEHR, T. A. and FRANZ, T. M. (1987) 'The current debate about the meaning of job stress', in J. M. Ivancevich and D. C. Ganster (eds), *Job Stress: from theory to suggestion*, New York: Haworth Press, 5–18.

BHAGAT, R. S., ALLIE, S. M. and FORD, D. L. (1991) 'An enquiry into the moderating role of styles of coping', in P. L. Perrewé (ed.) *Handbook on job stress* (special issue), *Journal of Social Behaviour and Personality* **6**: 163–185.

BHAGAT, R. S. and BEEHR, T. A. (1985) 'An evaluation summary and recommendations for future research', in T. A. Beehr and R. S. Bhagat (eds) *Human Stress and Cognition in Organisations: an integrated perspective,* New York: John Wiley and Sons, 417–431.

BILLINGS, A. and MOOS, R. (1984) 'Coping, stress and social resources among adults with unipolar depression', *Journal of Personality and Social Psychology* **46**: 877–891.

BRIEF, A. P. and ATIEH, J. M. (1987) 'Studying job stress: are we making mountains out of molehills?', *Journal of Occupational Behaviour* **8**: 115–126.

BRIEF, A. P. and GEORGE, J. M. (1991) 'Psychological stress and the workplace: a brief comment on Lazarus' outlook', in P. L. Perrewé (ed.) *Handbook on job stress* (special issue), *Journal of Social Behaviour and Personality* **6**: 15–20.

BUNCE, D. and WEST, M. A. (1996) 'Stress management and innovation interventions at work', *Human Relations* **49**: 209–231.

BURKE, R. J. (1971) 'Are you fed up with work? How can job tension be reduced?', *Personnel* **34**: 27–31.

BURKE, R. J. and BELCOURT, M. L. (1974) 'Managerial role stress and coping response', *Journal of Business Administration* **5**: 55–68.

BURKE, R. J. and WEIR, T. (1980) 'Coping with the stress of managerial occupations', in C. L. Cooper and R. Payne (eds) *Current Concerns in Occupational Stress*, Chichester: John Wiley, 299–335.

CARVER, C. S., SCHEIER, M. F. and WEINTRAUB, J. K. (1989) 'Assessing coping strategies: a theoretically based approach', *Journal of Personality and Social Psychology* **56**: 267–283.

COHEN, F. (1987) 'Measurement of coping', in S. V. Kasl and C. L. Cooper (eds) *Stress and Health: Issues in Research Methodology*, Chichester: John Wiley, 283–305.

COX, T. (1978) *Stress*, London: Macmillan.

COX, T. (1987) 'Stress, coping and problem solving', *Work and Stress* 1: 5–14.

COX, T. and FERGUSON, E. (1991) 'Individual differences, stress and coping', in C. L. Cooper and R. Payne (eds) *Personality and Stress: Individual Differences in the Stress Process*, Chichester: John Wiley and Sons, 7–30.

DECKER, P. J. and BORGEN, F. H. (1993) 'Dimensions of work appraisal: stress, strain, coping, job satisfaction and negative affectivity', *Journal of Counselling Psychology* 40: 470–478.

DEWE, P. J. (1987) 'Identifying strategies nurses use to cope with work stress', *Journal of Advanced Nursing* 12: 489–498.

DEWE, P. J. (1989) 'Examining the nature of work stress: individual evaluations of stressful experiences and coping', *Human Relations* 42: 993–1013.

DEWE, P. J. (1991) 'Primary appraisal, secondary appraisal and coping: their role in stressful work encounters', *Journal of Occupational Psychology* 64: 331–351.

DEWE, P. J. (1992) 'The appraisal process: exploring the role of meaning, importance, control and coping in work stress', *Anxiety, Stress and Coping* 5: 95–109.

DEWE, P. and GUEST, D. (1990) 'Methods of coping with stress at work: a conceptual analysis and empirical study of measurement issues', *Journal of Organisational Behaviour* 11: 135–150.

DEWE, P., COX, T. and FERGUSON, E. (1993) 'Individual strategies for coping with stress and work: a review', *Work and Stress* 7: 5–15.

DUCKWORTH, D. (1986) 'Managing without stress', *Personnel Management* 40-43.

EDWARDS, J. R. (1988) 'The determinants and consequences of coping with stress', in C. L. Cooper and R. Payne (eds) *Causes, Coping and Consequences of Stress and Work*, Chichester: John Wiley, 233–263.

EDWARDS, J. R. and BAGLIONI, A. J. (1993) 'The measurement of coping with stress: construct validity of the Ways of Coping Checklist and the Cybernetic Coping Scale', *Work and Stress* 7: 17–32.

EDWARDS, J. R. and COOPER, C. L. (1988) 'Research in stress, coping and health: theoretical and methodological issues', *Psychological Medicine* 18: 15–20.

EDWARDS, J. R., BAGLIONI, A. J. and COOPER, C. L. (1990) 'Stress, Type A, coping, and psychological and physical symptoms: a multi-sample test of alternative models', *Human Relations* 43: 919–956.

ENDLER, N. S. and PARKER, J. D. A. (1990) 'Multidimensional assessment of coping: a critical evaluation', *Journal of Personality and Social Psychology* 58: 844–854.

ERERA-WEATHERLY, P. I. (1996) 'Coping with stress: public welfare supervisors doing their best', *Human Relations* 49: 157–170.

EULBERG, J. R., WEEKLEY, J. A. and BHAGAT, R. S. (1988) 'Models of stress in organisational research: a metatheoretical perspective', *Human Relations* 41: 331–350.

FELDMAN, D. C. and BRETT, J. M. (1983) 'Coping with new jobs: a comparative study of new hires and job changers', *Academy of Management Journal* 26: 258–272.

FLEISHMAN, J. A. (1984) 'Personality characteristics and coping patterns', *Journal of Health and Social Behaviour* 25: 229–244.

FOLKMAN, S. (1982) 'An approach to the measurement of coping', *Journal of Occupational Behaviour* 3: 95–107.

FOLKMAN, S. (1984) 'Personal control and stress and coping processes: a theoretical analysis', *Journal of Personality and Social Psychology* 46: 839–852.

FOLKMAN, S. and LAZARUS, R. S. (1980) 'An analysis of coping in a middle aged community sample', *Journal of Health and Social Behaviour* 21: 219–239.

FOLKMAN, S., LAZARUS, R. S., DUNKEL-SCHETTER, C. A., DE LONGIS and GRUEN, R. J. (1986) 'Dynamics of a stressful encounter: cognitive appraisal, coping and encounter outcomes', *Journal of Personality and Social Psychology* 50: 992–1003.

GEORGE, J. M., BRIEF, A. P. and WEBSTER, J. (1991) 'Organisationally intended and unintended coping: the case of an incentive compensation plan', *Journal of Occupational Psychology* **64**: 193–205.

GREENGLASS, E. R. (1988) 'Type A behaviour and coping strategies in female and male supervisors', *Applied Psychology: An International Review* **37**: 271–288.

GREENGLASS, E. R. (1993) 'The contribution of social support to coping strategies', *Applied Psychology: An International Review* **42**: 323–340.

GREENGLASS, E. R., BURKE, R. J. and ONDRACK, M. (1990) 'A gender-role perspective of coping and burnout', *Applied Psychology: An International Review* **39**: 5–27.

GREENGLASS, E. R. and BURKE, R. J. (1991) 'The relationship between stress and coping among Type As', in P. L. Perrewé (ed.) *Handbook on job stress* (special issue), *Journal of Social Behaviour and Personality* **6**: 361–373.

HANDY, J. A. (1986) 'Considering organisations in organisational stress research', *Bulletin of the British Psychological Society* **39**: 205–210.

HARRIS, J. R. (1991) 'The utility of the transactional approach for occupational stress research', in P. L. Perrewé (ed.) *Handbook on job stress* (special issue), *Journal of Social Behaviour and Personality* **6**: 21–29.

HART, P. M., WEARING, A. L. and HEADEY, B. (1995) 'Police stress and well-being: integrating personality, coping and daily work experiences', *Journal of Occupational and Organisational Psychology* **68**: 133–156.

HAVLOVIC, S. J. and KEENAN, J. P. (1991) 'Coping with work stress: the influence of individual differences', in P. L. Perrewé (ed.) *Handbook on job stress* (special issue), *Journal of Social Behaviour and Personality* **6**: 199–212.

HOLROYD, K. A. and LAZARUS, R. S. (1982) 'Stress, coping and somatic adaptation', in L. Goldberger and S. Breznitz (eds) *Handbook of Stress: Theoretical and Clinical Aspects*, New York: Free Press, 21–35.

HOWARD, J. H., RECHNITZER, P. A. and CUNNINGHAM, D. A. (1975) 'Coping with job tension-effective and ineffective methods', *Public Personnel Management* **4**: 317–326.

KASL, S. V. (1983) 'Perusing the link between stressful life experiences and disease: a time for reappraisal', on C. L. Cooper (ed.) *Stress Research: Issues for the Eighties*, Chichester: John Wiley, 79–102.

KIEV, A. and KOHN, V. (1979) *Executive Stress*, New York: AMACOM.

KINICKI, A. J. and LATACK, J. C. (1990) 'Explication of the construct of coping with involuntary job loss', *Journal of Vocational Behaviour* **36**: 339–360.

KOESKE, G. F., KIRK, S. A. and KOESKE, R. D. (1993) 'Coping with job stress: which strategies work best?' *Journal of Occupational and Organisational Psychology* **66**: 319–335.

KUHLMANN, T. M. (1990) 'Coping with occupational stress among urban bus and train drivers', *Journal of Occupational Psychology* **63**: 89–96.

LATACK, J. C. (1986) 'Coping with job stress: measures and future directions for scale development', *Journal of Applied Psychology* **71**: 377–385.

LATACK, J. C. and HAVLOVIC, S. J. (1992) 'Coping with job stress: a conceptual evaluation framework for coping measures', *Journal of Organisational Behaviour* **13**: 479–508.

LAZARUS, R. S. (1991) 'Psychological stress in the workplace', in P. L. Perrewé (ed.) *Handbook on job stress* (special issue), *Journal of Social Behaviour and Personality* **6**: 1–13.

LAZARUS, R. S. and FOLKMAN, S. (1984) *Stress, Appraisal and Coping*, New York: Springer.

LAZARUS, R. S. and LAUNIER, R. (1978) 'Stress-related transactions between person and environment', in L. A. Pervin and M. Lewis (eds), *Perspectives in International Psychology*, New York: Plenum, 287–327.

LEANA, C. R. and FELDMAN, D. C. (1990) 'Individual responses to job loss: empirical findings from two field studies', *Human Relations* **43**: 1067–1083.

LEITER, M. P. (1990) 'The impact of family resources, control coping, and skill utilization on the development of burnout: a conceptualization study', *Human Relations* **43**: 1067–1083.

LEITER, M. P. (1991) 'Coping patterns as predictors of burnout: the function of control and escapist coping patterns', *Journal of Organizational Behaviour* **3**: 95–107.

LONG, B. C. (1993) 'Coping strategies of male managers: a prospective analysis of predictors of psychomatic symptoms and job satisfactions', *Journal of Vocational Behaviour* **42**: 184–199.

LONG, B. C., KAHN, S. E. and SCHUTZ, R. W. (1992) 'Causal model of stress and coping: women in management', *Journal of Counseling Psychology* **39**: 227–239.

McDONALD, L. M. and KORABIK, K. (1991) 'Sources of stress and ways of coping among male and female managers', in P. L. Perrewé (ed.) *Handbook on job stress* (special issue), *Journal of Social Behaviour and Personality* **6**: 163–185.

McGRATH, J. E. (1970) (ed.) *Social and Psychological Factors in Stress*, New York: Holt, Rinehart and Winston.

MOOS, R. H. and BILLINGS, A. G. (1982) 'Conceptualizing and measuring coping resources and processes', in L. Goldberg and S. Breznitz (eds) *Handbook of Stress: Theoretical and Clinical Aspects*, New York: Free Press.

MOOS, R. H., CRONKITE, R. C., BILLINGS, A. G. and FINNEY, J. W. (1987) *Health and Daily Living Form Manual*, revised version, Stanford CA: Social Ecology Laboratory.

MORLEY, N. L. and LUTHANS, F. (1984) 'An emic perspective and ethnoscience methods for organisational research', *Academy of Management Review* **9**: 27–36.

NELSON, D. L. and SUTTON, C. (1990) 'Chronic work stress and coping: a longitudinal study and suggested new directions', *Academy of Management Journal* **33**: 859–869.

NEWTON, T. J. (1989) 'Occupational stress and coping with stress: a critique', *Human Relations* **42**: 441–461.

NEWTON, T. J. (1995) *'Managing' Stress: Emotion and Power at Work*, London: Sage Publications.

NEWTON, T. J. and KEENAN, A. (1985) 'Coping with work-related stress', *Human Relations* **38**: 107–126.

NEWTON, T. J. and KEENAN, A. (1990) 'The moderating effect of the type A behaviour pattern and focus of control upon the relationship between change in job demands and change in psychological strain', *Human Relations* **43**: 1229–1255.

OAKLAND, S. and OSTELL, A. (1996) 'Measuring coping: a review and critique', *Human Relations* **49**: 133–155.

O'DRISCOLL, M. P. and COOPER, C. L. (1994) 'Coping with work-related stress: a critique of existing measures and proposal for an alternative methodology', *Journal of Occupational and Organisational Psychology* **67**: 343–354.

OSIPOW, S. H. and DAVIS, A. S. (1988) 'The relationship of coping resources to occupational stress and strain', *Journal of Vocational Behaviour* **32**: 1–15.

OSIPOW, S. H. and SPOKANE, A. R. (1984) 'Measuring occupational stress, strain, and coping', in S. Oskemp (ed.), *Applied Social Psychology Annual* **5**: 67–86.

PARASURAMAN, S. and CLEEK, M. A. (1984) 'Coping behaviour and managers' affective reactions to role stressors', *Journal of Vocational Behaviour* **24**: 179–183.

PARKER, D. F. and DE COTIIS, T. A. (1983) 'Organisational determinants of job stress', *Organisational Behaviour and Human Performance* **32**: 160–177.

PAYNE, R. A., JICK, T. D. and BURKE, R. J. (1982) 'Whither stress research? an agenda for the 1980s', *Journal of Occupational Behaviour* **3**: 131–145.

PEARLIN, L. I. and SCHOOLER, C. (1978) 'The structure of coping', *Journal of Health and Social Behaviour* **19**: 2–21.

PFEFFER, J. (1990) 'Management as symbolic action: the creation and maintenance of organisational paradigms', in L. L. Cummings and B. M. Staw (eds) *Information and Cognition in Organisations*, Greenwich: JAI Press, 1–52.

PUFFER, S. M. and BRAKEFIELD, L. T. (1989) 'The role of task complexity as a moderator of the stress and coping process', *Human Relations* 3: 199–217.

ROTH, S. and COHEN, L. J. (1986) 'Approach, avoidance and coping with stress', *American Psychologist* 41: 813–819.

SCHULER, R. S. (1984) 'Organisational stress and coping: a model and overview', in A. S. Sethi and R. S. Schuler (ed.), *Handbook of Organisational Stress Coping Strategies*, Cambridge, MA: Ballinger, 35–67.

SCHULER, R. S. (1985) 'Integrative transactional process model of coping with stress in organisations', in T. A. Beehr and R. S. Bhagat (eds) *Human Stress and Cognition in Organisations: an Integrative Perspective*, New York: John Wiley, 347–374.

SCHWARTZ, L. E. and STONE, A. A. (1993) 'Coping with daily work problems. Contributions of problem content, appraisals, and person factors', *Work Stress* 7: 47–62.

SHINN, M., ROSARIO, M., MORCH, H. and CHESTNUT, D. E. (1984) 'Coping with job stress and burnout in the human services', *Journal of Personality and Social Psychology* 46: 864–876.

STAW, B. M. (1990) 'Rationality and justification in organisational life', in L. L. Cummings and B. M. Staw, (eds) *Information and Cognition in Organisations*, Greenwich: JAI Press, 53–88.

STONE, I. A. and NEALE, J. M. (1984) 'New measure of daily coping: development and preliminary results', *Journal of Personality and Social Psychology* 46: 892–906.

STROMAN, C. A. and SELTZER, R. (1991) 'Racial differences in coping with job stress', in P. L. Perrewé (ed.) *Handbook on Job Stress* (special issue), *Journal of Social Behaviour and Personality* 6: 309–318.

STRUTTON, D. and LUMPKIN, J. (1992) 'Relationship between optimism and coping strategies in the work environment', *Psychological Reports* 71: 1179–1186.

THORNTON, P. I. (1992) 'The relation of coping, appraisal, and burnout in mental health workers', *The Journal of Psychology* 126: 261–271.

TRENBERTH, L. D., DEWE, P. J. and WALKEY, F. H. (1996) 'A factor replication approach to the measurement of coping', *Stress Medicine* 12: 71–79.

VAN MAANEN, J. (1979) 'Reclaiming qualitative methods for organisational research: a preface, *Administrative Science Quarterly* 24: 520–526.

VINGERHOETS, A. I. J. M. and MARCELISSEN, F. H. G. (1988) 'Stress Research: its present status and issues for future development', *Social Science and Medicine* 26: 279–291.

VITALIANO, P. P., RUSSO, L, CARR, J. E., MAIURO, R. D. and BECKER, J. (1985) 'The Ways of Coping Checklist: revision and psychometric properties', *Multivariate Behavioural Research* 20: 3–26.

WALSH, S. and JACKSON, P. R. (1995) 'Partner support and gender: contexts for coping with job loss', *Journal of Occupational and Organisational Psychology* 68: 253–268.

WOLFGANG, A. P. (1991) 'Job stress, coping and dissatisfaction in the health professions: a comparison of nurses and pharmacists', in P. L. Perrewé (ed.) *Handbook on job stress* (special issue), *Journal of Social Behaviour and Personality* 6: 213–226.

WONG, P. T. P. and REKER, G. T. (1984) 'Coping behaviours of successful agers', paper presented at the annual meeting of the Canadian Psychological Association, Ottawa.

CHAPTER TWO

Empirical versus theoretical approaches to the measurement of coping: a comparison using the ways of coping questionnaire and the cybernetic coping scale

JEFFREY R. EDWARDS AND A. J. BAGLIONI, JR

The purpose of this chapter is to compare empirical and theoretical approaches to the measurement of coping. To illustrate these approaches, we use the Ways of Coping Questionaire (WCQ; Folkman and Lazarus, 1988) and the Cybernetic Coping Scale (CCS; Edwards and Baglioni, 1993), which represent empirical and theoretical approaches to coping measurement, respectively. After reviewing the general context of research on coping, we describe the development of these measures and review available evidence regarding their psychometric properties. We then extend the confirmatory factor analyses reported by Edwards and Baglioni (1993) to examine the reliability, unidimensionality and other core aspects of the construct validity of these measures, based on a sample larger than that used by Edwards and Baglioni (1993). We conclude by comparing the performance of the two measures and by providing recommendations regarding the development of coping measures, with an emphasis on the relative merits of empirical versus theoretical approaches.

Research into coping with stress at work has proliferated in recent years (Coyne and Downey, 1991; Dewe, Cox and Ferguson, 1993; Edwards, 1988). This research has generated the need for valid and reliable measures of coping (Cohen, 1987; Endler and Parker, 1990; Latack and Havlovic, 1992; O'Driscoll and Cooper, 1994; Stone, Greenberg, Kennedy-Moore and Newman, 1991). Numerous coping measures have been proposed (e.g. Amirkhan, 1990;

Carver, Scheier and Weintraub, 1989; Dewe and Guest, 1990; Endler and Parker, 1990; Folkman and Lazarus, 1988; Latack, 1986; McCrae, 1984; Parasuraman and Cleek, 1984), representing a wide variety of definitions and proposed structures for coping.

Most of the available coping measures have been developed and validated using one of two general approaches (Carver *et al.*, 1989; Dewe and Guest, 1990; Parker and Endler, 1992). One approach involves assembling items from various sources (e.g. open-ended descriptions of coping episodes, existing coping measures) to represent a range of coping strategies and then empirically determining the structure underlying the items, usually based on exploratory factor analyses (e.g. Amirkhan, 1990; Dewe and Guest, 1990; Endler and Parker, 1990; Folkman and Lazarus, 1988). Although items constituting such measures are often selected on conceptual grounds, this approach is essentially empirically driven, given that empirical evidence is used to determine the number of dimensions underlying the items, the assignment of items to dimensions, and the meaning of the dimensions themselves, based on items with high loadings.

An alternative approach consists of defining a set of coping dimensions a priori, generating or adapting items to represent each dimension, and determining empirically the extent to which each item is associated with its intended dimension (e.g. Carver *et al.*, 1989; Edwards and Baglioni, 1993; Latack, 1986; Stone and Neale, 1984). This approach is primarily conceptually driven, in that theory is used to generate and define coping dimensions and to assign items to each dimension prior to analysis. Although the degree to which items represent their intended dimensions has often been evaluated using exploratory factor analysis (Carver *et al.*, 1989; Latack, 1986; Stone and Neale, 1984), some studies have employed confirmatory factor analysis (Edwards and Baglioni, 1993; Parker, Endler and Bagby, 1993), thereby providing a more rigorous and precise evaluation of the presumed factor structure.

The relative merits of these two approaches to the measurement of coping are a source of contention in the coping literature. Dewe and Guest (1990) argue in favour of the empirical approach, based on the limited empirical support for the validity of most coping frameworks and the lack of consensus among coping researchers as to which framework is superior. In contrast, Carver *et al.* (1989) and Parker and Endler (1992) advocate the theoretical approach, contending that the empirical approach results in scales constructed post hoc that are only loosely linked to theoretical coping dimensions.

The choice between empirical versus theoretical approaches to measurement depends to some extent on the stage of development of the research domain. At early stages, relevant constructs may not be clearly delineated, and an empirical approach may be appropriate. Imposing preconceived conceptual frameworks at this stage risks omitting relevant dimensions, creating unwarranted distinctions, and generating evidence that is divorced from phenomena as they naturally occur. At later stages, alternative conceptualizations of relevant constructs have often been advanced, and a theoretical

approach is needed to evaluate proposed operationalizations of each conceptualization and to test the relative merits of competing conceptualizations.

We believe that the conceptualization of coping is at a relatively mature
stage, thereby calling for a conceptual approach to the measurement of
coping. Our position is supported by the substantial accumulated evidence
documenting how people cope with acute and ongoing stressors (Coyne and
Downey, 1991; Dewe, Cox and Ferguson, 1993), and by the proliferation of
conceptual frameworks attempting to organize this evidence (Edwards, 1988;
Latack and Havlovic, 1992; Lazarus and Folkman, 1984; Silver and Wortman, 1980). Given this volume of research, we posit that the measurement of
coping will be advanced not by empirically generating additional coping
frameworks, but instead by comparing and evaluating frameworks that have
already been proposed (Carver et al., 1989).

In addition to these conceptual considerations, there are several methodological arguments in favour of a theoretical approach to the measurement of
coping. First, generating coping items without reference to a clearly defined
set of a priori coping dimensions provides little assurance that the resulting
items will represent a stable and interpretable factor structure or generate
measures with adequate reliability. To achieve this, a domain-sampling
procedure should be used, in which operational definitions of relevant coping
dimensions are constructed and multiple items are generated to tap each
dimension (Nunnally, 1978). Items generated for each coping dimension
should represent alternative indicators of the dimension, as opposed to
distinct facets of the dimension. Otherwise, the reliabilities of measures based
on the coping dimensions are likely to suffer (Stone and Neale, 1984), and the
interpretation of the measures will be ambiguous (Hattie, 1985; Wolins, 1982).

Second, empirically-driven procedures for developing measures capitalize
on sampling variability and may therefore yield results that do not generalize
beyond the data at hand (Campbell, 1976). This is illustrated by successive
exploratory factor analyses of responses to the WCQ (Folkman and Lazarus,
1988), which have yielded substantially different factor structures across
samples (Parker et al., 1993; Tennen and Herzberger, 1985). These differences
make it difficult to justify a single, general scoring procedure for the WCQ,
which in turn inhibits the accumulation of evidence regarding a replicable set
of coping dimensions.

Finally, studies that adopt an empirical approach to coping measurement
typically rely on exploratory factor analysis, which has several drawbacks in
terms of developing measures and their validation. For example, the criteria
used in exploratory factor analysis to determine the number of dimensions
(e.g. scree test, eigenvalues-greater-than-one rule) and to assign items to
dimensions (e.g. item loadings that exceed some fixed criterion of acceptability such as .30) are arbitrary and subjective (Cudeck and O'Dell, 1994;
Lambert, Wildt and Durand, 1990). Moreover, exploratory factor analysis
provides little statistical evidence regarding the correspondence between the
data and the estimated factor structure, as reflected in the overall model fit

and the fit of various components of the model (e.g. the degree to which covariation among items assigned to a scale is adequately explained by their common factor). These shortcomings are overcome by methods of analysis consistent with the theoretical approach to coping measurement, most notably confirmatory factor analysis (Bollen, 1989b; Hunter and Gerbing, 1982; Long, 1983).

Origins and development of measures

The Ways of Coping Questionnaire

The WCQ is based on Lazarus' transactional model of stress and coping (Lazarus, 1966; Lazarus and Folkman, 1984; Lazarus and Launier, 1978). Briefly, this model views stress as a relationship between the person and the environment that taxes or exceeds the person's resources and endangers his or her well-being. Coping is defined as the 'constantly changing cognitive and behavioural efforts to manage specific external and/or internal demands that are appraised as taxing or exceeding the resources of the person' (Lazarus and Folkman, 1984: 141). Two basic categories of coping include efforts to alter the troubled person-environment relationship (i.e. problem-focused coping) and efforts to regulate emotional distress (i.e. emotion-focused coping). Problem-focused coping encompasses numerous specific coping strategies, such as defining the problem, generating, evaluating, and selecting potential solutions, and attempting to cognitively reappraise the situation by shifting level of aspiration, reducing ego involvement, finding alternative channels of gratification, or developing new standards of behaviour. Emotion-focused coping includes strategies such as minimization, positive comparisons, seeking positive value from negative events, selective attention, distancing, avoidance, exercise, meditation, use of alcohol, venting anger, and seeking emotional support.

The first version of the WCQ, labelled the Ways of Coping Checklist (WCC; Folkman and Lazarus, 1980), consisted of sixty-eight binary items drawn from existing measures (Sidle, Moos, Adams, and Cady, 1969; Weisman and Worden, 1976) and derived from the transactional model (Lazarus and Folkman, 1984). These items were classified into two broad scales representing problem- and emotion-focused coping, based on judges' evaluations of item content (Folkman and Lazarus, 1980). However, subsequent factor analyses of these data revealed that these scales combined and collapsed multiple coping methods (Aldwin, Folkman, Schaefer, Coyne and Lazarus, 1980). Based on this, the two-factor conceptual structure was abandoned in favour of empirically-derived structures containing from three to eight factors (Aldwin et al., 1980; Parkes, 1984; Vingerhoets and Flohr, 1984; Vitaliano, Russo, Carr, Maiuro and Becker, 1985). Reliability estimates for scales corresponding to these factors have averaged .77 and ranged from .56 to .91.

The current sixty-seven-item WCQ was developed by Folkman and Lazarus (1985), who deleted or reworded WCC items that were redundant or unclear, added items suggested by respondents, and changed the binary response format to a four-point Likert scale ranging from 'does not apply and/or not used' to 'used a great deal.' Factor analyses of the WCQ (Aldwin and Revenson, 1987; Atkinson and Violato, 1993; Folkman and Lazarus, 1985; Folkman, Lazarus, Dunkel-Schetter, DeLongis and Gruen, 1986; Mishel and Sorenson, 1993; Parker et al., 1993; Smyth and Williams, 1991) have yielded four to eight factors. Scales based on these factors have exhibited reliabilities comparable to those for the WCC, averaging .73 and ranging from .56 to .85. The factor structure identified by Folkman et al. (1986) is currently advocated as the scoring protocol for the WCQ (Folkman and Lazarus, 1988).

A closer examination of successive factor analyses of the WCQ reveals that its factor structure is rather unstable. This is partly evidenced by the varying number of factors extracted. However, solutions with the same number of factors have often placed the same items on conceptually different factors or omitted some items entirely (e.g. Aldwin and Revenson, 1987; Folkman and Lazarus, 1985; Folkman et al., 1986). This is partly attributable to the content of the WCQ items, some of which confound different coping methods (e.g. 'Didn't let it get to me; refused to think too much about it') or are inherently ambiguous, describing a coping behaviour without specifying its focus or intent (e.g. 'I got professional help', 'I changed something about myself', 'I prayed', 'I did something I didn't think would work, but at least I was doing something'). This instability also reflects the process of conducting successive exploratory factor analyses on separate samples, which may yield different results simply due to sampling variability. Such analyses provide weak evidence regarding the generalizability of a given factor structure across samples.

The Cybernetic Coping Scale

The CCS was derived from Edwards' cybernetic theory of stress, coping and well-being (Edwards, 1988, 1992; Edwards and Cooper, 1988). This theory views stress as a discrepancy between the individual's perceived state and desired state that is considered important by the individual. The impact of this discrepancy on well-being is moderated by duration, or the amount of time the person spends thinking about the discrepancy. Coping is conceptualized as attempts to reduce or eliminate the negative effects of stress on well-being. Five forms of coping are identified, including attempts to bring the situation into conjunction with desires, adjust desires to meet the situation (i.e. accommodation), reduce the importance associated with the discrepancy (i.e. devaluation), direct attention away from the situation (i.e. avoidance), and improve well-being directly (i.e. symptom reduction). Hence, stress and coping are viewed as critical components of a negative feedback loop, in which stress damages well-being and activates coping; coping may improve well-being directly and indirectly, through the perceived and desired states comprising the discrepancy, the level of importance associated with the discrepancy, and the amount of attention directed towards the discrepancy.

The first version of the CCS was based on items drawn from existing coping measures (e.g. Aldwin *et al.*, 1980; Billings and Moos, 1984; Latack, 1986; Pearlin and Schooler, 1978; Sidle *et al.*, 1969), which were substantially revised and supplemented in accordance with the five dimensions indicated by the cybernetic theory (i.e. changing the situation, accommodation, devaluation, avoidance, symptom reduction). The eight items that were considered most representative of each dimension were combined and administered to samples of MBA students, executives and psychiatric in patients. Three confirmatory factor analyses were performed, based on a measurement model specified by assigning each item a priori to its intended factor and fixing loadings on all other factors to zero. Results were similar across samples, but indicated that certain factors were poorly represented, particularly accommodation and symptom reduction. Based on these results, additional items were written, and all items were evaluated for content and clarity by five judges, who were provided with definitions of the five coping dimensions and asked to rate the degree to which each item represented its intended dimension. The eight items that best represented each dimension, based on both conceptual and statistical criteria, were retained to form the second version of the CCS, which was used in this study.

The present investigation

This investigation reports confirmatory factor analyses of the WCQ and CCS. These analyses evaluate measurement models for the empirically derived eight-factor structure currently recommended for scoring the WCQ (Folkman *et al.*, 1986; Folkman and Lazarus, 1988) and for the theoretically derived five-factor structure underlying the CCS. By employing confirmatory factor analysis, this study avoids shortcomings of previous studies of the WCQ based on exploratory factor analysis (Aldwin and Revenson, 1987; Atkinson and Violato, 1993; Folkman and Lazarus, 1985; Folkman *et al.*, 1986; Mishel and Sorenson, 1993; Smyth and Williams, 1991). Moreover, this study extends previous confirmatory factor analyses of the WCQ (Parker *et al.*, 1993) by providing a more detailed assessment of model fit, and by reporting fit indices that avoid problems with previous indices, such as susceptibility to sample size (Gerbing and Anderson, 1992). Finally, this study augments analyses reported by Edwards and Baglioni (1993) by drawing from a larger sample and using additional criteria for evaluating model fit.

Methodology

Surveys were distributed to 982 MBA students at a large business school in the eastern US. A total of 230 surveys were returned, with 181 providing usable responses on all measures. The response rate (23 per cent) was probably limited due to the competing demands facing the students (e.g. classes, job search) and to the lack of an incentive (e.g. payment, course credit) for completing the survey. The final sample was predominantly male (67 per cent), averaged 27 years of age (range = 22 to 34 years), and averaged 4.27 years of work experience (range = 0 to 13 years). Career intentions of respondents included consulting (21 per cent), finance (21 per cent), marketing (16 per cent), production (6 per cent), and corporate planning (6 per cent), with the remaining 30 per cent distributed among real estate development, human resource management, investments, sales management, systems, law, and other specific job types indicated by the respondents. Respondents did not differ from non-respondents in terms of age, gender, or years of work experience.

Questionnaires included the sixty-seven-item WCQ presented in Lazarus and Folkman (1984: 328–333) and the forty-item second version of the CCS. Following Lazarus and Folkman (1984), a four-point response scale was used for the WCQ ('Not used' to 'Used a great deal'), whereas a seven-point response scale was used for the CCS ('Did not use at all' to 'Used very much'). To reduce order effects, half of the surveys presented the WCQ first, and the other half presented the CCS first. Questionnaires were distributed near the conclusion of the recruiting season, when most respondents were near the end of their job search. For both measures, respondents were asked to indicate how they coped with problems in locating what they viewed as the ideal job. Following Folkman *et al.* (1986) and Folkman and Lazarus (1988), eight WCQ scales were created, representing confrontative coping (6 items), distancing (6 items), self-control (7 items), seeking social support (6 items), accepting responsibility (4 items), escape-avoidance (8 items), 'planful' problem-solving (6 items), and positive reappraisal (7 items). Five eight-item CCS scales were constructed, representing changing the situation, accommodation, devaluation, avoidance, and symptom reduction.

Data analysis

To facilitate comparisons with earlier research, initial analyses consisted of calculating reliability estimates (Cronbach's alpha) and intercorrelations among the coping scales. Next, confirmatory factor analyses were conducted, examining item loadings, residuals, modification indices, correlations among the latent factors, and overall model fit. Analyses were performed using LISREL 8 (Joreskog and Sorbom, 1993). Criteria for assessing overall model fit included the chi-square test statistic, the goodness-of-fit index (GFI) and adjusted goodness-of-fit index (AGFI) (Joreskog and Sorbom, 1993), the Tucker-Lewis Index (TLI; Tucker and Lewis, 1973), the Incremental Fit Index (Bollen, 1989a), and the Comparative Fit Index (CFI; Bentler, 1990). Despite the widespread use of the chi-square, GFI, and AGFI for assessing overall model fit, these indices have important drawbacks, most notably their dependence on sample size (Gerbing and Anderson, 1992). The TLI, IFI and CFI are independent of sample size, although the IFI and CFI exhibit less variance than the TLI and are therefore currently preferred as indices of fit (Gerbing and Anderson, 1992).

In addition to overall model fit, several specific aspects of the measurement models were examined. First, the magnitude and significance of each item loading on its assigned factor was assessed. The magnitude of this relationship provides a direct indication of the construct validity of the item in question (Bollen, 1989b; Schwab, 1980). Second, the product rule for internal consistency was examined by determining whether the correlation between any two items assigned to the same factor was equal to the product of their respective loadings.[1] This rule reflects the principle that, if two items represent the same construct, their relationship should be completely determined by their associations with that construct (Danes and Mann, 1984; Gerbing and Anderson, 1988). If this rule is satisfied, then the residual for the correlation between items assigned to the same factor should not differ from zero. Third, the product rule for external consistency was applied by determining whether the correlation between two items assigned to different factors was equal to the product of the item loadings times the correlation between the two factors involved. This rule embodies the notion that the relationship between two items representing different constructs should be completely determined by the relationship between those constructs and the association of each item with its assigned construct (Danes and Mann, 1984; Gerbing and Anderson, 1988). If this rule is satisfied, then the residual for the correlation between items assigned to different factors should not differ from zero. Fourth, modification indices for item loadings on factors other than the assigned factor were examined. A significant modification index signifies that model fit would be improved if an item were allowed to load on additional factors (Sorbom, 1975). Such evidence indicates that the item in question represents the influence of multiple factors and is therefore a poor indicator of its intended factor. Finally, modification indices for correlations between item measurement errors were examined. If a measurement model is correctly specified (i.e., the hypothesized factors and their intercorrelations account for the covariation among the items), then measurement errors should be random (i.e., uncorrelated), indicating that no omitted factors induce systematic covariation between items (Gerbing and Anderson, 1984; Lord and Novick, 1968). This would be evidenced by non-significant modification indices for correlations between measurement errors.

Findings

Table 2.1 presents reliability estimates and intercorrelations of the WCQ and CCS scales. Of the eight WCQ scales, only one (escape-avoidance) exhibited a

Table 2.1. Descriptive statistics, correlations and reliability estimates for the Cybernetic Coping Scale and Ways of Coping Questionnaire.

			CCS					WCQ							
	M	sd	1	2	3	4	5	1	2	3	4	5	6	7	8
CCS															
1 Changing the situation	36.46	10.11	(.90)	.09	-.16*	-.18*	.26**	.64**	-.31**	-.02	.28**	-.08	-.06	.76**	.43**
2 Accommodation	30.43	8.46	.04	(.79)	.65**	.48**	.21*	.10	.57**	.40**	.15	.43**	.31**	-.01	.27**
3 Devaluation	26.79	12.32	-.15*	.62**	(.94)	.55**	.14	-.10	.62**	.31**	-.05	.37**	.30**	-.21*	.17*
4 Avoidance	23.46	10.51	-.17*	.44**	.58**	(.93)	.33**	-.11	.75**	.44**	-.03	.51**	.54	-.36**	.10
5 Symptom reduction	31.30	9.67	.25**	.19*	.15*	.30**	(.86)	.65**	.21*	.22*	.53**	.30**	.39**	.11	.28**
WCQ															
1 Confrontive coping	5.08	2.78	.38**	.04	.03	.02	.40**	(.49)	-.08	.17	.74**	.20	.29**	.71**	.55**
2 Distancing	7.01	3.42	-.24**	.46**	.52**	.57**	.15*	.09	(.66)	.78**	.02	.67**	.60**	-.25*	.20*
3 Self-control	9.85	3.63	.12	.29**	.22**	.29**	.24**	.27**	.44**	(.60)	.14	.59**	.58**	.31**	.42**
4 Seeking social support	8.34	3.17	.32**	.09	-.07	-.05	.37**	.38**	-.01	.28**	(.62)	.42**	.43**	.36**	.32**
5 Accepting responsibility	3.54	2.55	-.06	.26**	.27**	.39**	.24**	.24**	.40**	.33**	.30**	(.60)	.82**	-.01	.22*
6 Escape-avoidance	6.45	4.25	-.04	.19*	.26**	.44**	.33**	.35**	.36**	.34**	.29**	.57**	(.73)	-.17	.11
7 Planful problem-solving	9.66	3.33	.56**	.01	-.11	-.23**	.07	.32**	-.12	.36**	.02	-.02	-.04	(.62)	.59**
8 Positive reappraisal	7.32	3.98	.34**	.23**	.16*	.12	.30**	.40**	.20**	.38**	.31**	.18*	.20**	.41**	(.70)

Note

N = 181.

Reliabilities (Cronbach's alpha) are reported along the diagonal (numbers in brackets). Values below the diagonal are simple scale correlations; values above the diagonal are correlations from a 13-factor multi-item measurement model estimated using LISREL.

* p < .05. ** p < .01.

reliability greater than .70, six ranged between .60 and .70, and one (confrontative coping) was less than .50. In contrast, all five CCS scales exhibited reliabilities greater than .70, ranging from .79 to .94. For the WCQ, the highest interscale correlation was between accepting responsibility and escape-avoidance which, based on interitem correlations, indicated a shared emphasis on putting the situation behind oneself. For the CCS, the highest intercorrelation was between the accommodation and devaluation scales, which was attributable to two accommodation items that conveyed elements of minimizing the situation (i.e. 'I told myself the situation was okay after all', 'I tried to convince myself that the way things were was, in fact, acceptable'). The devaluation and avoidance scales were also highly corre-lated, suggesting that avoidance may be accompanied by deciding that the problem is unimportant. The WCQ distancing scale was highly correlated with the CCS accommodation, devaluation and avoidance scales. Further inspection revealed that these correlations were attributable to specific items in the distancing scale that suggested accommodation ('Looked for the silver lining, so to speak; tried to look on the bright side of things'), devaluation ('Made light of the situation; refused to get too serious about it'), and avoidance ('Tried to forget the whole thing'). The WCQ escape/avoidance scale was highly correlated with the CCS avoidance scale, which was expected due to their similar emphases. Likewise, the WCQ 'planful' problem-solving scale was highly correlated with the CCS changing the situation scale, due to a shared emphasis on directing efforts towards the situation rather than the person.

Standardized item loadings and fit indices are reported in Tables 2.2 and 2.3, and correlations among the latent factors obtained from a full thirteen-factor measurement model are reported in Table 2.1. Both models yielded a significant chi-square, indicating that neither model fits the data. GFI and AGFI values were .68 and .64 for the WCQ and were .71 and .67 for the CCS. Although these values suggest that the CCS model fits slightly better than the WCQ model, none of the values approached the .90 criterion suggested by Benter and Bonnet (1980). However, as noted previously, the chi-square, GFI, and AGFI have known drawbacks (e.g. sensitivity to sample size), and indices less prone to these problems, such as the TLI, IFI and CFI, should be examined. Values for these indices ranged from .52 to .57 for the WCQ but ranged from .82 to .83 for the CCS. This indicates that the CCS model yielded a substantially better fit than the WCQ model, although the fit of the CCS model remained somewhat below the .90 criterion.

Correlations among the thirteen latent factors (see Table 2.1) paralleled those among the CCS and WCQ scales. However, the absolute magnitudes of the factor correlations were generally larger than those for the scales, due to the fact that the former are corrected for measurement error. Factor correlations for the CCS ranged from −.18 to .65, whereas factor correlations for the WCQ ranged from −.25 to .82, with four correlations exceeding .70. Although all correlations for both measures were significantly less than 1.0 ($p < .05$), thereby providing evidence for discriminant validity (Singh, 1991), the high correlations for the WCQ suggest some redundancy among its factors, particularly the distancing and self-controlling factors and the accepting responsibility and escape-avoidance factors.

As shown in Table 2.2, two of the eight WCQ factors contained items with non-significant loadings ($p > .05$), indicating that these items did not adequately represent the intended underlying factor. In contrast, all CCS items loaded significantly on the intended factor ($p < .05$; see Table 2.3). Tables 2.2 and 2.3 also show considerable variation in item loadings for the WCQ factors, whereas all but the accommodation factor for the CCS exhibited fairly consistent loadings. Nonetheless, tests for tau equivalence (Joreskog and Sorbom, 1988) yielded similar results for the WCQ and the CCS, in that within-factor item loadings did not significantly differ for four of the eight WCQ factors (distancing, self-controlling, accepting responsibility, escape-avoidance) and for two of the five CCS factors (avoidance, symptom reduction) factors ($p > .05$). Further inspection revealed that standard errors for item loadings were notably higher for the WCQ than for the CCS items, making it more difficult to demonstrate tau equivalence for the CCS.

Product rules for internal and external consistency were tested by examining standardized residuals. All eight WCQ factors yielded significant within-factor residuals ($p < .05$), with proportions ranging from 1/15 for planful problem-solving to 7/21 for self-controlling.[2] All eight factors also exhibited significant between-factor residuals, ranging from 47/264 for planful problem-solving to 69/264 for confrontative coping. As expected, the largest residuals were found for items sharing similar content that were assigned to different factors, such as items 28 ('I let my feelings out somehow') and 45 ('Talked to someone about how I was feeling') and items 40 ('Avoided being with people in general') and 43 ('Kept others from knowing how bad things were'). Like the WCQ, all five CCS scales exhibited significant within-factor residuals, ranging from 6/28 for changing the situation to 19/28 for accommodation. The residuals for the accommodation factor corresponded to several item pairs sharing specific content that was not explained by the common underlying factor (e.g. 'I tried to accept the situation as it was' and 'I tried to just accept things as they were'). All five CCS factors also exhibited significant between-factor residuals, ranging from 39/256 for changing the situation to 75/256 for accommodation. Again, these residuals represented items with similar content assigned to different factors, such as items 6 ('I told myself the situation was okay after all') and 20 ('I told myself the problem wasn't so serious after all'). In total, of the 1225 residuals tested for the WCQ, 276 were significant (22.5 per cent), whereas 184 of the 780 residuals tested for the CCS were significant (23.6 per cent).

Overall, forty of the fifty WCQ items yielded significant modification indices ($p < .05$) for loadings on at least one other factor, twenty-six items yielded significant modification indices on three or more factors, and six

Table 2.2. Confirmatory factor analysis of the WCQ.

Item		Item loading
1	*Confrontive coping*	
6	I did something which I didn't think would work, but at least I was doing something. [2,3,5,6,7]	.06
7	Tried to get the person(s) responsible to change his or her mind.	.29**
17	I expressed anger to the person(s) who caused the problem.	.16
28	I let my feelings out somehow. [4,5,6,7]	.50**
34	Took a big chance or did something very risky. [4]	.37**
46	Stood my ground and fought for what I wanted. [2,4,5,6,7]	.68**
2	*Distancing*	
12	Went along with fate; sometimes I just have bad luck. [4,5,6,8]	.46**
13	Went on as if nothing had happened.	.68**
15	Looked for the silver lining, so to speak; tried to look on the bright side of things. [1,3,4,7,8]	.34**
21	Tried to forget the whole thing. [5,6]	.54**
41	Didn't let it get to me; refused to think too much about it. [1,4,5,6]	.46**
44	Made light of the situation; refused to get too serious about it. [1,4,5,6]	.42**
42	I asked a relative or friend I respected for advice. [1]	.66**
45	Talked to someone about how I was feeling. [2,3,5,6,8]	.65**
5	*Accepting responsibility*	
9	Criticized or lectured myself. [4,6,8]	.51**
25	I apologized or did something to make up.	.42**
29	Realized I brought the problem on myself. [6]	.53**
51	I made a promise to myself that things would be different next time.	.65**
6	*Escape-avoidance*	
11	Hoped a miracle would happen. [1,4,7]	.57**
16	Slept more than usual. [8]	.32**
33	Tried to make myself feel better by eating, drinking, smoking, using drugs or medication, and so forth.	.47**
40	Avoided being with people in general. [7]	.51**
47	Took it out on other people. [2,3]	.51**
50	Refused to believe that it had happened. [2,3,5,8]	.40**
58	Wished that the situation would go away or somehow be over with. [1,7,8]	.65**
59	Had fantasies or wishes about how things might turn out.	.60**

3 Self-controlling

Item		Loading
10	Tried not to burn my bridges, but leave things open somewhat.[1,4,7]	.28**
14	I tried to keep my feelings to myself.[1,4,7,8]	.61**
35	I tried not to act too hastily or follow my first hunch.[1,7,8]	.28**
43	Kept others from knowing how bad things were.[1,4,7,8]	.62**
54	I tried to keep my feelings from interfering with other things too much.[2,7]	.47**
62	I went over in my mind what I would say or do.[1,2,4,7,8]	.34**
63	I thought about how a person I admire would handle this situation and used that as a model.[1,2,4,7,8]	.23**

4 Seeking social support

Item		Loading
8	Talked to someone to find out more about the situation.[1,7,8]	.37**
18	Accepted sympathy and understanding from someone.[2,5,6,7]	.61**
22	I got professional help.[5]	.13
31	Talked to someone who could do something concrete about the problem.[1,7,8]	.36**

7 Planful problem-solving

Item		Loading
1	Just concentrated on what I had to do next: the next step.	.32**
26	I made a plan of action and followed it.[5,6]	.56**
39	Changed something so things would turn out all right.[1,5,8]	.46**
48	Drew on my past experiences; I was in a similar situation before.	.32**
49	I knew what had to be done, so I doubled my effort to make things work.	.71**
52	Came up with a couple of different solutions to the problem.[5,6]	.47**

8 Positive reappraisal

Item		Loading
20	I was inspired to do something creative.[1,7]	.36**
23	Changed or grew as a person in a good way.[2,3,5,6]	.74**
30	I came out of the experience better than when I went in.[1]	.79**
36	Found new faith.[2,5,6,7]	.41**
38	Rediscovered what is important in life.[1,5,6]	.37**
56	I changed something about myself.[1,4,6]	.54**
60	I prayed.[2,7]	.29**

$\chi^2 = 2169.49^{**}$ df = 1147 GFI = .68 AGFI = .64 TLI = .52 CFI = .55 IFI = .57

Note

N = 181.

Table entries are standardized factor loadings. For each time, numerical superscripts indicate the factor(s) for which that item yielded a significant modification index ($p < .05$). GFI = Goodness of Fit Index; AGFI = Adjusted Goodness of Fit Index; TLI = Tucker-Lewis Index; CFI = Comparative Fit Index; IFI = Incremental Fit Index.

* $p < .05$. ** $p < .01$.

Table 2.3. Confirmatory factor analysis of the Cybernetic Coping Scale.

	Item loading		Item loading
Changing the situation		20 I told myself the problem wasn't so serious after all.	.88**
1 I tried to change something about the situation so things would turn out.	.66**	24 I told myself the problem wasn't worth worrying about.	.82**
10 I tried to bring about what I thought should happen.[2,4]	.67**	28 I told myself the problem wasn't such a big deal after all.	.91**
14 I made a plan of action to change the situation and followed it.	.54**	32 I tried to convince myself that there were other things in life that were more important.[2,4,5]	.63**
18 I tried to change the situation to get what I wanted.	.79**	36 I told myself the problem was not very important in the grand scheme of things.	.78**
22 I tried to change the things about the situation that were bothering me.[4,5]	.72**	*Avoidance*	
26 I focused my efforts on changing the situation.	.81**	4 I tried to keep myself from thinking about the problem.	.72**
35 I worked on changing the situation to get what I wanted.	.80**	8 I tried to turn my attention away from the problem.	.76**
39 I tried to fix what was wrong with the situation.	.84**	12 I tried to just forget the whole thing.[1,5]	.82**
Accommodation		16 I tried to think about other things.[5]	.74**
6 I told myself the situation was okay after all.[3,5]	.47**	25 I refused to think about the problem.	.77**
15 I tried to adapt to the situation.[1,3,5]	.35**	29 I tried to keep my mind off the problem.	.81**
17 I tried to accept the situation as it was.[3,5]	.48**	33 I tried to simply ignore the problem.[2,3,5]	.81**
		37 I tried to avoid thinking about the problem.[1,2]	.86**

Item	Factor loading
19 I tried to just accept things as they were.[1,3,5]	.46**
23 I made an effort to change my expectations.	.66**
27 I tried to convince myself that the way things were was, in fact, acceptable.[3]	.60**
31 I tried to adjust my expectations to meet the situation.[1,3,4]	.76**
40 I tried to convince myself that the problem was not very important after all.[3]	.68**
Devaluation	
3 I tried to convince myself that the problem was not very important after all.[2]	.81**
7 I told myself the problem was unimportant.	.84**
11 I tried to convince myself that the problem was, in fact, pretty insignificant.[2,5]	.82**
	.82**
Symptom Reduction	
2 I tried to just let off steam.	.67**
5 I tried to let my feelings out somehow.	.51**
9 I did something that I thought would soothe my nerves.	.69**
13 I tried to relieve my tension somehow.	.76**
21 I did something I enjoyed, just to make myself feel better.	.63**
30 I tried to just get it off my chest.	.71**
34 I tried to just calm down.[4]	.61**
38 I just tried to relax.	.69**

$\chi^2 = 1495.27**$ df = 730   GFI = .71   AGFI = .67   TLI = .32   CFI = .83   IFI = .83

Note

Table entries are standardized factor loadings. For each item, numerical superscripts indicate the factor(s) for which that item yielded a significant modification index ($p < .05$). Item numbers in bold indicate items forming the 20-item CCS (See 'Discussion').

$* p < .05. ** p < .01.$

items yielded significant modification indices on five factors. For example, items 62 ('I went over in my mind what I would say or do') and 63 ('I thought about how a person I admire would handle the situation and used that as a model') not only loaded on the self-controlling factor, but also yielded significant modification indices for the confrontative coping, distancing, social support, planful problem-solving, and positive reappraisal factors. This apparently reflects the inherent ambiguity of these items, which do not specify the content of what was mentally rehearsed or how the admired person would handle the situation. In contrast, 17 of the 40 CCS items yielded significant modification indices for other factors, with five items yielding significant modification indices on three factors. Again, items with a larger number of significant modification indices were stated in more ambiguous terms (e.g. item 22, 'I tried to change the things about the situation that were bothering me').

For both models, modification indices for within-factor measurement error correlations corresponded closely to the within-factor residuals, meaning that nearly every item pair that yielded a significant residual also produced a significant modification index ($p < .05$). This is not surprising, because the residual between any within-factor item pair can be eliminated by allowing their measurement errors to correlate. All eight WCQ factors yielded significant modification indices for between-factor correlated errors, ranging from 18/264 for planful problem-solving to 30/184 for accepting responsibilitiy. For the CCS, significant modification indices were found for all five factors, ranging from 20/256 for avoidance to 26/256 for changing the situation and symptom reduction. Although not identical, these results paralleled the pattern of residuals for both models, such that a significant modification index was usually accompanied by a significant residual.

Discussion

The results of the preceding analyses provide little support for the validity of the WCQ but provide moderate to strong support for the validity of the CCS. Reliability estimates for the WCQ scales ranged from .49 to .73, with only one exceeding .70, whereas reliabilities for the CCS scales ranged from .79 to .94. The measurement model for the WCQ provided a poor fit to the data, and two of the eight factors contained items with non-significant loadings. In contrast, the CCS yielded substantially better fit, and all item loadings were significant and, for the most part, large in magnitude. Furthermore, significant modification indices for item loadings on other than the assigned factor were obtained for 80 per cent of the WCQ items but for only 43 per cent of the CCS items.

Other evidence provided comparable support for the WCQ and the CCS. In particular, product rules for internal and external consistency for the WCQ were satisfied in 79.3 per cent and 77.3 per cent of the tests performed, respectively. For the CCS, the product rule for internal consistency was satisfied for 63.6 per cent of the tests performed, somewhat lower than the percentage obtained for the WCQ, whereas the product rule for external consistency was met for 79.2 per cent of the tests performed, slightly higher than that for the WCQ. Likewise, for the WCQ, modification indices were significant for 22.1 per cent of the within-factor measurement error correlations and 12.3 per cent of the between-factor measurement error correlations, while the corresponding figures for the CCS were 36.4 per cent and 9.2 per cent, respectively.

Given that the CCS was superior to the WCQ in terms of reliability, item loadings, and overall fit, it was surprising that the CCS did not yield fewer significant residuals and modification indices for measurement error correlations than the WCQ. One potential explanation for this anomaly is that the proportion of significant interitem correlations was substantially larger for the CCS than for the WCQ (60.6 per cent vs 38.9 per cent). Because a residual is typically smaller in absolute magnitude than the corresponding interitem correlation, it follows that a residual is unlikely to be significant when the items themselves are not significantly correlated. Likewise, correlations between measurement errors are usually smaller than the correlation between the items involved, thereby implying that items with a non-significant correlation are unlikely to yield a significant modification index for the correlation between their measurement errors. Hence, the WCQ may have exhibited a modest proportion of significant residuals and modification indices for measurement errors correlations simply because the WCQ items themselves were not highly correlated.

Based on these results and those reported by Parker et al. (1993), there are difficulties with the use of the WCQ in its current form (Folkman and Lazarus, 1988). The limited support for the WCQ may be attributed largely to the empirical approach used for its development. As noted previously, accounts of the development of the WCQ (e.g. Folkman and Lazarus, 1980, 1988) suggest that, although the selection of the WCQ items was guided by Lazarus' theory, multiple items were not generated to represent a priori dimensions specified by the theory. Moreover, the WCQ scales were constructed not by grouping items that shared similar content, but rather by performing exploratory factor analyses, thereby requiring the use of subjective criteria regarding the number of dimensions represented by the WCQ and the assignment of items to dimensions. As a result, the WCQ contains items that are conceptually heterogeneous, do not yield a stable factor structure, and provide an incomplete representation of the coping dimensions specified by Lazarus' theory (cf. Parker and Endler, 1992; Parker et al., 1993).

Unlike the WCQ, the CCS was developed by generating items intended to represent a priori coping dimensions specified by Edwards' (1992) cybernetic theory and employing confirmatory factor analysis combined with evaluations of item content to determine the degree to which items corresponded to their assigned dimensions. As a consequence of this procedure, the CCS has been shown to exhibit psychometric properties that, with few exceptions, were superior to those found for the WCQ. Furthermore, scales based on the CCS correspond directly and unambiguously to the coping dimensions associated with the cybernetic theory, thereby enhancing the utility of the measure for empirical investigations based on the theory.

Despite these strengths, the current version of the CCS has several drawbacks, such as highly specific content shared by some items assigned to the same scale and redundancy in certain items constituting the accommodation and devaluation scales. However, by dropping the flawed items from these two scales and retaining the four best items from each scale (see Table 2.3), a 20-item CCS may be formed. Confirmatory factor analysis of the associated measurement model indicated somewhat better fit than for the full 40-item CCS (TLI, IFI and CFI values were .94, .95, and .95, respectively), and reliabilities of all scales were .78 or higher. However, because the 20-item CCS was derived empirically, this evidence should be considered tentative, pending cross-validation. Furthermore, a third version of the CCS is currently under development. Until this version is available, the 20-item CCS may be used to provide a parsimonious representation of the coping dimensions outlined by the cybernetic theory with little loss of information over the full 40-item CCS.

Conclusion

This chapter has compared empirical and theoretical approaches to the measurement of coping, as represented by the WCQ and CCS, respectively. The findings reported here provide information regarding the validity of the WCQ and CCS that may be useful to researchers engaged in empirical coping research. At a more general level, the comparison of the WCQ and the CCS has demonstrated several important advantages of the theoretical approach to coping measurement, as evidenced by the superior psychometric properties of the CCS over the WCQ and the close correspondence between the CCS factors and a priori coping dimensions of theoretical interest. Although generalizations based on the measures and sample used here are necessarily limited, the results of this investigation clearly favour the theoretical approach to coping measurement over the empirical approach (cf. Carver et al., 1989). Consequently, future coping measures should be developed by generating multiple indicators of coping dimensions corresponding to relevant theory, and these measures should be empirically evaluated using confirmatory procedures such as those employed here. By following this approach, the psychometric properties of the resulting measures will be enhanced, information will be accumulated regarding the relative merits of the various coping frameworks proposed in the literature (Edwards, 1988; Latack and Havlovic, 1992; Lazarus and Folkman, 1984; Silver and Wortman, 1980), and the linkages between coping theory and measurement will be strengthened.

Notes

[1] Coefficient alpha is often considered to be an index of internal consistency. This is partly a misnomer, because alpha depends solely on the number of items in a scale and the average interitem correlation (Nunnally, 1978), neither of which indicates that internal consistency has been established. For this reason, alpha should not be considered to be an index of unidimensionality (Green, Lissitz and Mulaik, 1977; Hattie, 1985).

[2] The denominators for these ratios indicate the total number of residuals tested for a given scale. For internal consistency, this represents the total number of correlations among the items constituting a scale, whereas for external consistency, this represents the total number of correlations between the items constituting a scale and the remaining items in the measure.

References

ALDWIN, C., FOLKMAN, S., SCHAEFER, C., COYNE, J. C. and LAZARUS, R. S. (1980) 'Ways of coping: a process measure', paper presented at the meeting of the American Psychological Association, Montreal, September, 1980.

ALDWIN, C. M. and REVENSON, T. A. (1987) 'Does coping help? A re-examination of the relation between coping and mental health', *Journal of Personality and Social Psychology* **53**: 337–348.

AMIRKHAN, J. H. (1990) 'A factor analytically derived measure of coping: the coping strategy indicator', *Journal of Personality and Social Psychology* **59**: 1066–1074.

ATKINSON, M. and VIOLATO, C. (1993) 'A factor analysis of the ways of coping questionnaire based on data from saddening experiences', *Psychological Reports* **72**: 1159–1164.

BENTLER, P. M. (1990) 'Comparative fit indexes in structural models', *Psychological Bulletin* **107**: 238–246.

BENTLER, P. M. and BONETT, D. G. (1980) 'Significance tests and goodness of fit in the analysis of covariance structures', *Psychological Bulletin* **88**: 588–606.

BILLINGS, A. G. and MOOS, R. H. (1984) 'Coping, stress, and social resources among adults with unipolar depression', *Journal of Personality and Social Psychology* **46**: 877–891.

BOLLEN, K. A. (1989a) 'A new incremental fit index for general structural equation models', *Sociological Methods and Research* **17**: 303–316.

BOLLEN, K. A. (1989b) *Structural Equations with Latent Variables*, New York: Wiley.

CAMPBELL, J. P. (1976) 'Psychometric theory', in M. Dunnette (ed.) *Handbook of Industrial and Organizational Psychology*, Chicago: Rand McNally, 185–222.

CARVER, C. S., SCHEIER, M. F. and WEINTRAUB, J. K. (1989) 'Assessing coping strategies: a theoretically based approach', *Journal of Personality and Social Psychology* **56**: 267–283.

COHEN, F. (1987) 'Measurement of coping', in S. V. Kasl and C. L. Cooper (eds) *Stress and Health: Issues in Research Methodology*, Chichester: John Wiley, 283–305.

COYNE, J. C. and DOWNEY, G. (1991) 'Social factors and psychopathology: stress, social support, and coping processes', *Annual Review of Psychology* **42**: 401–425.

CUDECK, R. and O'DELL, L. L. (1994) 'Applications of standard error estimates in unrestricted factor analysis: significance tests for factor loadings and correlations', *Psychological Bulletin* **115**: 475–487.

DANES, J. E. and MANN, O. K. (1984) 'Unidimensional measurement and structural equation models with latent variables', *Journal of Business Research* **12**: 337–352.

DEWE, P., COX, T. and FERGUSON, E. (1993) 'Individual strategies for coping with stress at work: a review', *Work and Stress* **7**: 5–15.

DEWE, P. J. and GUEST, D. E. (1990) 'Methods of coping with stress at work: a conceptual analysis and empirical study of measurement issues', *Journal of Organizational Behavior* **11**: 135–150.

EDWARDS, J. R. (1988) 'The determinants and consequences of coping with stress', in C. L. Cooper and R. Payne (eds) *Causes, Coping, and Consequences of Stress at Work*, New York: Wiley, 233–263.

EDWARDS, J. R. (1992) 'A cybernetic theory of stress, coping, and well-being in organizations', *Academy of Management Review* **17**: 238–274.

EDWARDS, J. R. and BAGLIONI, A. J. (1993) 'The measurement of coping with stress: construct validity of the Ways of Coping Checklist and the Cybernetic Coping Scale', *Work and Stress* **7**: 17–31.

EDWARDS, J. R. and COOPER, C. L. (1988) 'The impacts of positive psychological states on physical health: a review and theoretical framework', *Social Science and Medicine* **27**: 1447–1459.

ENDLER, N. S. and PARKER, J. D. A. (1990) 'Multidimensional assessment of coping: a critical evaluation', *Journal of Personality and Social Psychology* **58**: 844–854.

FOLKMAN, S. and LAZARUS, R. S. (1980) 'An analysis of coping in a middle-aged community sample', *Journal of Health and Social Behavior* **21**: 219–239.

FOLKMAN, S. and LAZARUS, R. S. (1985) 'If it changes, it must be a process: a study of emotion and coping during three stages of a college examination', *Journal of Personality and Social Psychology* **48**: 150–170.

FOLKMAN, S. and LAZARUS, R. S. (1988) *Manual for the Ways of Coping Questionnaire*, Palo Alto, CA: Consulting Psychologists Press.

FOLKMAN, S., LAZARUS, R. S., DUNKEL-SCHETTER, C., DELONGIS, A. and GRUEN, R. J. (1986) 'Dynamics of a stressful encounter: cognitive appraisal, coping, and encounter outcomes', *Journal of Personality and Social Psychology* **50**: 992–1003.

GERBING, D. W. and ANDERSON, J. C. (1984) 'On the meaning of within-factor correlated measurement errors', *Journal of Consumer Research* **11**: 572–580.

GERBING, D. W. and ANDERSON, J. C. (1988) 'An updated paradigm for scale development incorporating unidimensionality and its assessment', *Journal of Marketing Research* **25**: 186–192.

GERBING, D. W. and ANDERSON, J. C. (1992) 'Monte Carlo evaluations of goodness of fit indices for structural equation models', *Sociological Methods and Research* **21**: 132–160.

GREEN, S. B., LISSITZ, R. W. and MULAIK, S. A. (1977) 'Limitations of coefficient alpha as an index of test unidimensionality', *Educational and Psychological Measurement* **37**: 827–838.

HATTIE, J. (1985) 'Methodology review: assessing unidimensionality of tests and items', *Applied Psychological Measurement* **9**: 139–164.

HUNTER, J. E. and GERBING, D. W. (1982) 'Unidimensional measurement, second order factor analysis, and causal models', in B. M. Staw and L. L. Cummings (eds) *Research in Organizational Behavior*, Greenwich, CT: JAI Press, 267–320.

JORESKOG, K. G. and SORBOM, D. (1988) *LISREL 7*, Chicago: SPSS, Inc.

JORESKOG, K. G. and SORBOM, D. (1993) *LISREL 8*, Hillsdale, NJ: Erlbaum.

LAMBERT, Z. V., WILDT, A. R. and DURAND, R. M. (1990) 'Assessing sampling variation relative to number-of-factors criteria', *Educational and Psychological Measurement* **50**: 33–48.

LATACK, J. C. (1986) 'Coping with job stress: measures and future directions for scale development', *Journal of Applied Psychology* **71**: 377–385.

LATACK, J. C. and HAVLOVIC, S. J. (1992) 'Coping with job stress: a conceptual evaluation framework for coping measures', *Journal of Organizational Behavior* **13**: 479–508.

LAZARUS, R. S. (1966) *'Psychological Stress and the Coping Process*, New York: McGraw-Hill.

LAZARUS, R. S. and FOLKMAN, S. (1984) *Stress, Coping, and Adaptation*, New York: Springer.

LAZARUS, R. S. and LAUNIER, R. (1978) 'Stress-related transactions between person and environment', in L. A. Pervin and M. Lewis (eds) *Perspective in Interactional Psychology*, New York: Plenum, 287–327.

LONG, J. S. (1983) *Confirmatory Factor Analysis: A Preface to LISREL*, Beverly Hills, CA: Sage.

LORD, F. M. and NOVICK, M. R. (1968) *Statistical Theories of Mental Test Scores*, Reading, MA: Addison-Wesley.

McCRAE, R. R. (1984) 'Situational determinants of coping responses: loss, threat, and challenge', *Journal of Personality and Social Psychology* **46**: 919–928.

MISHEL, M. H. and SORENSON, D. S. (1993) 'Revision of the ways of coping checklist for a clinical population', *Western Journal of Nursing Research* **15**: 59–76.

NUNNALLY, J. C. (1978) *Psychometric Theory*, New York: McGraw-Hill.

O'DRISCOLL, M. P. and COOPER, C. L. (1994) 'Coping with work-related stress: a critique of existing measures and proposal for an alternative methodology', *Journal of Occupational and Organizational Psychology* **67**: 343–354.

PARASURAMAN, S. and CLEEK, M. A. (1984) 'Coping behaviors and managers' affective reactions to role stressors', *Journal of Vocational Behavior* **24**: 179–193.

PARKER, J. D. A. and ENDLER, N. S. (1992) 'Coping with coping assessment: a critical review', *European Journal of Personality* **6**: 321–344.

PARKER, J. D. A., ENDLER, N. S. and BAGBY, R. M. (1993) 'If it changes, it might be unstable: examining the factor structure of the ways of coping questionnaire', *Psychological Assessment* **5**: 361–368.

PARKES, K. R. (1984) 'Locus of control, cognitive appraisal, and coping in stressful episodes', *Journal of Personality and Social Psychology* **46**: 655–668.

PEARLIN, L. I. and SCHOOLER, C. (1978) 'The structure of coping', *Journal of Health and Social Behavior,* **19**: 2–21.

SCHWAB, D. P. (1980) 'Construct validity in organizational behavior', in L. L. Cummings and B. M. Staw (eds), *Research in Organizational Behavior* (vol. 2), Greenwich, CT: JAI Press.

SIDLE, A., MOOS, R. H., ADAMS, J. and CADY, P. (1969) 'Development of a coping scale', *Archives of General Psychiatry* **20**: 225–232.

SILVER, R. L. and WORTMAN, C. B. (1980) 'Coping with undesirable life events', in J. Garber and M. E. P. Seligman (eds) *Human Helplessness: Theory and Application*, New York: Academic Press, 279–340.

SINGH, J. (1991) 'Redundancy in constructs: problem, assessment, and an illustrative example', *Journal of Business Research* **22**: 255–280.

SMYTH, K. A. and WILLIAMS, P. D. (1991) 'Patterns of coping in black working women', *Behavioral Medicine* **17**: 40–46.

SORBOM, D. (1975) 'Detection of correlated errors in longitudinal data', *British Journal of Mathematical and Statistical Psychology* **28**: 138–151.

STONE, A. A., GREENBERG, M. A., KENNEDY-MOORE, E. and NEWMAN, M. G. (1991) 'Self-report, situation-specific coping questionnaires: what are they measuring?', *Journal of Personality and Social Psychology* **61**: 648–658.

STONE, A. A. and NEALE, J. M. (1984) 'New measure of daily coping: development and preliminary results', *Journal of Personality and Social Psychology* **46**: 892–906.

TENNEN, H. and HERZBERGER, S. (1985) 'Ways of coping scale', in D. J. Keyser and R. C. Sweetland (eds) *Test Critiques* vol. 3, Kansas City, KS: Test Corporation of America, 686–697.

TUCKER, L. R. and LEWIS, C. (1973) 'The reliability coefficient for maximum likelihood factor analysis', *Psychometrika* **38**: 1–10.

VINGERHOETS, A. J. and FLOHR, P. J. (1984) 'Type A behaviour and self-reports of coping preferences', *British Journal of Medical Psychology* **57**: 15–21.

VITALIANO, P. P., RUSSO, J., CARR, J. E., MAIURO, R. D. and BECKER, J. (1985) 'The ways of coping checklist: revision and psychometric properties', *Multivariate Behavioral Research* **20**: 3–26.

WEISMAN, A. D. and WORDEN, J. W. (1976) 'The existential plight of cancer: significance of the first 100 days', *International Journal of Psychiatry in Medicine* **7**: 1–15.

WHEATON, B. (1987) 'Assessment of fit in overidentified models with latent variables', *Sociological Methods and Research* **16**: 118–154.

WOLINS, L. (1982) *Research Mistakes in the Social and Behavioural Sciences*, Ames, IA: The Iowa State University Press.

CHAPTER THREE

The self-regulation of experience: openness and construction

MICHAEL ROSENBAUM

> Openness and closure (or, more accurately, opening and closing) are more than interesting metaphors for developmental processes. As with all other expressions of life metabolism, it is the coordination of these processes that constitutes an ever-present challenge. (Mahoney, 1991: 343)

As living human organisms we are subject to a continuous flow of experience. Some experiences are pleasant and we would like to heighten and protract them; others are disturbing or obstructive, and we would like to ignore, suppress or change them. In most cases, we cannot alter our experiences directly. But we can control our response to our experiences and, thereby, indirectly influence subsequent experiences.

In this chapter I will discuss two major patterns of response to experience: opening responses and closing responses. Opening responses entail non-judgemental acceptance of one's experiences pretty much as they are. Closing responses aim to interpret or organize those experiences on the basis of one's existing theories and constructs, and thereby close off aspects of the experience that do not coincide with those theories and constructs. Because the term 'closing' may have a somewhat negative connotation in comparison to 'opening', I prefer to use the term 'constructive' to point to the active process by which people construe their responses to their current experiences on the basis of their past experiences and expectations of future events.

The chapter will start with a definition of what I mean by experience and a brief review of the literature related to opening and constructive responses. This will be followed by a discussion of the role of self-control skills in facilitating the two responses, and then by a look at the contribution of these responses to coping with issues related to health and illness. My basic premise is that since opening and constructive responses to experience play a significant role in coping with stress and, consequently, in people's physical and mental well-being, the self-control of these responses is of value and importance.

What is experience?

'Experience' is a formidably broad term with a range of meanings both as noun and verb. Here I am concerned with primary experience as defined fifty years ago by a dictionary of psychology (Harriman, 1947): '. . . a conscious response to stimuli devoid of interferences, expectations, and other distortions'. 'Experiencing' in this sense is '. . . a way of knowing that is immediate, embodied, holistic, and contextual,' to quote Bohart's description (Bohart, 1993: 2). Because experiencing is the primary way we relate to real, living situations, it is very closely connected to affect, but it is not the same thing. It includes a broader range of behaviours than just emotions or mood. It is a complex behaviour that has perceptual-sensory, action-motoric and visceral components.

An important feature of experience, as the term is used here, is that it constantly changes in concert with bodily and contextual fluctuations. No two 'experiences' can be identical. As Epstein (1991), discussing spontaneous behavior from an operant conditioning point of view, explains: 'You never brush your teeth exactly the same way twice, and even the rat's lever press varies in subtle ways with each occurrence. Variability is typical of all behaviour . . . Even if, somehow, you could repeat some response precisely, it would still be novel in the sense that each occurrence is the product of a changed organism' (Epstein, 1991: 362).

The inherent novelty of every experience is not always appreciated. Because every experience contains too much data to process effectively, persons often focus only on the data that are consistent with their expectations (Kelly, 1955) and construe similar experiences occurring at different times as identical, when they are not. Increased sensitivity to the variability and novelty of one's everyday experience is the essence of an opening response to experience. The tendency to search for common elements in different experiences is the essence of a constructive response to experience.

Opening and constructive responses to experience

The opening response (OR) to experience – often termed openness to experience (OTE) – has been treated fairly extensively in the literature, perhaps because of its centrality in humanistic psychology. Much of the psychological literature has tended to treat the OR as a personality trait or stable disposition. Humanistic and existential theories consider the OR to be the major characteristic of the 'fully functioning' individual (Maslow, 1970; Mittelman, 1994; Rogers, 1980). As Rogers (1980) sees it, persons who are open to experience have the stable ability to be aware of their deepest thoughts and feelings and to intensely experience the full range of visceral, sensory, emotional, and cognitive experiences within themselves without feeling threatened.

A number of measures have been developed to assess OTE as a personality trait (Coan, 1972; Fitzgerald, 1966; McCrae and Costa, 1983). Most recently, McCrae and Costa (1991) maintain that OTE, along with neuroticism, extraversion, agreeableness and conscientiousness, is a basic dimension of personality and, furthermore, argue that, of all of these dimensions, openness to experience is 'the most relevant to the study of imagination, cognition, and personality' (McCrae, 1993-94: 40). These authors define OTE in rather broad terms. In their view, 'Open individuals are characterized both by a broader and deeper scope of awareness and by a need to enlarge and examine experience; they are imaginative, aesthetically responsive, emphatic, exploring, curious, and unconventional' (McCrae and Costa, 1991: 228). OTE has also been treated as a basic personality trait in Tellegen's (1985) model of personality traits, albeit under the label of 'absorption'. Roche and McConkey (1990) maintain that Tellegen's absorption is a characteristic of individuals who are open 'to experience emotional and cognitive alterations across a variety of situations' (ibid., 92).

Some researchers, however, view OTE as a psychological state. For example, Marsha Linehan (1994) defines OTE as the acceptance of one's primary experience just as it is: 'without constrictions, and without distorting, without judgments, without evaluating, without trying to keep it, and without trying to get rid of it' (Linehan, 1994: 80). In this definition, OTE is similar to the state of mindfulness described by Teasdale, Sigal and Williams (1995) on the basis of the work of Kabat-Zinn (1990): 'The essence of this state is to "be" fully in the present moment, without judging or evaluating it, without reflecting backwards on past memories, without looking forward to anticipate the future, as in anxious worry, and without attempting to "problem-solve" or otherwise avoid any unpleasant aspects of the immediate situation' (ibid., 33).

Alternatively, both openness and closedness to experience have been described as ways by which individuals process information about the world around them. Johnston and Hawley (1994), for example, propose that 'the mind/brain can perceive familiar habitats in terms of conceptually driven processing and unexpected inputs in terms of data-driven processing' (ibid., 68). These ways of processing correspond to opening and closing responses to experience. In the words of these authors, a 'closed mind' processes data on the basis of pre-conceived ideas (cognitive constructs, schema), while an 'open mind' employs data-driven processing. Their view is supported by a similar distinction suggested in some social psychology literature, which has proposed that information about other people's psychological conditions can be processed either directly and immediately without any inferential processes (i.e. openness) or on the basis of inferences from past learning (i.e. closedness) (see Asch, 1952; Baron and Misovich, 1993).

In contrast to the opening response, the constructive response (CR) to experience has neither been previously named nor dealt with explicitly, even though constructivistic assumptions dominate current psychological thinking. These assumptions, however, have led to the unnamed, implicit treatment of the CR in the literature dealing with stress. Many personality traits or stable dispositions that have been hypothesized to be instrumental in coping with stress are in fact CRs to stressful experiences. This can be seen, for example, in Kobasa's (1979) concept of 'hardiness,' which, according to its author, moderates the stress-illness relationship. In this conception, hardy people cope well with stress because they judge their activities as being valuable and important ('commitment'), controllable and challenging (Orr and Westman, 1990). The evaluation and judgement of the experience are the hallmark of the constructive or closing response. The CR can also be seen in Antonovsky's (1990) 'sense of coherence', a personality disposition to construe stressful experiences as comprehensible, manageable and meaningful. Here, the CR seems to be inherent in or very close to the cognitive interpretation of experience that is entailed in making it coherent. Similarly, the constructive response to experience (CTE) can be discerned in Seligman's distinction between persons with an 'optimistic' explanatory style, who cope well with stress, and those with a 'pessimistic' explanatory style, who cope less well (Seligman *et al.*, 1988). The very possession of an explanatory style implies the evaluation of current experience on the basis of habitual ways of information processing that characterize a constructive approach to experience.

Both opening and constructive responses to experience are complex reactions whose precise boundaries are often less clear-cut than one would wish. In reality there is neither a purely opening response nor a purely constructive response, because every response to experience is based to some extent on one's conceptual system and to some extent on the 'objective' elements of the experience. The question is the extent to which any given behaviour can be characterized as opening or constructive. The following definitions of the OR and CR are of their 'ideal' state.

As I define it, the opening response is largely a psychological state that is marked by acceptance and mindfulness and characterized by:

1 awareness of one's current experiences;
2 recognition of the flow and continuously changing features of one's experience;
3 acceptance of one's experiences as they are without any attempt to judge or to evaluate them, whether on the basis of logical rules or existent personal constructs, and without comparison to past experiences or anticipation of the future;
4 ability to respond to experiences in maximal congruence with them, with no attempt to deny or transform them;
5 full absorption and involvement in the occurring experience.

My concept of a constructive response (CR) to experience is derived from the constructivist's assertion that 'humans actively create and construe their personal realities' (Mahoney and Lyddon, 1988: 200). That is, people interpret, explain, and so build, or 'construct,' their experience on the basis of their personal theories. A CR may be characterized by the following cognitive activities:

1 interpretation of the experience in line with one's personal constructs;
2 rational analysis of the causes and likely outcomes of the experience (e.g. 'Why did this happen to me?' and 'What can be the consequences of my present experience?');
3 avoidance of the experience via a range of mechanisms, including distraction, suppression, denial, and the like;
4 use of problem-solving techniques.

Much of the literature tends to treat OTE and CTE as opposing responses and to pit one against the other. The humanistic psychology literature clearly regards OTE as a desirable quality and CTE as an impediment to a full life. Some of the literature of the cognitive and behavioural schools, on the other hand, tends to value constructive responses and, by implication, de-emphasize opening responses. As used here, neither of the two types of response is morally or psychologically better or worse than the other. They simply have different uses, different functions. An open approach to experience enables change and plasticity, a constructive approach promotes self-perpetuation and stability. To cope effectively with the vagaries of daily life, people must strike a balance between the two. Something of this need is acknowledged in the psychotherapeutic procedures aimed at helping individuals to cope with stress. While the literature on coping with stress focuses on constructive responses to experience, and not on opening responses, various of the actual procedures emphasize both ORs and CRs (Ellis, 1994; Linehan, 1994).

Change and stability

In the larger context of human behaviour, it can be argued ORs and CRs reflect a basic dialectic in the interaction of human beings with their environment. On the one hand, the organism strives towards stability, predictability, equilibrium, homeostasis, and self-organization. Constructive reactions help achieve that stability by enabling individuals to interpret their primary experience in accord with their previously established conceptual system. On the other hand, since the environment is in perpetual flux, there is no such thing as permanent stability (Lewis, 1995) and the organism is forever driven towards change and development. In a major way, change and development are the products of disequilibrium, heterostasis and instability.

Yet, paradoxically, change occurs as the organism tries to create order out of disorder through the process of self-organization (Prigogine and Stengers,

1984). Self-organization and the quasi-stability that goes with it are achieved through cognitive appraisal and interpretive schema (Lewis, 1995). Conversely, once order is achieved, change occurs only when the organism returns to a temporary state of disequilibrium as a consequence of its opening up to experience. New behaviours may spontaneously appear precisely when people do not expend mental effort to evaluate and judge their experiences (Epstein, 1991).

Self-control as a constructive process

Although some individuals may be more prone to CRs and others to ORs, both responses can be facilitated by self-control. Self-control may be defined as a conscious and volitional behaviour aimed at overcoming any of a wide range of self-generated impediments to a desired target behaviour (Kanfer and Gaelick-Buys, 1991). All self-control behaviours are non-automatic responses to experience. The self-control literature has so far focused mainly on self-control behaviours that are, by my present definition, constructive responses to experience, which entail the construction of experience so as to prevent its interfering with the performance of a desired behaviour. Two functions of self-control fit the constructive mould quite well.

One is *redressive* self-control (Rosenbaum, 1988), which is applicable in situations in which certain experiences, namely emotions, visceral reactions or intrusive thoughts, may interfere with the smooth execution of a target behaviour. We may take, for example, persons who suffer from speech anxiety but wish to speak in public. Their anxiety – that is, their experiential reaction – naturally interferes with their target behaviour. To cope with their anxiety, they may employ any of various CRs. They may try to transform their anxiety to something positive, either by construing the need to cope with the feeling as a challenge or by using encouraging self-statements. They may also try to ignore their anxiety and suppress anxiety-related cognitions.

Most self- and stress-management therapies (e.g. Kanfer and Schefft, 1988; Meichenbaum, 1985) focus on training clients in self-control skills that are directed at the adoption of effective coping responses. These skills include such things as cognitive reframing of the experience, disputing irrational beliefs (Ellis, 1973), creating positive expectancies (Bandura, 1977), and producing contrasting new experiences (e.g. distraction, positive self-statements, etc.). I have labelled the use of self-control to regulate one's emotional responses to experience so that they do not interfere with the performance of a target behaviour the redressive function of self-control (Rosenbaum, 1988).

The other function of self-control that entails a constructive approach to experience is what I have termed its reformative function (Rosenbaum, 1988). The reformative function of self-control is called for when individuals decide to change current habits (for example, smoking or overeating) so as to avert future disruptions to their well-being. Its role is to help persons to alter or reform their habitual or long-term behaviour. Whereas redressive self-control is directed at enabling the person to resume normal functioning that had been temporarily disrupted, reformative self-control is directed at enabling the person to cease a previous pattern of behaviour and replace it with a different pattern. For all practical purposes, the two self-control functions usually go hand in hand. For example, whether or not a person perseveres in a self-change programme (such as dieting, exercising and so forth) depends to a considerable extent on whether he or she simultaneously employs redressive self-control to reduce the pain or discomfort that the change may entail (for a fuller discussion of the interrelationship between the two self-control functions, see Rosenbaum, 1990a, 1993).

Redressive self-control is based mainly on the cognitive control of emotions, thoughts and visceral reactions; reformative self-control is based mainly on problem solving, planning and delay of gratification. Their differences notwithstanding, both the redressive and reformative functions of self-control are achieved through constructive responses to experience. Both functions require a variety of self-control skills to carry out. I have termed this assortment of self-control skills 'learned resourcefulness' (Rosenbaum, 1983). The term is borrowed from Meichenbaum (1985), who described the graduates of his stress-inoculation training programme as having acquired a sense of resourcefulness in coping with stress. It emphasizes the learnability of the skills.

To assess individual differences in learned resourcefulness, I have developed the Self-Control Schedule (SCS; Rosenbaum, 1980). This schedule consists of items drawn from the following content areas: (a) use of cognitions and self-instructions to cope with emotional and physiological responses, (b) application of problem-solving strategies (e.g. planning, problem definition, evaluating alternatives, and anticipation of consequences), (c) ability to delay gratification, and (d) a general belief in one's ability to self-regulate internal events. The self-control skills assessed by this scale all fall under the concept of the constructive response to experience. The reliability and validity of the SCS is well established (e.g. Rosenbaum, 1990a; 1993), and the scale has been translated into various languages and used in a number of countries. Factor analytical studies of the SCS provide support for the distinction between skills needed for the redressive function of self-control and those needed for the reformative function (Gruber and Wildman, 1987; Rude, 1989; Sugiwaka and Agari, 1993, 1995). Both types of self-control skill are subsumed under the concept of learned resourcefulness, which is a subcategory of the constructive response to experience.

Opening responses through experiential self-control

Although self-control behaviour has thus far been treated largely as a constructive response to experience, it can also be applied to the process by which people open themselves to experience. In this case, openness to experience is the target behaviour and self-control behaviours are directed at helping individuals to be OTE.

Humanistic psychology, in which OTE plays a central role, assumes that the barrier to OTE consists of psychological defences, and it works to help clients achieve OTE through the provision of interpersonal conditions (e.g. unconditional positive regard and empathy) which encourage them to abandon their defences and become more open. My point of departure is that OTE is often impeded by the extensive use of constructive responses, that is by the appraisal of current experiences on the basis of one's pre-conceived personal constructs. I would like to argue, from a cognitive-behavioural point of view, that OTE can be facilitated by the use of a specific repertoire of self-control skills that consists of opening responses. These skills are inherent in what I have termed experiential self-control (E-SC; Rosenbaum, 1993). They include the ability to focus on one's experiential reactions, to be aware of the various components of these reactions, to attend to bodily sensations and feelings and to accept them as they are, to relinquish analytical thinking, to create images that are congruent with the experience, and to be receptive of uncertain, unfamiliar or paradoxical experiences.

Training in techniques such as relaxation (J. C. Smith, 1988), hypnosis (Spanos and Flynn, 1989), mindfulness (Kabat-Zinn, 1990; Marlatt, 1994), and a variety of 'acceptance' skills (Linehan, 1994) is in fact training in the experiential self-control skills that both constitute and facilitate OTE. For example, in relaxation exercises, clients learn how to let go of analytical processes and to focus on current experiences. J. C. Smith (1988) suggested three processes that are basic to all relaxation techniques: '*focusing*, the ability to identify, to differentiate, maintain attention on, and return attention to simple stimuli for an extended period; *passivity*, the ability to stop unnecessary goal directed and analytic activity; and *receptivity*, the ability to tolerate and accept experiences that may be uncertain, unfamiliar, or paradoxical' (ibid., 312). All three behaviours – focusing, passivity, and receptivity – are part of the experiential self-control (E-SC) repertoire. OTE requires and entails focus on the moment to moment experience with limited cognitive activity, a passive, 'letting go' stance which avoids mental exertion, and receptivity to the experience without simultaneous self-evaluation and thinking about consequences. Indeed, learning to relax can be seen as a self-control process that requires many of the skills of E-SC.

People who are capable of opening responses when they wish to make them have a rich repertoire of experiential self-control skills. On the basis of this assumption we have developed the Experiential Self-Control Scale (E-SCS; Hefer, 1991; Rosenbaum, 1995) to assess individual differences in experiential self-control. The scale is in the early stages of development and will probably undergo a number of modifications. Initially it was named the Releasive Self-Control Scale (Hefer, 1991). Only later was the name changed to the Experiential Self-Control Scale (Rosenbaum, 1993). At this stage, the

E-SCS consists of twenty-nine items, each describing a common everyday experience in which experiential self-control is likely to be called for. Most of the items refer to personal and interpersonal activities that we assume that the average individual would like to fully experience, such as being at a party, having a love relationship, listening to music, reading a book, watching a film, or taking a bath. Subjects are asked to indicate the extent to which they can be open to these experiences.

The internal reliability of the scale is acceptable (α = .83). Factor analysis yielded four factors: (a) enjoyment of pleasant activities ('At a party with friends who know how to have fun, I allow myself to be carried along'); (b) emotional expressiveness ('When I get really good news, I get so carried away by excitement that I practically sing for joy'); (c) readiness to give up analytical thinking ('I very much enjoy losing my head whenever possible'); and (d) focusing on the current experience ('Listening to music I like makes me forget my surroundings').

As expected, the E-SCS positively correlated with the Tellegen Absorption Scale (TAS; Tellegen and Atkinson, 1974), which was found to be highly correlated to hypnotizability and associated with OTE (Roche and McConkey, 1990). It was not found to be correlated with Spielberger's Trait Anxiety Scale (Spielberger, Gorsuch and Lushene, 1970) or with the Social Desirability Scale (Crowne and Marlowe, 1960). Furthermore, the E-SCS was consistently found to be orthogonal to our measure of learned resourcefulness; the Self-Control Schedule. It clearly assesses a different dimension of self-control than redressive and reformative self-control.

Nevertheless both the SCS and the E-SCS were positively correlated with Burger and Cooper's (1979) Desirability of Control Scale and with King and Emmons' (1990) Emotional Expression Questionnaire and negatively related to their Ambivalence Over Emotional Expression Questionnaire (AEQ). In other words, individuals who have a tendency to use self-control behaviour, whether it is redressive, reformative or experiential, are likely to express a general desire for control, be emotionally expressive and be less ambivalent about overt emotional expression than those less inclined to self-control.

Preliminary data on the E-SCS are encouraging, though further experimental studies are required to establish better its validity as a measure of experiential self-control.

Health behaviour as a function of constructive and opening self-control skills

Behavioural medicine and health psychology provide excellent arenas for testing hypotheses derived from personality and self-control theories. The issues with which health psychologists are concerned, such as adherence to medical regimes, adoption of a healthful life style, coping with the stresses of illness and aversive medical procedures, and proneness to specific diseases, are also relevant to the present discussion of the self-control skills entailed in constructive and opening responses to experience.

Most of the research on health psychology to date has focused on constructive responses to experience and on the specific role of learned resourcefulness, that is the redressive and reformative self-control skills assessed by the Self-Control Schedule. It has been consistently found that highly resourceful subjects (who score high on the SCS) are better able to adopt healthful behaviour than low resourceful subjects. Such people were found to be more successful in giving up smoking on their own (Katz and Singh, 1986), in changing their eating habits (Leon and Rosenthal, 1984; T. Smith, 1979), and in curbing their intake of alcohol (M. P. Carey, K. B. Carey, Carnrike and Meisler, 1988). Highly resourceful dialysis patients demonstrated greater adherence to a strict fluid intake regimen than those less resourceful (Rosenbaum and Ben-Ari Smira, 1986) and adjusted better to epilepsy (Derry, Chovaz, McLachlan and Cummings, 1993; Rosenbaum and Palmon, 1984) and to pain in general (Rosenbaum, 1980b; Weisenberg, Wolf, Mittwoch and Mikulincer, 1990). Among chronic pain patients, Toomey, Seville, Mann, Abashian and Wingfield (1995) found an association between high resource-fulness and the number of planned pain-related physician consultations and between low resourcefulness and the number of pain-related emergency room visits (interpreted as a more impulsive utilization of health services).

In contrast to the considerable attention given to constructive responses in health psychology (Antonovsky, 1990; Kobasa, 1979; Rosenbaum, 1990a), the role of opening responses in health has received little study. Moreover, two recent studies found the Openness to Experience scale of the NEO-PI person-ality inventory (Costa and McCrae, 1992) to be related to high-risk health behaviour (Booth-Kewley and Vickers, 1994; Trull and Sher, 1994).

These findings notwithstanding, my basic proposition is that effective coping with stress in general and illness in particular requires both the open-ness to experience skills assessed by the Experiential Self-Control Scale and the construction of experience skills assessed by the Self-Control Schedule. It requires both the ability to take in and accept the stresses that one encounters and the ability to reframe and restructure the stressful experience in accord-ance with one's existing conceptual system.

This proposition was confirmed in an MA thesis by Agam (1995). The participants in this study were forty-nine end-stage renal disease patients in renal dialysis. They completed the SCS, the E-SCS and a self-efficacy measure asking them to assess their expected ability to effectively cope with their renal disease. When separately analysed, neither the E-SCS nor SCS scores were correlated with self-efficacy. However, a multiple regression analysis showed that, along with the patients' age and gender, the interaction of SCS and E-SCS did predict the patients' self-efficacy expectancies.

To understand the meaning of the interaction, the researcher divided the patients into four groups according to whether they were high or low resourceful (above or below the group's median score on the Self-Control

Schedule) and high or low on experiential self-control (above or below the group's median score on the Experiential Self-Control Scale). Findings showed that the subjects' experiential self-control moderated the association of learned resourcefulness and self-efficacy expectancies. Among low experiential self-control subjects, there was no difference in the self-efficacy of high and low resourceful subjects. However, among high experiential self-control subjects, there were large differences. In fact, subjects who were low on both learned resourcefulness and experiential self-control had the lowest self-efficacy expectancies, while those who were high on both expressed the strongest convictions that they could cope effectively with their renal disease. Those who could be open to their experiences while cognitively controlling them (that is, able to apply a constructive approach when needed) had the greatest confidence in their ability to cope with their disease.

The interaction of learned resourcefulness and experiential self-control was also found to predict the subjects' perceived quality of life. Here the effects of the two repertoires of self-control behaviours were additive. Namely, subjects high on both the SCS and the E-SCS reported the highest perceived quality of life, while subjects low on both reported the lowest.

The findings of another study (Yoschpe and Rosenbaum, 1995) suggest that the joint effects of resourcefulness and experiential self-control on individuals' well-being may be mediated by their ambivalence about emotional expression. King and Emmons (1990) reported that conflict over emotional expressiveness (as assessed by the Ambivalence Over Emotional Expression Questionnaire; AEQ) is related to measures of physical and psychological well-being. The conflict may be an outcome of the individual's efforts to control his or her emotions (Emmons, King and Sheldon, 1993), or of the fear of losing control once emotions are openly expressed. If the latter is the cause, then people who have the redressive and experiential self-control skills to manage their emotional responses can be expected to be less ambivalent about expressing their emotions. Yoschpe and Rosenbaum's findings lend support to this hypothesis: subjects' scores on the SCS and E-SCS were negatively related to their scores on the AEQ and positively related to their scores on King and Emmons' (1990) Emotional Expressiveness Questionnaire. Since the SCS and the E-SCS do not correlate with one another, each of these scales shares a unique variance with the AEQ. In addition, the study found that ambivalence about emotional expressiveness mediated the effects of learned resourcefulness and experiential self-control on subjects' subjective well-being.

Both of the above studies suggest that people who strongly believe they can control their responses to their experiences are less conflicted about expressing their emotions (and perhaps less conflicted about other aspects of their behaviour as well); are more confident in their ability to cope with stress; and feel a greater sense of well-being than persons with a weaker sense of self-control. This is important because the way that people deal with their emotional experience has been linked to both physical and psychological health. For example, it has been suggested that people who suppress their emotional reactions and maintain a seemingly stoic attitude are more prone to cancer (Eysenck, 1988; Levy, 1985) than those who express their feelings, and that people who have difficulty in dealing with anger and hostility are more prone to coronary heart disease (Friedman and Booth-Kewley, 1987; Williams *et al.*, 1980) than those who handle these emotions more easily. Moreover, the inhibition of emotional expressiveness has been associated with increased autonomic activity, which, if chronic, may lead to the development of physical illness (Pennebaker, 1985). All in all, numerous studies suggest that the twofold ability to accept and be in tune with one's emotional experiences – that is to be open to experience – and to reframe and restructure those experiences in accordance with a pre-existent conceptual system – that is to construct experience – may play a considerable role in the maintenance of physical and psychological health (see also Salovey and Mayer, 1989-90).

Conclusions and implications

In this chapter I have discussed constructive (closing) responses and opening responses to experience and their relationship to self-control behaviour. People employ self-control to overcome impediments to performing target behaviours. When the impediments derive from their own experiences (e.g. emotions, visceral reactions or intrusive thoughts), people may use various CRs to eliminate or reduce that interference. These CRs are the self-control skills entailed in learned resourcefulness (Rosenbaum, 1990a), and which manifest themselves in self-organized closure to experience. Yet there are circumstances in which the desired target behaviour is openness to experience (OTE), and a person's constructive responses interfere with the ability to be open to experience. In this case, they may use experiential self-control skills that both foster OTE and consist of a variety of ORs.

Cognitive psychotherapy emphasizes the use of constructive responses (i.e. learned resourcefulness) to overcome the obstructive effects of undesirable experiential reactions such as anxiety and depression (Beck, 1976; Ellis, 1973). The intervention promotes stability through fostering self-organized closure. Yet therapy also strives to produce change and promote growth and development. To achieve these goals, clients must be helped to become more OTE. Here experiential self-control skills may come into play. These skills can be acquired in therapy through training, for instance in relaxation, hypnosis, mindfulness (Kabat-Zinn, 1990; Marlatt, 1994), and a variety of 'acceptance' skills (Linehan, 1994). Recent research has shown that the addition of any of these 'opening' techniques to the standard procedures of cognitive therapy substantially enhances treatment outcome (Kirsch, Montgomery and Sapirstein, 1995; Linehan, 1994; Marlatt, 1994).

The intricate interplay between openness and closedness to experience, as defined in this chapter, characterizes many human endeavours. For example, a worker on an assembly line is expected to adhere closely to a well defined routine while being open and attentive to any changes on the line. His or her psychological well-being depends, to some extent, on the ability to be open to the experiences on the assembly line – that is, on the ability to recognize the challenges and pleasures involved even in work that may appear to be dry and boring (Csikszentmihalyi, 1990). Scientific work is also a good example of the interplay between opening and constructive responses. Most scientific work is guided by theory and an established conceptual system, yet serendipitous discoveries have often played a major role in advancing human knowledge. 'Part of the fun in research is in relishing the unexpected. Part of reality is also the recognition that great discoveries are rarely programmable' (Merbaum and Lowe, 1982: 121). Fully functioning persons are guided by established personal theories and at the same time are able to be open to their ever changing experiences and to the discovery of new things about themselves. Individuals who achieve an optimal balance between opening and constructive approaches to experience by the use of self-control skills are likely to enjoy a considerable sense of psychological and physical well-being.

References

AGAM, S. (1995) 'The impact of learned resourcefulness and experiential self-control on adjustment of dialysis patients', unpublished master's thesis, Tel Aviv University, Tel Aviv.

ANTONOVSKY, A. (1990) 'Pathways leading to successful coping and health', in M. Rosenbaum (ed.) *Learned Resourcefulness: On Coping Skills, Self-control and Adaptive Behaviour*, New York: Springer, 31–63.

ASCH, S. E. (1952) *Social Psychology*, New York: Prentice Hall.

BANDURA, A. (1977) 'Self-efficacy: toward a unifying theory of behavior change', *Psychological Review* **84**: 191–215.

BARON, R. M. and MISOVICH, S. J. (1993) 'Dispositional knowing from an ecological perspective', *Personality and Social Psychology Bulletin* **19**: 541–551.

BECK, A. T. (1976) *Cognitive Therapy and the Emotional Disorders*, New York: International University Press.

BOHART, C. A. (1993) 'Experiencing: the basis of psychotherapy', *Journal of Psychotherapy Integration* **3**: 51–67.

BOOTH-KEWLEY, S. and VICKERS, R. R. (1994) 'Association between major domains of personality and health behavior', *Journal of Personality* **62**: 282–298.

BURGER, J. M. and COOPER, H. M. (1979) 'The desirability of control', *Motivation and Emotion* **3**: 381–393.

CAREY, M. P., CAREY, K. B., CARNRIKE, C. L. M. and MEISLER, A. W. (1988) 'Learned resourcefulness, drinking, and smoking', paper presented at the annual meeting of the Association for the Advancement of Behavior Therapy, New York.

COAN, R. W. (1972) 'Measurable components of openness to experience', *Journal of Consulting and Clinical Psychology* **39**: 346.

COSTA, P. T. and McCRAE, R. R. (1992) *Revised NEO Personality Inventory (NEO-PI-R) and NEO Five-Factor Inventory (NEO-FFI) Professional Manual*, Odessa, FL: Psychological Assessment Resource.

CROWNE, D. P. and MARLOWE, D. A. (1960) 'A new scale of social desirability indepen-
dent of psychopathology', *Journal of Consulting Psychology* **66**: 547–555.
CSIKSZENTMIHALYI, M. (1990) *Flow: the Psychology of Optimal Experience*, New York:
Harper Perennial.
DERRY, P. A., CHOVAZ, C. J., MCLACHLAN, R. S. and CUMMINGS, A. (1993) 'Learned
resourcefulness and psychosocial adjustment following temporal lobectomy in
epilepsy', *Journal of Social and Clinical Psychology* **12**: 454–470.
ELLIS, A. (1973) *Humanistic Psychotherapy: the Rational-emotive Approach*, New York:
McGraw-Hill.
ELLIS, A. (1994) 'Acceptance in rational-emotive therapy', in S. C. Hayes, N. S.
Jacobson, V. M. Follette and M. J. Dougher (eds) *Acceptance and Change:
Content and Context in Psychotherapy*, Reno, NV: Context Press, 91–102.
EMMONS, R. A., KING, L. A. and SHELDON, K. (1993) 'Goal Conflict and the Self-
Regulation of Action', in D. M. Wegner and J. W. Pennebaker (eds.) *Handbook of
Mental Control*, Englewood Cliffs, NJ: Prentice Hall, 528–551.
EPSTEIN, R. (1991) 'Skinner, creativity, and the problems of spontaneous behavior',
Psychological Sciences **2**: 362–370.
EYSENCK, H. J. (1988) 'Personality, stress and cancer: prediction and prophylaxis',
British Journal of Medical Psychology **61**: 57–75.
FITZGERALD, E. T. (1966) 'Measurement of openness to experience: a study of regres-
sion in the service of the ego', *Journal of Personality and Social Psychology* **6**:
655–663.
FRIEDMAN, H. S. and BOOTH-KEWLEY, S. (1987) 'The "disease-prone personality": a
meta-analytic view of the construct', *American Psychologist* **42**: 539–555.
GRUBER, V. A. and WILDMAN, B. G. (1987) 'The impact of dysmenorrhea on daily
activities', *Behavior Research and Therapy* **25**: 123–128.
HARRIMAN, P. L. (1947) *Dictionary of Psychology*, New York: The Wisdom Library.
HEFER, C. (1991) 'Releasive Self-control', unpublished master's thesis, Tel Aviv
University, Tel Aviv.
JOHNSTON, W. A. and HAWLEY, K. J. (1994) 'Perceptual inhibition of expected inputs:
the key that opens closed minds', *Psychonomic Bulletin and Review* **1**: 56–72.
KABAT-ZINN, J. (1990) *Full Catastrophe Living*, New York: Delacorte.
KANFER, F. H. and GAELICK-BUYS, L. (1991) 'Self-management methods', in F. H.
Kanfer and A. P. Goldstein (eds) *Helping People Change: A Textbook of
Methods*, New York: Pergamon, 305–360.
KANFER, F. H. and SCHEFFT, B. K. (1988) *Guiding the Process of Therapeutic Change*,
Champaign, IL: Research Press.
KATZ, R. C. and SINGH, N. (1986) 'A comparison of current smokers and self-cured
quitters on Rosenbaum's self-control schedule', *Addictive Behaviors* **11**: 63–65.
KELLY, G. A. (1955) *The Psychology of Personal Constructs*, New York: Norton.
KING L. and EMMONS, R. A. (1990) 'Conflict over emotional expression: psychological
and physical correlates', *Journal of Personality and Social Psychology* **58**:
864–877.
KIRSCH, I., MONTGOMERY, G. and SAPIRSTEIN, G. (1995) 'Hypnosis as an adjunct to
cognitive-behavioral psychotherapy: a meta-analysis', *Journal of Consulting and
Clinical Psychology* **53**: 214–220.
KOBASA, S. C. O. (1979) 'Stressful life events, personality, and health: an inquiry into
hardiness', *Journal of Personality and Social Psychology* **37**: 1–11.
LEON, G. R. and ROSENTHAL, B. S., (1984) 'Prognostic indicators of success or relapse
in weight reduction', *International Journal of Eating Disorders* **3**: 15–24.
LEVY, S. M. (1985) *Behavior and cancer*, San Francisco: Jossey-Bass.
LEWIS, M. D. (1995) 'Cognition-emotion feedback and the self-organization of
developmental paths', *Human Development* **38**: 71–102.
LINEHAN, M. M. (1994) 'Acceptance and change: the central dialectic in psychotherapy',

in S. C. Hayes, N. S. Jacobson, V. M. Follette, and M. J. Dougher (eds) *Acceptance and Change: Content and Context in Psychotherapy*, Reno, NV: Context Press, 73–86.

McCRAE, R. R. (1993-94) 'Openness to experience as a basic dimension of personality', *Imagination, Cognition and Personality* **13**: 39–55.

McCRAE, R. R. and COSTA, P. T. (1983) 'Joint factors in self-reports and ratings: neuroticism, extraversion, and openness to experience', *Personality and Individual Differences* **4**: 245–255.

McCRAE, R. R. and COSTA, P. T. (1991) 'Adding Liebe und Arbeit: the full five-factor model and well-being', *Personality and Social Psychology Bulletin* **17**: 227–232.

MAHONEY, M. J. (1991) *Human Change Processes*, New York: Basic Books.

MAHONEY, M. J. and LYDDON, W. J. (1988) 'Recent developments in cognitive approaches to counseling and psychotherapy', *The Counseling Psychologist* **16**: 190–234.

MARLATT, G. C. (1994) 'Addiction, mindfulness, and acceptance', in S. C. Hayes, N. S. Jacobson, V. M. Follette, and M. J. Dougher (eds) *Acceptance and Change: Content and Context in Psychotherapy*, Reno, NV: Context Press, 175–197.

MASLOW, A. H. (1970) *Motivation and Personality*, 2nd edition, New York: Harper & Row.

MEICHENBAUM, D. (1985) *Stress Inoculation Training*, New York: Pergamon Press.

MERBAUM, M. and LOWE, M. R. (1982) 'Serendipity in research in clinical psychology', in P. C. Kendall (ed.) *Handbook of Research Methods in Clinical Psychology*, New York: Wiley and Sons Inc, 95–123.

MITTELMAN, W. (1994) 'Openness, optimal functioning, and final causes: a further reply to Tobacyk', *Journal of Humanistic Psychology* **34**: 100–107.

ORR, E. and WESTMAN, M. (1990) 'Does hardiness moderate stress, and how?: a review', in M. Rosenbaum (ed.) *Learned Resourcefulness: On Coping Skills, Self-control and Adaptive Behavior*, New York: Springer, 64–94.

PENNEBAKER, J. W. (1985) *The Psychology of Physical Symptoms*, New York: Springer-Verlag.

PRIGOGINE, I. and STENGERS, R. (1984) *Order out of Chaos*, New York: Bantam.

ROCHE, S. M. and McCONKEY, K. M. (1990) 'Absorption: nature, assessment, and correlates', *Journal of Personality and Social Psychology* **59**: 91–101.

ROGERS, C. R. (1980) *A Way of Being*, Boston: Houghton Mifflin.

ROSENBAUM, M. (1980a) 'A schedule for assessing self-control behaviors: preliminary findings', *Behavior Therapy* **11**: 109–121.

ROSENBAUM, M. (1980b) 'Individual differences in self-control behaviors and tolerance of painful stimulation', *Journal of Abnormal Psychology* **89**: 581–590.

ROSENBAUM, M. (1983) 'Learned resourcefulness as a behavioral repertoire for the self-regulation of internal events: issues and speculations', in M. Rosenbaum, C. M. Franks and Y. Jaffe (eds) *Perspectives on Behavior Therapy in the Eighties*, New York: Springer, 54–73.

ROSENBAUM, M. (1988) 'Learned resourcefulness, stress, and self-regulation', in S. J. Fisher and J. Reason (eds) *Handbook of Life Stress, Cognition and Health*, Chichester: John Wiley, 483–496.

ROSENBAUM, M. (1989) 'Self-control under stress: the role of learned resourcefulness', *Advances in Behavior Research and Therapy* **11**: 249–258.

ROSENBAUM, M. (1990a) 'The role of learned resourcefulness in self-control of health behavior', in M. Rosenbaum (ed.), *Learned Resourcefulness: On Coping Skills, Self-control and Adaptive Behavior*, New York: Springer, 3–30.

ROSENBAUM, M. (1990b) 'A model for research on self-regulation: reducing the schism between behaviorism and general psychology', in G. H. Eifert and I. M. Evans (eds) *Unifying Behavior Therapy: Contribution of Paradigmatic Behaviorism*, New York: Springer, 126–149.

ROSENBAUM, M. (1993) 'The three functions of self-control behavior: redressive, reformative and experiential', *Work and Stress* 7: 33–46.

ROSENBAUM, M. (1995) Unpublished data on the experiential self-control scale, Tel Aviv University.

ROSENBAUM, M. and BEN-ARI SMIRA (1986) 'Cognitive and personality factors in the delay of immediate gratification of hemodialysis patients', *Journal of Personality and Social Psychology* 51: 357–364.

ROSENBAUM, M. and PALMON, N. (1984) 'Helplessness and resourcefulness in coping with epilepsy', *Journal of Consulting and Clinical Psychology* 52: 244–253.

RUDE, S. S. (1989) 'Dimensions of self-control in a sample of depressed women', *Cognitive Therapy and Research* 13: 363–375.

SALOVEY, P. and MAYER, J. D. (1989–90) 'Emotional intelligence', *Imagination, Cognition, and Personality* 9: 185–211.

SELIGMAN, M. E. P., CASTELLON, C., CACCIOLA, J., SCHULMAN, P., LUBORSKY, L., OLLOVE, M. and DOWNING, R. (1988) 'Explanatory style change during cognitive therapy for unipolar depression', *Journal of Abnormal Psychology* 97: 13–18.

SMITH, J. C. (1988) 'Steps towards a cognitive-behavioral model of relaxation', *Biofeedback and Self-Regulation* 13: 307–329.

SMITH, T. V. G. (1979) 'Cognitive correlatives of response to a behavioral weight control program', unpublished doctoral dissertation, Queen's University, Kingston, Ontario, Canada.

SPANOS, N. P. and FLYNN, D. M. (1989) 'Simulation, compliance and skill training in the enhancement of hypnotizability', *British Journal of Experimental and Clinical Hypnosis* 6: 1–8.

SPIELBERGER, C. D., GORSUCH, R. L. and LUSHENE, R. E. (1970) *Manual for the State-Trait Anxiety Inventory*, Palo Alto, CA: Consulting Psychologists Press.

SUGIWAKA, H. and AGARI, I. (1993) 'Two kinds of self-control as effective stress-management technique: combination of redressive and reformative functions', paper presented at the International Congress of Health Psychology, Tokyo, Japan.

SUGIWAKA, H. and AGARI, I. (1995, July) 'The effects of controllability and predictability as situational factors on two types of self-control', paper presented at the World Congress of Behavioral and Cognitive Therapies, Copenhagen, Denmark.

TEASDALE, J. D., SEGAL, Z. and WILLIAMS, M. J. (1995) 'How does cognitive therapy prevent relapse and why should attentional control (mindfulness) training help?', *Behavior Research and Therapy* 33: 25–39.

TELLEGEN, A. (1985) 'Structures of mood and personality and their relevance to assessing anxiety with an emphasis on self-report', in A. H. Tuma and J. D. Messe (eds) *Anxiety and Anxiety Disorders*, Hillsdale, NJ: Erlbaum, 681–706.

TELLEGEN, A. and ATKINSON, G. (1974) 'Openness to absorbing and self-altering experiences ('absorption'), a trait related to hypnotic susceptibility', *Journal of Abnormal Psychology* 83: 268–277.

TOOMEY, T. C., SEVILLE, J. L., MANN, J. D., ABASHIAN, S. W. and WINGFIELD, M. S. (1995) 'Relationship of learned resourcefulness to measures of pain description, psychopathology and health behavior in a sample of chronic pain patients', *The Clinical Journal of Pain* 11: 259–266.

TRULL, T. J. and SHER, K. J. (1994) 'Relationship between the five-factor model of personality and axis I disorders in a non-clinical sample', *Journal of Abnormal Psychology* 103: 350–360.

WEISENBERG, M., WOLF, Y., MITTWOCH, T. and MIKULINCER, M. (1990) 'Learned resourcefulness and perceived control of pain: a preliminary examination of construct validity', *Journal of Research in Personality* 24: 101–110.

WILLIAMS, R. B., HANEY, T. L., LEE, K. L., KONG, Y., BLUMENTAL, J. and WHALEN, R. E. (1980) 'Type A behavior, hostility, and coronary atherosclerosis', *Psychosomatic Medicine* **42**: 539–550.

YOSCHPE, O. and ROSENBAUM, M. (1995) 'The impact of learned resourcefulness and experiential self-control on subjective well-being via their impact on ambivalence about emotional expression', unpublished manuscript, Tel Aviv University, Tel Aviv.

PART TWO

Work problems and coping

CHAPTER FOUR

Coping with acute workplace disasters

CATHERINE LOUGHLIN AND JULIAN BARLING

Coping with acute disasters at work

Workplace disasters are inevitable, and can produce tremendous stress for the individuals involved. While much is known about work stress, particularly chronic work stress, far less is known about acute workplace disasters. In this chapter we review the literature on the effects of acute workplace disasters in order to understand their effects on employee health and well-being. We then ascertain how individuals cope with these situations. Understanding how people cope with workplace (or technological) disasters is particularly important because their negative effects may exceed those of natural disasters (Baum, 1988).

Defining a 'disaster' is no easy task (see Quarantelli, 1985). Issues concerning both the severity (i.e. its effect or outcomes) and the breadth of the events (i.e. the number of people involved) have typically been considered (e.g. Baum, 1987). Quarantelli (1985) suggests that the degree of social disruption is also critical in defining a disaster, irrespective of the degree of physical destruction. Thus, consistent with the notion that both the physical environment and its inhabitants are to be considered, a disaster occurs when the demands posed by a crisis exceed the resources, preparation and capabilities of a group to respond (e.g. Quarantelli, 1985). Thus, a disaster is an event characterized by great power, sudden onset, excessive demands on individual coping, and large scope (affecting many people). These events are generally beyond both the realm of normal, everyday experience and the immediate control of victims, and as close to universally stressful as any event can be (Baum, 1987). By viewing a disaster as a collective stressor, we invoke theory and research on stress to aid us in understanding and predicting the effects of a disaster on individuals and organizations.

Although the definition of the term 'stress' is no less problematic, there seems to be some agreement about the following terms which we will use throughout. *Stressors* are the originating events or objective environmental characteristics in a given stressful situation; they are quantifiable and objectively verifiable. *Stress* reflects the subjective interpretation or experience of stressors, i.e. different people experiencing the same event may interpret or perceive it in different ways. *Strain* refers to the outcome of stress (e.g. psychosomatic complaints, depression, anxiety). Also, there are different types of stressor. *Acute stressors* have a clear and specific onset, are of short-term duration and have a low probability of occurring again. In contrast, *chronic stressors* last a long time, are highly repetitive and it is usually difficult to specify the exact time of their onset (Pratt and Barling, 1988).

The distinction between acute and chronic stress can be obscure. For example, although DSM-III-R (1987) defined an acute stressor as any event continuing for six months or less, some researchers have suggested that acute stressors have a far shorter duration than this (e.g. Pratt and Barling, 1988). DSM-IV (1994) now defines an acute stressor as lasting for less than one month. On the other hand, Baum (1987) considers acute effects to last for a year. Perhaps more importantly, some events which begin as acute stressors may become chronic and exert long-term effects (e.g. the Three Mile Island (TMI) nuclear disaster). Thus, just when acute stressors should be viewed as chronic remains unclear.

An important observation when studying acute workplace (i.e. techno-logical) disasters, is that most workplace or technological disaster research has focused on chronic events whose long-term consequences are uncertain, such as gradual toxic exposures to chemicals or radiation, as in TMI. In some cases, chronic disasters are not even marked by a clear point of onset (e.g. asbestos inhalation in Baum, 1988). In contrast, the majority of natural disasters studied have been acute (i.e. short-term) events with a recognizable *low point*, when the worst is perceived to have passed (e.g. hurricanes or floods). This leaves some uncertainty in drawing conclusions concerning acute technological or workplace disasters, which are the focus of this chapter.

Some unique aspects of workplace diasters should be noted. First, they may unleash tremendous destruction, yet are difficult to predict. This is important, as it reduces perceived and actual controllability. Second, they usually provide some focus for blame (e.g. the organization). Third, there may be some confusion about the actual degree of damage done in workplace disasters (e.g. Three Mile Island, where spouses could not even agree on whether or not to evacuate). By contrast, natural disasters generally seem to trigger more social cohesion, with communities working together to overcome the adversities of nature (Handford, Martin and Kales, 1988).

There are several characteristics inherent in a disaster that are typically associated with subsequent pathology. When considering the trauma poten-tial of disasters in general (i.e. regardless of whether they are acute or chronic, natural or human made) there seem to be certain event and victim characteristics that are consistently associated with subsequent disruption and pathology (Bolin, 1985; Baum, 1987). According to Baum (1987), the extent of horror (witnessing death or dealing with dead or dying victims) or terror (related to the proximity of victims to the raw physical events of the disaster, e.g. collapse of buildings or bridges) to which victims are exposed appears to affect the degree of trauma: the suddenness or scope (i.e. impact ratio), and intensity of the event are important. Also, the degree of preparedness of those affected, their familiarity with the given type of disaster event and the extent of warnings are all important in determining the effects of the event. Finally, low points appear to be important (i.e. if there is no clear low point, consequences may become chronic or may increase).

One additional point warrants clarification. There are three separate groups of workers that can be considered. First, and perhaps most obvious, are on-site workers who are employed in the organization experiencing the disaster. However, there are two additional groups of workers directly involved in the consequences of a disaster, namely on-site rescue workers (i.e. those who aid those affected at the scene of the event), and off-site rescue workers (e.g. medical personnel receiving victims at a nearby hospital). We will focus primarily on on-site workers and on-site rescuers, due to the minimal information available on off-site rescuers. The disaster site becomes the place of work for on-site rescue workers as well as on-site employees. Although it would be helpful to consider on-site employees and on-site rescue workers separately, the state of the literature does not permit this.

We will also attempt to limit our discussion to *acute* workplace disasters (i.e. events with a clear and specific onset, short-term duration, and a low probability of recurrence). Where the literature is scarce, we will incorporate findings from disasters of an acute/chronic or chronic nature.

The effects of acute workplace disasters on employee health and performance

The most common response to an acute natural disaster is the disaster syndrome, in which 'dazed behaviour and psychic detachment, is frequently followed by shock, a sense of loss, anxiety, and in some cases, activity directed towards saving lives or restoring property to its former state' (Baum, 1987: 18). Contrary to the media image of disasters, people tend not to panic or compete for scarce resources, but rather proceed in a rather organized fashion and often engage in prosocial behaviour (e.g. rescue attempts and group formation). Natural disasters are typically associated with acute (i.e. short-term) strain (e.g. disruption of normal activities, anxiety) (Baum, 1987). Although it is possible that this might simply reflect a paucity of studies addressing long-term effects, when long-term effects do occur, they are usually a function of the severity of interference with individuals and social functioning.

Smith, North and Price (1988) summarize typical emotional response to technological disasters. They include repeated recall of the event, reliving the trauma, 'psychic numbing', searching for a scapegoat, searching for meaning, demoralization, somatic symptoms, hostility, distrust of authorities, alienation, increased alcohol consumption, sleep disturbances, and recurrent dreams or nightmares. The notion of 'survivor guilt' (i.e. the feeling of responsibility felt by those who escape death in a disaster in which others perish) has also been identified, and arose from studies of the survivors of the Coconut Grove fire in 1942 (Handford et al., 1988). In that incident a violent fire swept through Coconut Grove, a popular Boston nightclub, causing the death of nearly 500 people.

In contrast to focusing on individual symptoms, some researchers have concentrated on psychiatric illnesses or syndromes in those affected by disasters. Lifton and Olson (1976) identified a survivor syndrome in response to disasters, that includes death anxiety, death guilt, 'psychic numbing', impaired human relationships, and a struggle to give significance and meaning to the disaster experience. Others refer to 'reactive disaster syndrome' characterized by the expression of anxiety, depression, sadness and loss of concentration (e.g. Lopez-Ibor et al., 1985). Post-traumatic stress disorder (PTSD) is also specifically trauma-related and its study is most useful in understanding a variety of mental health reactions to acute stress resulting from disaster situations (Handford et al., 1988). Indeed, some authors conclude that PTSD is the psychiatric disorder most closely associated with the experience of disaster (Smith et al., 1988).

A few additional comments concerning PTSD are necessary. Some researchers suggest that pre-existing personality factors are more likely to be responsible for PTSD following a disaster than the event itself (e.g. McFarlane, 1988). Others argue that the nature of the disaster is a more significant determinant of PTSD than prior personality (e.g. Leopold and Dillon, 1963; Hoiberg and McCaughey, 1984). Some authors caution that there is a tendency for clinicians to over-diagnose PTSD when there is ongoing litigation (Rosen, 1995). Leopold and Dillon (1963) stated that if PTSD is untreated it tends to worsen with time. Finally, some authors feel that PTSD is attracting too much attention as a negative psychological consequence of exposure to disasters, and are working on the development of a measure to assess both the positive and negative responses to disaster experiences (Joseph, Williams and Yule, 1993).

On-site workers who are present during the disaster and may be responsible for bringing it under control are sometimes referred to as 'first party victims'; in industrial plants these would be, for example, plant workers, maintenance personnel and security personnel. They are usually the first people exposed to the effects of the disaster and are at greatest risk of severe injury or even loss of life as a result of the event. They are also most likely to be exposed to scenes of death, decomposing bodies, dismemberment, disfigurement, or other severe injury. Handford et al. (1988) consider these individuals to be a special group that may be uniquely affected by technological and workplace disasters, and subsequently require specialized mental health services to assist them in coping.

Some studies have considered the well-being of workers on the job at the time of an acute workplace disaster. Findings suggest that the strain experienced by workers following an acute workplace disaster dissipates quickly after the event. For example, Barling, Bluen and Fain (1987) studied an explosion at the world's largest dynamite factory in which fourteen people were killed and fourteen others injured. Two weeks after the event, no significant effects existed in either the personal functioning (marital satisfaction and psychological distress) or organizational functioning (organizational commitment and job satisfaction) of workers physically exposed to the explosion or those performing a different job function at the same plant. Furthermore, in another study the responses of police officers to an acute stressor (viz. retrospective reports of having been involved in a shooting in the previous twenty years) indicated that the strain associated with this acute stressor dissipated within a week of the event (Loo, 1986). This research suggests that for these workers the strain following an acute stressor may dissipate fairly quickly after the event.

Barling et al. (1987) did raise the possibility that effects not seen immediately after the disaster may become apparent later. Theorell et al.'s (1994) findings support this. In a study of train drivers who had experienced a 'person under train' incident, sickness absence was elevated for three weeks following the disaster. There were no effects for the following two months. Thereafter, however, there was again a significant increase in sickness absence from three to twelve months following the incident. The possibility that acute events exert long-term effects, therefore, cannot be excluded.

Unfortunately, not enough is known about on-site workers experiencing acute disasters. Obtaining information from employees who were on the job when an acute disaster occurred can be difficult. For example, Wilkinson (1983) surveyed official, unofficial and off-site rescuers, as well as victims and observers after the collapse of the Hyatt Regency Hotel's Skywalks in Kansas City in 1981. He was prevented from obtaining data from hotel employees who were on the job at the time of the accident, possibly because of the company's fear of legal reprisals.

Before definitive conclusions can be reached, other factors that might moderate the psychological effects of the workplace should be considered. For example, the degree of terror and horror experienced by individuals in Barling et al.'s (1987) study might be considered relative to the Coconut Grove fire, which was also an acute disaster event but which had a greater degree of terror and horror associated with it (and higher subsequent degrees of pathology). Markowitz et al. (1987) studied firefighters and found short-term psychological effects one to two months after responding to a chemical fire (e.g. demoralization, specific emotional distress). But can a chemical fire really be considered an 'acute stressor', given the 'persistent threat' that would follow (e.g. uncertainty regarding how harmful the chemicals really were)?

Such questions make it difficult to determine whether or not there are long-term effects following an acute disaster.

What happens when an acute workplace disaster turns into a chronic situation? Koscheyev et al. (1993) studied chief operators at the Chernobyl power plant at four time periods following the nuclear disaster which occurred in 1986. They found that health concerns, depression and other indicators of strain increased significantly over time. They suggest that this may be due to the continuing uncertainty concerning the safety of working at the Chernobyl plant (i.e. no 'low point' was reached). Kasl, Chisholm and Eskenazi (1981) found that six months after the TMI accident, nuclear workers at the plant differed from a matched sample of controls in their reported exposure to radiation at the time of the accident and in their feeling that their health had thus been endangered. They had lower job satisfaction and reported more uncertainty about their occupational futures. TMI workers also reported greater frequency of anger, anxiety and various psychophysiological symptoms six months after the accident. In contrast, twelve months after the accident, using different measures, Bromet et al. (1980) found that there was only a small difference between TMI and control workers, although TMI workers reported feeling more rewarded by their jobs.

In 1983, Raphael et al. noted the paucity of research on the aftermath of disasters for rescue workers. By 1988, Dunning acknowledged that the psychological well-being of emergency services personnel sent to disasters was at risk, and noted that there was little empirically-based data that could serve as a guide for prevention. Dunning (1988) argued, from a review of the limited research examining on-site rescuers, and from anecdotal reports of emergency workers after involvement in disasters, that participation causes psychological, physical and/or behavioural impairment that is usually temporary but sometimes permanent. Although there has been some movement towards filling this gap in the literature in the past decade, we still know very little about the emotional, cognitive and behavioural responses, coping strategies and long-term effects of disaster work on these rescue workers. We will present what is known at this stage.

In 1977 a crowded train in Granville, near Sydney, Australia was derailed and hit the supporting pillars of a bridge, which collapsed, killing eighty-three passengers and injuring over 200 others. In their study of the rescue workers involved in the disaster Raphael et al. (1983) showed that most rescue workers found the event to be personally stressful. They identified five main stressors: helplessness, the magnitude of the destruction, the sight and smell of mutilated dead bodies, witnessing the anguish of the relatives and the suffering of the injured, and working under time pressure (because the collapsed bridge had not stabilized by the time of rescue). One month after the disaster about 25 per cent of the rescue workers still had symptoms of anxiety, depression and insomnia. Certainly, involvement in graves registration duty (for on-site rescuers) was found to be associated with subsequent PTSD for almost half the rescue workers involved (Sutker, Uddo, Brailey, Vasterling and

Errera, 1994). However, working in a disaster situation can also have positive effects. Three-and-a-half per cent of the rescue workers in Raphael *et al.*'s (1983) study felt more positively about their lives as a result of their involvement in the disaster. Thus, it should also be noted that there is considerable variability in the experiences of workers both during and following a disaster. The major sources of stress for nurses (i.e. on-site personnel) are the urgency of the situation and the desire to perform at maximum levels, problems of organizing disaster care, identification with victims, the need to avoid conflict with other workers, and conflicts between their nursing and family roles (Miles *et al.*, 1984).

Wilkinson (1983) found that after the collapse of the Hyatt Regency Hotel Skywalks in 1981, the most common symptoms among official, unofficial or off-site rescuers, and victims and observers were repeated recollections of the disaster, sadness, fatigue, recurring feelings of anxiety and depression, dreams of the disaster, and sleep disturbances. Wilkinson (1983) also recognized the possibility of feelings of guilt after the accident. However, the reasons for the guilt feelings were diverse. For instance, several rescuers wished that they had done more to relieve pain and suffering, or death; others expressed guilt because they were alive.

Coping with acute workplace disasters

Once an understanding of the psychological consequences of workplace disasters is achieved, the next logical step involves moving towards a consideration of how any such effects can be alleviated. When we speak of 'coping' we typically mean the cognitive and behavioural efforts a person makes to manage those internal and external demands that exceed personal resources (Lazarus and Folkman, 1984). In the area of work stress, researchers have often relied on Lazarus and his colleagues' concepts of emotion- versus problem-focused coping. Emotion-focused coping involves strategies that an individual uses to control emotional distress resulting from a stressful incident. Problem-focused coping reduces the actual stressor by attempting to modify the environment or the individual's behaviour within that environment. A third type of coping, also considered in the literature, is that of appraisal-focused coping. In this case, cognitive strategies such as denial or redefinition of the stressful situation are used in order to cope (Moos and Billings, 1982).

Baum, Fleming and Singer (1983) found that in coping with chronic stress following the Three Mile Island incident, both emotion-focused coping and self-blame were associated with less stress than were problem-focused coping and denial. Given that individuals are far more likely to gain control over their emotions than the events at hand, emotion-focused coping may be most effective in this type of situation and lead to the greatest sense of control. Also, emotional regulation and assuming responsibility for difficulties were related to each other and to perceived control. They concluded that a 'control-oriented' coping style (in which the perception of control is actively created and maintained by whatever means necessary) can be effective in reducing the distress associated with technological catastrophes. Baum, Fleming and Davidson (1983) concluded that technological or workplace catastrophes are more likely than natural disasters to cause chronic stress and widespread effects: technology is supposed to be under our control. However, incidents like TMI, the leak of toxic gas at Bhopal and Chernobyl show that this is not always the case. Where people are unable to regain a sense of control in their lives, chronic stress seems to be the result.

One method of coping with work stress that has received considerable attention is social support (House, 1981). McCarroll et al. (1993) discuss the coping strategies of personnel dealing with violent deaths. They found it important that the personnel received social support and full disclosure of gruesome details before they participated in body handling. During the work, the following strategies were used to maximize coping: concealing odours in any way possible, wearing gloves, viewing remains as non-human, avoiding emotional identification with the deceased, humour, professionalizing the job, and work-group support. Finally, after exposure, debriefings with the work group (professional counselling was seen as unacceptable), social support from family, alcohol, and in some cases memorial services, were all seen as helping workers cope with this unique stressor. Although these appear to be effective in the short term, their long-term consequences (e.g. use of alcohol) are unclear.

The family can also be an important source of support for rescue workers (e.g. Paton and Kelso, 1991). Wilkinson (1983) found that after the Hyatt Skywalks disaster, the family was considered to be an important resource for post-disaster coping for rescuers. Friends were the second most important source of support. Paton and Kelso (1991) also suggest that spouses could be a valuable resource for the organization responsible for the relief workers. Wives' perceptions of their husbands' reactions to a disaster (all relief workers were male in this study) could limit the use of denial, which is important as these individuals may be at risk for developing longer term problems (e.g. Horwitz et al., 1980).

Wilkinson (1983) found that after the Hyatt Regency disaster, the most common coping strategy among those interviewed (half of whom were official, unofficial or off-site rescuers) was talking about the event (for 37 per cent of the sample this was their principal outlet), although whom they were talking to is not identified. Thus, whether from family, friends or co-workers, social support is an important part of the coping process.

It should be noted that coping can be occupation specific. Employees within occupations have unique ways of coping with their work stress. Palmer (1983) studied emergency medical technicians who are in constant contact with people who are injured and/or dying, and found unique and apparently functional ways of coping. For example, the use of humour (using comedy as an escape or safety-valve), psychological distancing through technical language (e.g. death becomes a 'signal 27') and rationalization (e.g. you 'lose' a few, but without us none would survive) are used within this occupation but may be of little use in other occupations.

The role of the organization in facilitating personal coping

Despite the obvious need for coordination in successful disaster management, organizations are sometimes hesitant to assume responsibility for overall coordination, usually because of ambiguity about who has legitimate authority in a disaster (Baum, 1987).

Disaster-related interventions can be implemented before, during or after the disaster under the advice of mental health professionals. The most appropriate and cost-effective role for the mental health worker during times of disaster is that of a consultant to primary caregivers such as family physicians, clergy and full-time disaster personnel. Handford et al. (1988) state that to help the victims of technological accidents or disasters cope successfully, mental health systems must be ready to change their normal procedures and reach out to the community. This can be achieved by mental health professionals training temporary emergency mental health workers (e.g. college students) in how to deal with disaster-related human reactions, as was the case after Hurricane Agnes in Pennsylvania, in 1972. The services provided by the mental health system can include individual supportive counselling, working with other social service agencies to arrange for secure shelter and food, establishing 'crises lines', training volunteers to interact with victims, offering advice to planners, monitoring the psychological well-being of rescue workers, advising the media (e.g. about the nature of responses to a disaster and the availability of resources), and promoting public education (e.g. through offering immediate one-day seminars for community mental health workers, clergy and other community leaders). The underlying goal is to provide a social support network until people can re-establish their own support networks.

The importance of early intervention cannot be overstated. In their study of the rescue workers involved in the Granville rail disaster, Raphael et al. (1983) advocated early intervention after stressful rescue experiences and also highlighted the importance of debriefing sessions for rescue workers. In debriefing sessions, experiences, fears and triumphs are shared, and horrifying experiences discussed in the safety of the group. Both empirical findings and personal reports from participants forcefully support the necessity of including programmes in organizational interventions following disaster rescue work. This would assist with personal coping and returning the individual to normal functioning as soon as possible.

Group information sessions or therapeutic approaches may also be particularly useful. Wilkinson (1983) found that after the Hyatt Regency disaster, 30 per cent of those interviewed had attended group sessions after the disaster and found them helpful. Members of such support groups often continue to support one another after formal therapy sessions have stopped (Handford *et al.*, 1988). The clergy and other community leaders also fulfil an important role in interpreting the meaning of disastrous events. They can help victims and rescue workers put the event in a larger context (Handford *et al.*, 1988), and be an important source of social support. However, to facilitate this process, any stigma associated with participation in support or therapeutic groups must be diminished.

One problem that can reduce the chances for successful coping is the 'trauma membrane': the tendency for well-meaning relatives, friends, therapists, and/or whole communities to shield victims from memories of the event. This might encourage denial, and decrease the likelihood of seeking help when in need (Lindy, 1985). In contrast, Handford *et al.* (1988) suggest that the value of mental health intervention is that individuals soon reach the point where remembering the disastrous event becomes tolerable rather than a cause for concern or despair.

With respect to emergency workers, there is sometimes a reluctance by management to acknowledge the effects of duty-based disaster work because it is believed that the selection process should have excluded the most vulnerable individuals, and also because disasters are perceived to bring out the best in rescue workers (Dunning, 1988). Further, formally recognizing the possibility of duty-related trauma reactions occurring might increase the likelihood of successful workers' compensation claims and duty disability requirements (Dunning, 1988).

Where an organization's culture encourages denial of psychological strain following disasters, problems will be concealed and people will suffer. Even if potential decrements in organizational functioning and productivity caused by workers suffering from work-related trauma was not an adequate incentive for management to implement programmes to help them deal with work-related trauma, growing legal concerns would provide a sufficient impetus concerning organizations' responsibility for the health and safety of their employees (Dunning, 1988; Sells, 1994).

In an attempt to understand how organizations can facilitate the coping of rescue workers, it is helpful to look at one specific organization. Dover Air Force Base in Delaware is the designated East Coast mortuary for the US Military and has handled several disasters through the years (e.g. the Jonestown massacre, the bombings in Beirut, and the plane crash in Gander, Newfoundland in 1985). They customarily deal with about sixty deaths a month. In the case of the plane crash in Gander, the base was forced to handle the remains of nearly five times that number at one time. Consequently, they engaged the help of volunteers from the military. Thus, these volunteers were on-site workers, almost none of whom had been exposed to such a traumatic experience previously, but who were expected to be involved with the ill, the dead and the dismembered, and then return to and continue their normal job assignments.

The following observations and recommendations are gleaned from the extensive debriefings of rescue workers involved in the Gander disaster (Ursano and Fullerton, 1988):

- There are good reasons for using on-site volunteers in the case of mass tragedies (it is difficult to force someone to take part in body identification when this not part of their normal duties; large numbers of people are necessary and volunteers will often go the extra mile).
- Although some volunteers could be seeking the recognition that typically follows assisting in such a crisis (e.g. medal, promotion), and others may be curious, most seem to do it because of camaraderie (i.e. if they were the one who had been killed, they would want someone to do it for them and their families).
- Communication is important throughout the disaster, including an acknowledgement of the horrors that might be encountered and the symptoms that workers might experience.
- The primary stressors for workers seem to be the intensity of the sensory experience, (e.g. viewing, smelling and touching human remains, and identifying with the victims as human beings).
- Working with faulty equipment can be a significant stressor in addition to the human aspects of the tragedy.
- Some administrators felt that it was critical that the commanders and supervisors should assist in the mortuary to gain exposure to what was experienced by the volunteers. This would make them more sensitive to resultant problems in volunteers should they appear later. However, this suggestion was somewhat controversial.
- Policies and procedures for these types of event need to be documented. Many workers commented on 'reinventing the wheel' each time a disaster struck.
- There should be compulsory breaks, and frequent rotation of volunteers. (In the disaster following the bombing of a federal building in Oklahoma City in 1995, rescuers worked for no more than two hours before they were given a compulsory break).
- Rescue workers or body-handlers need a transition place to go to before returning to the outside world: e.g. a place where they can sit with other volunteers after their shift, have something to eat and drink and talk about the gruesome nature of the event. (In the Oklahoma disaster, a specific area was set up for rescue workers, and at the end of every two-hour work shift they attended mandatory debriefing sessions.)
- Minute details must be attended to in order to minimize the trauma on workers. For instance, regarding food preparation, it would not be appropriate to serve barbecued ribs to rescue workers who had been identifying charred bodies all day.
- It can be 'therapeutic' to get into another work setting immediately after the disaster (assuming there has been closure on the tragedy).

- Deprogramming sessions can be very important, not only in terms of the information they provide, but also in terms of giving workers access to mental health professionals, opening the lines of communication and encouraging a feeling of community.
- Chaplains are often the primary mental health workers in disaster situations. They are used as confidants by leaders and workers alike.
- Awards of recognition given to workers after the disaster can acknowledge the value the organization places on the efforts of these workers. However, there are often problems in ensuring that everyone is recognized.
- Individuals who had been involved in rescue work before the Gander disaster stated that it 'only gets worse' the more often you do it. Thus, whether rescue workers ever become accustomed to the stressors is questionable.

Studies on acute disasters: some methodological problems

There are some methodological problems that commonly apply to studies on acute workplace disasters. First, problems emerge because of the sudden onset of events and the difficulty in planning studies beforehand. Second, because the precise occurrence of a disaster is virtually impossible to predict, it is unlikely that non-retrospective baseline data will be available. Another challenge emerges from the difficulty associated with choosing the appropriate temporal lag between measuring the stressor and the strain associated with it (Barling, 1990). Moreover, the duration of an acute stressor may be extended beyond its physical life by virtue of victims re-experiencing the event or associated events; intrusive imagery or secondary events can keep the trauma alive (Baum, O'Keefe and Davidson, 1990). Another problem emerges because the literature on the psychological consequences of disasters has proceeded without a formal theory, making it difficult to organize research findings that might guide future research (Green, 1985). Finally, studies providing information on rates of occurrence of symptoms yield much higher rates of psychopathology than those studies reporting data on psychiatric syndromes or illnesses (Smith *et al.*, 1988). Thus, although there can be large differences in the estimated rates of occurrence of survivor impairment from one study to another, these differences are often more attributable to how, when and from whom data were collected than to genuine differences in impairment. In general, adverse mental health consequences are found in much higher rates in studies using in-depth, open-format interviews, or self-report than those using standardized structured research instruments (Smith *et al.*, 1988).

Conclusion

The reality is that, as much as we attempt to prevent them, disasters will continue to happen. It is our hope that even if we cannot stop them from occurring, we can at least minimize their negative effects on the victims and on-site and off-site rescue workers.

Much is left to be learnt, and critical questions remain. Just some examples of these questions are: How long can volunteers engage in acute disaster work before they should take time off? Are there any positive effects following successful disaster work? What are the effects on mental health professionals and chaplains who counsel on-site and off-site rescue workers? When do acute disasters become chronic? What predicts whether a workplace disaster will exert a short-term or long-term effect (Theorell *et al.*, 1994)?

Having researched the literature, our overriding conclusion is that the various groups of disaster workers provide an invaluable service to others in time of need. It is now time for researchers to provide sufficient information and offer viable strategies to enable these workers to cope more effectively when they perform this important function.

Note

Preparation of this chapter was supported by separate grants from the Social Sciences and Humanities Research Council of Canada to both authors.

References

BARLING, J. (1990) *Employment, Stress and Family Functioning*, Toronto: John Wiley & Sons.

BARLING, J., BLUEN, S. D. and FAIN, R. (1987) 'Psychological functioning following an acute disaster', *Journal of Applied Psychology* **72(4)**: 683–690.

BAUM, A. (1987) 'Toxins, technology, and natural disasters', in G. R. and B. K. Bryout (eds.) *Cataclysms, Crises, and Catastrophes: Psychology in Action*, Washington, DC: American Psychological Association.

BAUM, A. (1988) 'Disasters, natural and otherwise', *Psychology Today*: 57–60.

BAUM, A., FLEMING, R. and DAVIDSON, L. M. (1983) 'Natural disaster and technological catastrophe', *Environment and Behavior* **15(3)**: 333–354.

BAUM, A., FLEMING, R. and SINGER, J. E. (1983) 'Coping with victimization by technological disaster', *Journal of Social Issues* **39(2)**: 117–138.

BAUM, A., O'KEEFE, M. K. and DAVIDSON, L. M. (1990) 'Acute stressors and chronic response: the case of traumatic stress', *Journal of Applied Social Psychology* **20(20)**: 1643–1654.

BOLIN, R. (1985) 'Disaster characteristics and psychosocial impacts', in B. J. Sowder (eds) *Disasters and Mental Health: Selected Contemporary Perspectives*, Rockville, MD: National Institute of Mental Health, US Department of Health and Human Services.

BROMET, E. J., PARKINSON, D. K. and SCHULBERG, H. C. (1980) *Three Mile Island: Mental Health Findings*, Pittsburgh: Western Psychiatric Institute and Clinic and the University of Pittsburgh.

DIAGNOSTIC AND STATISTICAL MANUAL OF MENTAL DISORDERS (DSM-IV) (1994) 4th Edition, Washington, DC: American Psychological Association.

DUNNING, C. (1988) 'Intervention strategies for emergency workers', in M. Lystad (ed.) *Mental Health Response to Mass Emergencies: Theory and Practice*, New York: Brunner/Mazel.

GREEN, B. L. (1985) 'Conceptual and methodological issues in assessing the psychological impact of disaster', in B. J. Sowder (ed.) *Disasters and Mental Health: Selected Contemporary Perspectives*, Rockville, MD: National Institute of Mental Health, US Dept of Health and Human Services.

HANDFORD, H. A., MARTIN, E. D. and KALES, J. D. (1988) 'Clinical interventions: technological accidents', in M. Lystad (ed.) *Mental Health Response to Mass Emergencies: Theory and Practice*, New York: Brunner/Mazel.

HOIBERG, A. and McCAUGHEY, B. G. (1984) 'The traumatic after effects of collision at sea', *American Journal of Psychiatry* **141(1)**: 70–73.

HORWITZ, M. J. (1985) 'Disasters and psychological responses to stress', *Psychiatric Annals* **15(3)**: 161–167.

HORWITZ, M. J., WILNER, N., KALTREIDER, N. and ALVAREZ, W. (1980) 'Signs and symptoms of post-traumatic stress disorder', *Archives of General Psychiatry* **37**: 85–92.

HOUSE, J. S. (1981) *Work Stress and Social Support*. Reading, MA: Addison-Wesley.

JOSEPH, S., WILLIAMS, R. and YULE, W. (1993) 'Changes in outlook following disaster: the preliminary development of a measure to assess positive and negative responses', *Journal of Traumatic Stress* **6(2)**: 271–279.

KASL, S. V., CHISHOLM, R. F. and ESKANAZI, B. (1981) 'The impact of the accident at the Three Mile Island on the behavior and well-being of nuclear workers', *American Journal of Public Health* **71**: 472–495.

KOSCHEYEV, V. S., MARTENS, V. K., KOSENKOV, A. A., LARTZEV, M. A. and LEON, G. R. (1993) 'Psychological status of Chernobyl nuclear power plant operators after the nuclear disaster', *Journal of Traumatic Stress* **6(4)**: 561–568.

LAZARUS, R. and FOLKMAN, S. (1984) *Stress, Appraisal and Coping*, New York: Springer Publishing.

LEOPOLD, R. L. and DILLON, H. (1963) 'Psycho-anatomy of a disaster: a long term study of post-traumatic neuroses in survivors of a marine explosion', *The American Journal of Psychiatry* **119**: 913–921.

LIFTON, R. J. and OLSON, E. (1976) 'The human meaning of total disaster: the Buffalo Creek experience', *Psychiatry* **39**: 1–18.

LINDY, J. D. (1985) 'The trauma membrane and other clinical concepts derived from psychotherapeutic work with survivors of natural disasters', *Psychiatric Annals* **15(3)**: 153–160.

LOO, R. (1986) 'Post-shooting stress reactions among police officers', *Journal of Human Stress* **12**: 27–31.

LOPEZ-IBOR Jr, J. J., Canas, S. F. and RODRIGUEZ-GAZAMO, M. (1985) 'Psychopathological aspects of the toxic oil syndrome catastrophe', *British Journal of Psychiatry* **147**: 352–365.

McCARROLL, J. E., URSANO, R. J., WRIGHT, K. M. and FULLERTON, C. S. (1993) 'Handling bodies after violent death: strategies for coping', *American Journal of Orthopsychiatry* **63(2)**: 209–215.

McFARLANE, A. C. (1988) 'The aetiology of post-traumatic stress disorder following a natural disaster', *British Journal of Psychiatry* **152**: 116–121.

MARKOWITZ, J. S., GUTTERMAN, E. M., LINK, B. G. and RIVERA, M. (1987) 'Psychological response of firefighters to a chemical fire', *Journal of Human Stress* **13(2)**: 84–93.

MILES, M. S., DEMI, A. S. and MOSTYN-AKER, P. (1984) 'Rescue workers' reactions following the Hyatt Hotel disaster', *Death Education* **8**: 315–331.

MOOS, R. and BILLINGS, A. (1982) 'Conceptualizing and measuring coping resources and processes', in L. Goldberger and S. Breznitz (eds), *Handbook of Stress: Theoretical and Clinical Aspects*, New York: Free Press.

PALMER, C. E. (1983) 'A note about paramedics' strategies for dealing with death and dying', *Journal of Occupational Psychology* **56**: 83–86.

PATON, D. and KELSO, B. A. (1991) 'Disaster rescue work: the consequences for the family', *Counselling Psychology Quarterly* **4(2/3)**: 221–227.

PRATT, L. I. and BARLING, J. (1988) 'Differentiating between daily events, acute, and chronic stressors: a framework and its implications', in J. R. Hurrell, L. R. Murphy, S. L. Sauter, and C. L. Cooper (eds), *Occupational Stress: Issues and Developments in Research*, London: Taylor and Francis, 41–53.

QUARANTELLI, E. L. (1985) 'What is a disaster? The need for clarification in definition and conceptualization in research', in B. J. Sowder *Disasters and Mental Health: Selected Contemporary Perspectives*, Rockville, MD: National Institute of Mental Health; US Dept. of Health and Human Services.

RAPHAEL, B., SING, B., BRADBURY, L. and LAMBERT, F. (1983) 'Who helps the helpers? The effects of a disaster on the rescue workers', *Omega* **14(1)**: 9–20.

ROSEN, G. M. (1995) 'The *Aleutian Enterprise* sinking and post-traumatic stress disorder: misdiagnosis in clinical and forensic settings', *Professional Psychology: Research and Practice* **26(1)**: 82–87.

SELLS, B. (1994) 'What asbestos taught me about managing risk', *Harvard Business Review*, March-April, 76–90.

SMITH, E. M., NORTH, C. S. and PRICE, P. C. (1988) 'Response to technological accidents', in M. Lystad (ed.) *Mental Health Response to Mass Emergencies: Theory and practice*, New York: Brunner/Mazel.

SUTKER, P., UDDO, M., BRAILEY, K., VASTERLING, J. J. and ERRERA, P. (1994) 'Psychopathology in war-zone deployed and non-deployed operation desert storm troops assigned graves registration duties', *Journal of Abnormal Psychology* **103**: 383–390.

THEORELL, T., LEYMANN, H., JODKO, M., KONARSKI, K. and NORBECK, H. E. (1994) '"Person under train" incidents from the subway driver's point of view – a prospective 1-year follow-up study: the design, and medical and psychiatric data', *Social Science and Medicine* **38**, 471–475.

URSANO, R. and FULLERTON, C. (eds) (1988) *Exposure to Death, Disasters, and Bodies*, Bethesda, MD: F. Edard Herbert School of Medicine, Uniformed Services University of the Health Sciences.

WILKINSON, C. B. (1983) 'Aftermath of a disaster: the collapse of the Hyatt Regency Hotel skywalks', *American Journal of Psychiatry* **140(9)**: 1134–1139.

CHAPTER FIVE

Work, family and psychological functioning: conflict or synergy?

ESTHER R. GREENGLASS

This chapter addresses the relationship between the spheres of work and family and its implications for psychological functioning. In the past, work and family were regarded as separate spheres. However, in recent research these spheres have been regarded as interrelated and capable of affecting each other in both positive and negative ways. Originally, research in the area of work was conducted primarily with men, whose employment tended to be seen as a source of stress and the family as a source of emotional support. However, with increasing numbers of women entering the labour force, it is clear (mainly from research on women) that work often functions as a buffer for stress emanating from familial roles. Contrary to traditional assumptions, the family can often serve as a source of considerable stress, and with so many dual-earner families having children, both women and men are attempting to juggle work and family roles.

In this chapter, discussion is focused on the interaction between work and family roles, including synergistic effects, and inter-role conflict. For both men and, women coping with role conflict is an increasingly significant issue. Discussion here focuses on coping with work-family conflict from both the individual and the organizational perspective.

Work and family spheres

When one income could support the family and men spent their time and energy generating that income, women's roles focused on taking care of life in all other spheres. It tended to be a world where the public sphere of economic activity was separate from the private one of family care, and a world in which organizations could assume that their employees had no responsibilities outside of their paid employment. The assumption was that unless one kept the two spheres separate, family matters could intrude on work, thus making one less efficient in one's job. This assumption derived from a paradigm that posited work as the most important pursuit and non-work activities as detracting from the main sphere of activity. The process of segmentation – the view that work and family are independent, and that they do not affect each other – was held by employees (mainly male) and researchers alike. However, research suggests that if segmentation does occur, it does not happen naturally. Workers actively try to separate work and family life as a means of dealing with work-related stress (Piotrkowski, 1979).

Organizations have tended to reinforce the idea of segmentation, assuming that their employees were workers only. Organizational demands had to be met and work had to come first. Companies have assumed that all workers define success similarly, and wish to follow one dominant career path. As argued by Bailyn (1993), organizations continue to reward the full commitment of their employees on the basis of the amount of visible time spent at the workplace. Separation of work and family has been supported and encouraged by such policies which, at the same time, tended to penalize workers who sought flexibility in working hours.

Theories of stress and coping in the workplace have been developed in studies on men, and are consistent with their lifestyle. One of the basic tenets of these theories has been that a person's work outside the home is one's primary or sole source of fulfilment. An individual's self-esteem is typically seen as depending largely on their performance on the job, often to the exclusion of how they behave in other roles (in the family, for example). The family tends to be seen primarily in terms of the benefits it provides to the worker: in particular, social support. Few discussions focus on the family as a source of stress for the worker. A person defined as 'successful' at coping is one who can distance him or herself from familial relationships, which are seen as distractions from the job. Both employees and researchers alike have tended to value the ability to keep work and familial roles separate, particularly on the way up the corporate ladder (Bartolemé and Evans, 1979; Burke and Greenglass, 1987).

Implications of women's increasing employment

There is increasing recognition that the relationship between work and family is an important one for understanding not only the psychological functioning of the individual but also the functioning of the organization. Although it has been assumed that the two spheres were guided by different values and standards and dominated by persons of different gender, demographic changes and new life-styles are calling for a greater integration of the spheres. As increasing numbers of women seek to combine work and family roles, organizations are being challenged to respond to changing circumstances in the work force.

Interest in the relationship between the two spheres was spurred by the steady increase in the rate of participation of women in the labour force and, in particular, of married women.

Given the persistence of gendered occupational ghettos and the fact that men earn significantly more than women, women still do not experience equality with men in the workplace. While women comprise 45 per cent of the labour force, 80 per cent of all wives in dual-earner families earn less than their husbands (Gilbert, 1993). Women and men between the ages of 25 and 29 are equally likely to have college degrees, but the same degrees buy higher salaries for men, and a woman is unlikely to achieve parity. Women continue to be under-represented in nearly all professional areas and less than 3 per cent of officer positions in Fortune 500 companies are held by women (Gilbert, 1993). Moreover, many women occupy clerical roles and thus are especially likely to experience lack of autonomy and control over job tasks, under-utilization of skills and lack of recognition for accomplishments (Haynes and Feinleib, 1980).

All of these developments have made more difficult the task of providing family care, and to date the institutional response has been inadequate to meet the needs of individuals and families. With the influx of married women into the paid labour force, there was a corresponding increase in studies that examined whether employment was good for women. In general, studies comparing physical and psychological health status between housewives and employed wives and mothers have shown that employed wives have better health (Merikangas, 1985; Verbrugge, 1982; Waldron and Herold, 1984). Bringing in a wage is a tangible symbol of competence, and employed wives have been found to have higher self-esteem, more self-confidence and a greater sense of personal competence and autonomy (Baruch, Biener and Barnett, 1987).

An examination of studies on dual-earner families over the last twenty years reveals a greater trend to focus on women than on men. In these studies, issues centre around whether women who expand their traditional role (wife, mother) to include employment can do so without harming themselves, their children and/or their spouses. One approach employed to articulate the diversity in professional women's involvement in career and family over a lifespan is seen in the work of Lee (1994). She developed six models based on a review of the literature and designed to represent the variation in women's career patterns on three dimensions: timing of children, level of involvement in career over time and level of involvement with family over time. One of these models, 'Early Career Orientation Sustained', posits a high degree of involvement in a woman's career sustained over time, and a low level of involvement with family sustained over time. In this model the timing of children does not interfere with the woman's career. Personal characteristics of the women in this model included high levels of energy and drive, profound passion for their work and a sense of identity that was strongly tied to their professional accomplishments. Not highly invested in their role as a parent, they were equanimous about delegating primary responsibility for childcare to someone else. In another model, 'Early Career and Family Orientation', career and family are launched early (in the woman's twenties) and a pattern of mutual accommodation of career to family and family to career is established from the start. Women adhering to this model have a moderate level of involvement in their career in their twenties and thirties, increasing to high as children become more independent. Their involvement in family is moderate in their twenties and thirties, decreasing to low as children become more independent. Using these models, Lee (1994) put forth a series of hypotheses consisting of individual and family outcomes of various ways of combining career and family. The hypotheses were tested through interviews with professional women who fitted into each of the six models. Findings indicated that certain factors were likely to influence women's emergent patterns of combining career and family over time. These include degree of salience of the parental role, needs for independence versus needs for interdependence, and avail-ability of spousal and surrogate care for children.

In a study that examined the importance of family structure to the careers of managers, Schneer and Reitman (1993) studied the interactive effects of marital status, parental status, spousal employment status and gender on income and career satisfaction in a sample of Masters in Business Adminis-tration degree holders. The findings indicated that men who had children and employed wives earned less than men whose wives were not employed. Further, the results revealed that members of post-traditional families (families with children and employed wives) were more satisfied with their careers than members of 'traditional' families. This may be because their career paths have enabled them to fulfil the multiple roles of spouse, parent and worker (Sieber, 1974), thus resulting in greater career satisfaction. This research demonstrates the importance of family structure (i.e. traditional versus post-traditional) for analyses of career satisfaction and income.

Interaction between work and family roles

Because the majority of employed women also occupy family roles, assess-ments of the effects of their employment cannot be made in isolation. The quality of their familial relationships also have to be taken into account. Negative job conditions, both directly and in combination with difficult family situations, may mean that for some women the costs of employment exceed the benefits. Difficult family situations include having very young children, having a large number of children, and being a single parent. Haynes and Feinleib (1982) found that although employment per se, even in a low prestige job, was not an additional risk factor for coronary heart disease (CHD) in women, certain role combinations increased this risk. For clerical workers who had three or more children or were married to a blue collar husband, there was a two or threefold increase in risk. Clerical workers who had non-supportive bosses and who suppressed their anger were also at significantly higher risk. For these women, familial roles were associated with extensive responsibility and husbands who tended not to share much in family work. Given these conditions at home, clerical work can become especially stressful for women (Haynes and Feinleib, 1982).

Inter-role Conflict and its Effects

One element of the work-family interface that has been extensively studied is the conflict a person may experience between the two roles. Role conflict may be seen as the simultaneous occurrence of two or more sets of pressures, such that compliance with one would make more difficult compliance with the other (Kahn, Wolfe, Quinn, Snock and Rosenthal, 1964). Inter-role conflict is a form of role conflict in which the sets of opposing pressures arise from participation in different roles. An examination of the literature on role conflict suggests that work-family conflict exists when time devoted to the requirements of one role makes it difficult to fulfil the requirements of another (time-based conflict), and when strain from participation in one role makes it difficult to fulfil requirements of another (strain-based conflict) (Greenhaus and Beutell, 1985). Using these forms of role conflict, Greenhaus and Beutell proposed a model of work-family conflict in which any role characteristic that affects a person's time involvement or strain within a role can produce conflict between that role and another role. Research findings on work-family conflict are generally consistent with the notions of time-based and strain-based conflict.

Buffering Effects of Work

The interaction between work and family roles is further demonstrated by research showing that challenging work can mitigate the effects of troubled familial relationships. In their research with Massachusetts women employed in social work or practical nursing, Barnett and Marshall (1991) showed that for employed mothers, high rewards from challenge at work buffered the impact of distress of a difficult parenting experience. In fact, their research showed that the psychological distress of women with troubled parent-child relationships was no worse than that of women with good relationships with their children, provided that they were in challenging jobs. In other words, if there are problems with children, a job with rewards compensates for the effects of a stressful parenting experience. The existence of interactive effects raises the possibility of rewards in one sphere mitigating or exacerbating the negative effects of experience of stress in the other.

The Scarcity Hypothesis and the Enhancement Hypothesis

In examining the relationship between multiple roles and mental and physical health, researchers have subscribed to one of two major views. The *scarcity hypothesis* (Goode, 1974) assumes that human energy is fixed and limited in quantity. Because any social role draws on this energy pool, a greater number of roles creates a greater likelihood of stress, overload and conflict, with negative consequences for health. On the other hand, the *enhancement hypothesis* (Marks, 1977; Sieber, 1974) does not see energy as fixed: role involvement increases one's energy reservoir. According to this view, the more roles one has, the more potential sources one has of self esteem, stimulation and social status. A person involved in several roles can trade off or bargain with respect to the less valued aspects of each role. Both hypotheses are incomplete since they focus on the quantity or number of roles rather than on their quality. Further, neither hypothesis acknowledges that being both a parent and a paid worker is usually experienced differently by men and women.

The Rational Model

In general, researchers have assumed that having roles in more than one sphere necessarily leads to conflict. One such view that is widely held and has significantly influenced the direction of research is the rational model of work-family conflict (Greenhaus, Bedeian and Mossholder, 1987; Keith and Schafer, 1980; Staines, Pleck, Sheppard and O'Connor, 1978). According to this model, the amount of conflict one perceives rises in proportion to the number of hours one expends in both work and family domains. It is further argued that the more hours one spends in work activities, the more one experiences interference from work to family. The more time spent in family activities, the more interference is experienced from family to work. The rational view predicts that the total amount of time spent performing work and family roles is positively related to role overload. Some empirical support has been reported for the rational model of work-family conflict. In a study of male and female psychologists and managers, Gutek, Serle and Klepa (1991) reported a high correspondence between hours spent in either the work or family domain and conflict originating in that domain. This relationship was stronger for women than for men. Greenhaus *et al.* (1987) also found that extensive time commitment to work was positively related to work-family conflict.

The Gender-Role Perspective

Another model of work-family conflict takes the gender-role perspective (Gutek *et al.*, 1991) in which gender both directly influences perceived work-family conflict and moderates the relationship between hours spent in paid work and family work as well as perceived work-family conflict. Additional hours of work in one's gender-prescribed domain (such as more housework hours for women) should be felt as less of an imposition by the role holder than additional hours of work in the domain traditionally associated with the other gender. Research supports this proposition. Women often report no more work-family conflict than men, despite the fact that their combined work and family hours often exceed those of men. Women spend more hours in housework than men (Denmark, Shaw and Ciali, 1985) and more hours in combined work and family activity (Pleck, 1985; Berk and Berk, 1979). In their study of women and men professionals, Gutek, Serle and Klepa (1991) found that although women spent more hours than men in family work, they reported the same level of family-work conflict. These data suggest that although women may put in many hours at the workplace and in unpaid work at home, this does not necessarily translate into the experience of role conflict, especially if the hours are expended in 'gender-appropriate' domains.

Spillover

When the positive or negative effects of one domain carry over to or affect the other, spillover is said to occur. This prevalent view suggests that workers carry emotions, attitudes and behaviours from work into their family life, and vice versa (Belsky, Perry-Jenkins and Crouter, 1985; Crouter, 1984; Piotrkowski, 1979). When negative spillover is discussed, it refers to the deleterious effects of stressors from one domain affecting the other domain and increasing strain (Gutek, Repetti and Silver, 1988). High levels of work role stressors may either contribute to decreased family satisfaction directly, or contribute indirectly through their effects on job satisfaction (Gutek, Repetti and Silver, 1988; Staines and O'Connor, 1980). Stressors that arise when the two domains interact, i.e. as in work-family conflict, may further detract from the well-being of two-career partners and affect outcomes in multiple domains (Kopelman, Greenhaus and Connoly, 1983).

Most of the research on spillover concentrates on negative outcomes when nonwork interferes with performance on the job (e.g. Greenhaus and Beutell, 1985). This emphasis on the negative effects of family participation on the work domain cannot help but hurt the acceptance of women in non-traditional fields (Kirchmeyer, 1993). Moreover, research that emphasizes negative spillover from non-work to work being dysfunctional often supports a rationalization for preventing the full participation of women at all levels of the workforce. Increasingly, however, research has supported the idea of positive outcomes of multiple domain participation (Thoits, 1983). A closer examination of research on work-family interaction indicates evidence of *positive* spillover despite the presence of the negative kind, and suggests instead that the two kinds represent separate dimensions of experience rather than opposite ends of the same continuum (Crouter, 1984). Overall, the findings support a more balanced view of work-nonwork relationships and suggest that any comprehensive theory should encompass both positive and negative aspects of spillover. Positive non-work-to-work spillover involves non-work supporting or enhancing work, whereas the negative kind involves non-work making work difficult or unsatisfactory (Kirchmeyer, 1993). In a study examining positive and negative spillover in a sample of male and female Canadian managers with families, Kirchmeyer (1993) found that the benefits of multiple domain participation seemed to outweigh the negative consequences. Women and men were almost identical in the extent to which they perceived participation in non-work activities as supporting their work. As in the results reported by Gutek et al. (1991), women were less likely to agree with statements about negative spillover from non-work domains. Since family work represents a traditional gender role domain for women, they would be less likely to experience family-to-work interference than men and thus would be less likely to feel that non-work was an imposition on their work activities. Research of this kind challenges outmoded stereotypes about family responsibilities functioning as a barrier to work participation by women to a greater extent than men. Moreover, the research has demonstrated that multiple role participation can be a benefit to well-being in women as well as men.

Compensation and Accommodation

Other views of the processes linking work and non-work spheres invoke the concepts of compensation and accommodation. When individuals compensate they seek satisfaction in one sphere to compensate for lack of satisfaction in another. The idea of compensation originated from Dubin (1967), who concluded that employees perceive their lives as centring outside work, where they seek intimate relationships and enjoyment. The concept of compensation provides an explanation of why some employees become more involved in their work when they are experiencing problems at home. In general, the idea of compensation views employees as seeking greater satisfaction from the sphere of either work or non-work spheres as a result of being dissatisfied with the other.

The term 'accommodation' has also been used to explain some of the processes involved in the work-non-work relationship. Individuals may limit their involvement in work or family life so that they can accommodate the demands of the other sphere. This process may apply more to female employees than to their male counterparts. Many women are more involved with their families than their work, and may limit their involvement at work in order to better accommodate their family obligations. The limited hours worked by women as compared to men reflect their efforts to accommodate their family roles (Hughes and Galinsky, 1994). Thus, the process of accommodation suggests a causal order that is the reverse of compensation – higher involvement in one sphere (family) leads to low involvement in the other (work). This is in contrast to the concept of compensation, which was developed primarily with male employees. For men, work has traditionally been considered to be their most important realm of activity. Thus, a man who is more oriented towards his family than his work may be seen as compensating for a job that is unfulfilling. Thus, greater involvement in the non-work sphere is seen as an effect of dissatisfaction with work.

This illustrates the importance of considering the processes underlying the relationship between work and family, and not just the end result. The outcomes of compensation and accommodation may be similar (an imbalance between the spheres of work and non-work), but the two processes are quite different (Lambert, 1990). The reasons why they occur differ as well. Compensation occurs when workers respond to unsatisfactory job or familial conditions by becoming more involved in the other sphere in the hope of obtaining greater satisfaction there. Accommodation occurs when an individual responds to an overly-involving family or work situation by limiting involvement in the other sphere.

The importance of examining the processes linking the spheres of work and non-work becomes evident when one considers current trends in managerial policy in the workplace (Lambert, 1990). While employer-supported childcare may facilitate the combination of the two spheres, the kind of process by which this occurs may differentially affect workers' satisfaction with family or work life. Day care, for example, may operate according to the principle of accommodation, whereby it may make it easier for workers to accommodate workplace demands (i.e. an eight-hour day), but at the expense of family satisfaction. While children are young, working an eight-hour day may result in feelings of disillusionment because one is not able to participate more fully in the child's care. This underlines the importance of examining the underlying processes and their implications for the individual and the organization.

Utilitarian approach to role conflict

Another means of understanding role conflict employs the utilitarian approach which explains competition between roles using the concepts of rewards and costs in determining levels of role investment (Homans, 1976). Individuals are seen as investing in roles that provide a favourable balance of rewards to costs. Farrell and Rusbult (1981) found a significant positive relationship between perceived reward for work and job commitment. They also noted a significant negative relationship between perceived costs of work and job commitment.

According to the utilitarian approach, the competition between roles is inevitable and often one role will gain acceptance at the expense of the other. Thus, there is a close link between the utilitarian approach and role conflict models of effects of work on family life and vice versa (e.g. Greenhaus and Beutell, 1985; Holahan and Gilbert, 1979).

A utilitarian approach has been employed to develop hypotheses regarding expectations that partners bring to their marriage, particularly at the interface between work and family. According to Major (1993) the concept of distributive justice may be applied to marriage partners' sense of personal entitlement regarding what they should put into and receive from marriage. Moreover, the idea of distributive justice may also moderate partners' perceptions of equity and satisfaction within the marital relationship. Perceptions of entitlement, in turn, appear to be tied to interpersonal processes that justify and legitimize an unequal division of family labour. Despite the fact that increasing numbers of women are employed outside the home, they continue to be assigned the primary role of parenting and in fact do most of the work within the home (Greenglass, 1995).

The differential assignment of familial work is also sanctioned and promoted by employers and managerial policy. Employees with family needs (i.e. picking up a child early from daycare) may be treated differently. Sometimes companies create a 'mommy track', a secondary track for employees who do not fit the mould for traditional top performance. However, the 'mommy track' works against women and the goal of gender equity by reinforcing a differential perception of women and men with children (Aldous, 1990). Through managerial policy, gender becomes translated into a structural barrier to women's equal participation with men in career development. Thus, gender as an organizer can affect personal lives of spouses in dual-earner families by legitimizing unequal division of labour. In fact, the more employers' policy reflects a traditional gendered workplace culture in which women with children leave the workplace and men with children are basically unencumbered by familial responsibilities, the more difficult it is for both partners to achieve equity in their own relationship.

Social-exchange Approach

A social exchange approach may also be applied to the understanding of the work-non-work interface from the perspective of the partners in a dual earner family. The family can be seen as a unit of social exchange in which there may be inequities between partners in resources such as power, obligation and benefits (Granrose, Parasuraman and Greenhaus, 1992). In this exchange, the spouse may be seen as serving many functions for the partner. A man who subscribes to traditional gender role expectations in marriage could experience relative deprivation in the amount of support he perceives he is getting from his employed partner, particularly if she is experiencing role overload. As a result, he may be less motivated to provide services and support in the home, leading to even greater role overload for the wife in this marriage (Granrose, Parasuraman and Greenhaus, 1992).

At the same time, when an employed married woman contributes financially to a marriage, thus sharing the traditional 'breadwinner' role with her partner, she may be expecting greater support and service within the home from her spouse. If, however, he subscribes to traditional masculine gender role norms, he may be reluctant to engage in work that he regards as inappropriate to his role. This will likely contribute to increased perceptions of inequity in both partners, thus resulting in lower motivation for each to support the other.

Multiple roles and stress

Multiple roles can be viewed as sources of stress, satisfaction or both (Repetti, 1987; Valdez and Gutek, 1987). Multiple roles are more often associated with positive than with negative effects (Repetti and Crosby, 1984). There is empirical evidence to suggest that they have benefits that can be seen not only in improved mental health (Repetti and Crosby, 1984) but also in occupational life. Crosby (1982) found that women who occupied roles of employee, spouse and parent were more satisfied with their jobs than women who were only employees and spouses. Russell, Altmeier and VanVelzen (1987) found that teachers who were married reported less burnout than teachers who did not have a spousal role. Compared with married teachers, the unmarried teachers reported less personal accomplishment in their jobs. Having a family provides one with a sense of purpose, commitment and raison d'être: feelings which may then generalize to other spheres of life, including work. It is possible that this commitment may act as a buffer against burnout, despite the long hours of work involved in occupying multiple roles. The single person, on the other hand, lacking social support at home, may be more likely to seek support at work. Perceiving their jobs as their social life, single individuals may become more involved with people at work and are thus more vulnerable to burnout (Greenglass, 1991). Pietromonoaco, Manis and Frohardt-Lane (1986) found that women with three, four or five roles had higher self-esteem and were more satisfied with their jobs than women with one or two roles. When asked to discuss the major source of stress, women with families typically mentioned problems associated with family members while women without partners or children mentioned absence of partners or children as their chief source of stress. Multiple roles can provide buffers against failure or frustrations in any given role (Pietromonoaco et al., 1986).

In men there are asymmetrically permeable boundaries between work and family, which means that work is allowed to infringe on men's family time since the most important thing a man can do to demonstrate that he is a good husband and father is to be a good provider (Barnett and Baruch, 1987). In contrast, the traditional family role identity and the work identity of women entail no such overlap in role behaviours.

There are also gender differences in the acceptability of segmenting work and family roles. Current norms require men to leave their familial obligations at home. Women in general do not have this choice.

While men may be doing more work within the home than in the past, domestic chores and childcare are far from equally shared (Hanna and Quarter, 1992). Research shows that women continue to undertake the bulk of domestic labour in five countries including Canada, the US, Australia, Sweden, and Norway (Baxter, 1997). Frankenhauser (1991) found that women's averge total workload (paid and unpaid work) was 78 hours per week, whereas men's total workload was 68 hours per week. She also found that the increased workload interfered with women's ability to wind down effectively, with resultant negative effects on health. Additional research has shown that women had a higher total workload than men, that work stress peaked at ages 35–39, and that total workload increased with the number of children in the household (Lundberg et al., 1994).

Given gender-related differences in identity role relationships, it is not surprising that men and women differ in their vulnerability to stress. Studies have examined the relative contribution of family-related versus work-related stress to negative mental and physical health outcomes. Among women, family role stressors have been found to be more strongly tied to psychological distress and physical illness than the work-related sources of stress (Kandel, Davies and Raveis, 1985). For men, workplace stressors were more strongly related to symptoms of psychological distress than were family role stressors. In general, women appear to be especially vulnerable to the negative effects of family stress (Kessler, Price and Wortman, 1985).

At the same time, research findings suggest that work may serve as a buffer against stress arising from familial roles (Baruch, Biener and Barnett, 1987). The trend in the literature to overemphasize work as a source of stress, as is the case when men are studied, has tended to obscure the benefits of employment. In the study of women's lives, the workplace has been seen as offering many women an arena in which the balance of psychological demands and control is more favourable than at home (Baruch et al., 1987). The workplace, more often than not, offers benefits such as challenge, control, structure, positive feedback and self-esteem, as well as a valued set of social ties.

Social support and the work-non-work interface

Increasingly, research is demonstrating the importance of social support as a contributor to psychological and physical health, as well as general well-being (Dignam, Barrera Jr. and West, 1986; Greenglass, 1993; Hobfoll, 1988). It may also moderate the effects of work-family role conflict on the well-being of individuals in two-earner relationships. Data indicate that social support from friends, spouse and family reduces role conflict between work and non-work spheres (Greenglass, Pantony and Burke, 1988). Spouse support in particular has been negatively correlated with role overload and work-family role conflict (Burke, Weir and Duwors, 1980; Holahan and Gilbert, 1979). Social support systems may help individuals mobilize their psychological resources, master strain, share tasks and obtain information, skills and practical advice (e.g. Caplan, 1974).

Granrose *et al.* (1992) developed a conceptual model that delineates the variables that influence the provision of maintenance levels of support within two-earner families. The model posits that support provided is influenced by three broad conceptual categories: the support environment, the provider's resources, and the provider's willingness to provide support. The support environment is represented by personal characteristics of the provider and the recipient, the social support network outside the marriage and the organization and physical characteristics of the family setting. The provider's resources include money, knowledge and time. Willingness to provide support includes: gender role and family role norms, marital rewards perceived by the providers, and equity perceptions of relative resources, relative work and family commitment and support received. Because gender and family role norms differ for each spouse, gender differences are likely to exist in the pattern of two-earner relationships and provision of support.

Using the proposed model along with a review of existing literature on work-family dynamics, Granrose *et al.* (1992) proposed testable hypotheses related to their model. They argued that partners would give more support if they were in a supportive environment, if their personal characteristics increased the probability that support needs were perceived and are able to be answered, and if they had sufficient time, money or information as is gained from being in the same field or at a later career stage than their spouses. Spouses provide more support if they value the family or if they have children in the home and if relative positions in the family and work do not violate gender or family role norms or perceptions of equity.

Granrose *et al.* (1992) point out that there are some relationships which cannot be clarified on the basis of previous research. For example, they raise the issue of the relationship between involvement in work and exchange of support. If individuals are highly involved in their career, they may have time conflicts which make it difficult to devote time to support another person. If their work lives are much more rewarding than their family life, they may be unwilling to expend effort to give support. However, individuals who are highly committed to work and *also* to the family may go out of their way to make time and effort to be supportive because of the value they put on family life. In this case, the absolute levels of family and work commitment and the absolute levels of resources available may not be adequate predictors of the amount of support given. If willingness is strong, more of the available resources may be devoted to providing support.

Individual coping and role conflict

Managing work-non-work conflict also involves the use of individual coping strategies. Typically, coping is classified into strategies that attempt to alter stressful encounters between the person and the environment – problem-focused strategies – and those that attempt to regulate stressful emotions – emotion-focused strategies (Lazarus and Folkman, 1984). Generally, problem-focused coping is associated with positive outcomes (Etzion and Pines, 1986; Greenglass, 1988; Greenglass, Burke and Ondrack, 1990) and emotion-focused strategies, especially wishful thinking or denial of a problem, are associated with poor outcomes (Greenglass, 1988).

In an early study of coping with work-home role conflict, Hall (1972) surveyed women and developed a model in which coping behaviour was classified into three types. Type I involved influencing the role environment by changing expectations of others or by hiring outside help. Type II involved changing one's own attitudes in order to reduce conflict by prioritizing activities or by lowering one's standards, and Type III involved trying to meet the expectations of others by working harder than before. The Type I coping style was hypothesized to be the most satisfying. However, results showed that how one coped was less important than the fact that one was at least attempting to cope.

Amatea and Fong-Beyette (1987) surveyed professional women and classified their inter-role coping efforts into a 2 × 2 matrix of emotion- versus problem-focused coping by active versus passive strategies. The active-emotion square includes such coping behaviour as seeing the positive side and talking with friends, whereas the passive-emotion square includes denial and suppression. The active-problem square includes prioritizing activities, hiring outside help and negotiating with others. The passive-problem square includes strategies such as working harder to meet expectations and eliminating roles. Results showed that the women used problem-focused solutions more frequently than emotion-focused solutions.

In a study of teachers, Bhagat, Allie and Ford (1991) investigated the role of both organizational stress (as measured by stressful events that often occur in schools) and life stress (as measured by the life experiences survey of Sarason, Johnson and Siegel, 1978) in the prediction of seven indicators of life strains and the moderating roles of two personal styles of coping: problem-focused coping and emotion-focused coping. When the relationships were analysed in terms of the two components of coping, as recommended by Folkman and Lazarus (1980), problem-focused coping emerged as a more effective moderator compared to emotion-focused coping. Problem-focused coping moderated seven out of the fourteen hypothesized relationships, while emotion-focused coping moderated only one. These results indicated that problem-solving coping strategies moderated organizational stress/life strain and personal life stress/life strain relationships to a greater extent than did emotion-focused coping.

Kirchmeyer (1993) examined successful strategies employed by male and female managers in coping with demands of multiple domains. Coping was assessed using Hall's (1972) strategies. Also employed were measures of positive and negative spillover from non-work to work (Kirchmeyer, 1992). Results indicated higher levels of agreement with statements about positive non-work to work spillover than those about negative spillover. For these managers, the benefits of multiple domain participation outweighed the burdens. Men and women were almost identical in the extent to which they perceived non-work participation as supporting their work. In keeping with previous research (Gutek et al., 1991), women were less likely to agree with statements about negative spillover from non-work (family) domains to work. Because family work represents a traditional gender-role domain for women, they would find non-work or family activities less of an imposition.

Further, results from this study suggest that it was not simply more active coping that enhanced the net spillover effect, rather it was greater use of certain strategies. Most effective in helping individuals cope with demands from multiple life domains were those strategies aimed at altering one's own attitudes as opposed to altering those of others, and increasing one's own personal efficiency as opposed to decreasing one's activity level or relying on others.

Other researchers have emphasized the need for developing strategies to prevent undue stress while attempting to cope with demands from multiple domains. For example, training programmes are needed to prevent excessive demands of work from affecting the well-being of the family. Hall and Ritcher (1988) emphasize the need for helping individuals develop effective coping styles throughout their careers. Individual workers should be given appropriate training in recognizing the danger signals of excessive role encroachment, and trained to manage the diverse effects of stress from either domain. Emphasis should be placed on developing effective social support systems at home and at work.

In a study that examined the effects of individual and collective efforts to reduce the stresses of combined paid work and parenting, Shinn, Wong, Simko and Ortiz-Torres (1989) compared the relative impact of three levels of coping on parental well-being. These were individual coping strategies, social support from various sources and flexibility in job scheduling. Coping was classified into strategies that attempt to alter stressful encounters between person and environment (problem-focused) and those that attempt to regulate stressful emotions (emotion-focused) (Lazarus and Folkman, 1984). Social support systems may aid employees to mobilize their psychological resources. Organizations can undertake activities to reduce stressors at work. Flexitime can potentially reduce schedule conflict for parents at all stages of their career and has beneficial effects for job satisfaction, absenteeism and ease of child care (Cummings and Molloy, 1977). The study was conducted with 644 full-time employees with children. Respondents, who completed a self-adminis-tered questionnaire, included 208 married fathers, 287 married mothers and 149 single mothers (Shinn et al., 1989). Results showed that individual coping was the most powerful predictor of outcomes, with problem-focused coping being associated with positive outcomes and emotion-focused coping associated with negative ones.

Contrary to their expectations, individual coping efforts were more important than both social support and flexibility of job schedules (one to two hours' discretion in starting time) in accounting for parents' well-being. In this study, flexible job schedules were only weakly related to outcome measures (e.g. family satisfaction, family distress, job satisfaction, job distress and physical health). This is consistent with Bohen and Viveros-Long's (1981) findings that flexitime made little difference to working parents. However, the schedule flexibility may have been too weak to have much impact on the well-being of the respondents. It may be that an hour or two of discretion about when one comes to work is not the type of flexibility that parents need most. Perhaps they need to have the flexibility to take an afternoon off to take a child to the doctor or visit a teacher. Also, respondents cannot always take advantage of even the most flexible flexitime programme. Respondents reported that their childcare responsibilities constrained their use of flexitime.

Organizational initiatives and work-family interface

In response to the influx of women into the workforce, there is greater pressure on organizations to implement policy changes that create new ways of thinking about work, careers, family and the combination of these. In particular, organizations need to recognize that there are specific barriers which make advancement of women in the organization difficult if not impossible. Often these barriers include organizational inflexibility in defining work schedules and work sites, as well as an absence of programmes to enable employees to balance work and family responsibilities.

A number of work/family initiatives have been implemented in American and Canadian companies. These include parental leave, family care leave, sick leave for dependent care, adoption assistance, flexible spending accounts, domestic partner benefits, childcare centres, family daycare networks, emergency childcare, pre-school programmes, after-school programmes, training and support groups, dependent care resource and referral, relocation assistance and elder care programmes (Mattis, 1994). Research on the impact of work/family programmes in the Johnson and Johnson company showed that between 1990 and 1992, supervisors became significantly more supportive of employees when work/family problems arose. Supervisors were also seen as more supportive of flexible time and leave policies. Contrary to some expectations, there was no impact on absenteeism or tardiness among their employees (Families and Work Institute, 1993).

The use of flexitime as a way of coping with family and work responsibilities underlines how the issue of time is regarded within the organization. Traditionally, career roles within organizations favour those employees who put most of their time into work at the office. Managers expect to see their employees at work during a certain part of the day and often use time at work as one of the criteria for performance evaluation (Bailyn, 1993). Given their economic and psychological position, women tend to be caretakers of the family and in the process are forced to retreat from the core positions in organizational life (Bailyn, 1993). Assumptions that long hours at the office equate with commitment and productivity need to be re-examined. In an era of high tech information communication, visible time at work is no longer a valid basis for judgement of high performance. As Bailyn (1993) argues, what is needed is a different view of career development based on combining the legitimacy of private needs with a rethinking of the conditions of work.

An Alternative Career Model

The need for an alternative concept of a career model was raised by Greenglass (1973) who argued that the narrowness and inflexibility of most careers create unintended limits and, in many cases, exclude creative, energetic and intelligent persons. An alternative career model is needed for individuals who wish to combine career and family and are prepared to make a dual commitment. There are individuals who think of themselves as a permanent part of the labour force who also have a serious commitment to family. This dual commitment on the part of either male or female may last five years for some, and a lifetime for others.

The alternative career model proposed by Greenglass (1973) is more diversified and differentiated than the accepted pattern, which assumes uniformity of motives and degree of career commitment. As Sanders (1965) discussed, the professional career model could be represented by an ascending spiral movement indicating upward career choices but paced more slowly than the standard one, with longer horizontal stop-overs. In short, the person would have the option of pursuing his or her career either full-time or part-time while explicitly stating the rate at which he or she intends to progress. Instead of suffering the negative consequences so often experienced by those who deviate from the acceptable rate of progress through the ranks, individuals wishing to devise a career pattern more congruent with their own unique lifestyle should be able to do so.

Traditional careers are based on long-term planning using linear continuity as a key construct. It is presumed that one's career direction is up and that what happens at the outset of one's career determines the future career progression. What is needed is a system that assumes not a continuous progression towards some predetermined job but alternating times of low and high involvement in work. This permits employees to exit and re-enter the position without elimination in earlier rounds. Such a system would depend on individual negotiation built around discontinuous career planning. The onus would be on organizations to adjust job requirements to varying degrees of employee involvement (Bailyn, 1993). In short, organizations need to be attuned to the diverse needs of their employees.

Summary

This review of research and theory examining the relationship between work and family spheres has shown that more than one paradigm is needed to increase our understanding in this area. The focus has been on the conflict that is often observed between work and family role obligations and the deleterious effects of this conflict on the individual man or woman. At the same time, the discussion has indicated that we should focus on the ways in which multiple role occupancy can enhance well being. Often, challenging work can mitigate the effects of troubled familial relationships. Rewards in one sphere can lessen the negative effects of experience in another. This synergy between roles is often overlooked in paradigms that focus exclusively on the conflict between them. It is, however, recognized that demands associated with work and home can lead to conflict between the two spheres.

The effective management of work and family cannot be achieved through individual effort alone. Organizational initiatives are required to achieve a harmony between the demands of work and family. Redefinitions of work progress need to be incorporated into managerial policy and employee relations. Stereotypes of gender roles need to be replaced by the acknowledgement of the diversity of skills and needs of individual workers. It is only in this way that change can be brought about which will result in a significant improvement in workers' quality of life.

Note

Thanks are due to Lisa Fiksenbaum for her assistance in preparing this manuscript and to Ronald Burke and Michael Leiter for their comments on earlier drafts.

References

ALDOUS, J. (1990) 'Specifications and speculation concerning the politics of workplace family policies', *Journal of Family Issues* **11**: 355–367.

AMATEA, E. S. and FONG-BEYETTE, M. L. (1987) 'Through a different lens: examining professional women's interrole coping by focus and mode', *Sex Roles* **17**: 237–252.

BAILYN, L. (1993) *Breaking the Mould: Women, Men and Time in the New Corporate World*, New York: The Free Press.

BARNETT, R. and BARUCH, G. (1987) 'Determinants of fathers' participation in family work', *Journal of Marriage and the Family* **49**: 29–40.

BARNETT, R. C. and MARSHALL, N. L. (1991) 'The relationship between women's work and family roles and their subjective well-being and psychological distress', in M. Frankenhaeuser, J. Lundberg and M. Chesney (eds), *Women, Work and Health: Stress and Opportunities*, New York: Plenum, 111–137.

BARTOLEMÉ, F. and EVANS, P. L. (1979) 'Professional lives versus private lives – shifting patterns of managerial commitment', *Organization Dynamics* **3**: 3–29.

BARUCH, G. K., BIENER, L. and BARNETT, G. R. (1987) 'Women and gender in research on work and family stress', *American Psychologist* **42**: 130–136.

BAXTER, J. (1997) 'Gender equality and participation in housework: a cross-national perspective', *Journal of Comparative Family Studies* **28**: 220–247.

BELSKY, J., PERRY-JENKINS, M. and CROUTER, A. (1985) 'The work-family interface and marital change across transition to parenthood', *Journal of Family Issues* **6**: 205–220.

BERK, R. and BERK, S. F. (1979) *Labor and leisure at home*, Beverly Hills, CA: Sage.

BERNARD, J. (1981) 'The good provider role: its rise and fall', *American Psychologist* **36**: 1–12.

BHAGAT, R. S., ALLIE, S. M. and FORD, D. L. (1991) 'Organizational stress, personal life stress and symptoms of life strains: an inquiry into the moderating role of styles of coping', *Journal of Social Behavior and Personality* **6**: 163–184.

BOHEN, H. H. and VIVEROS-LONG, A. (1981) *Balancing Jobs and Family Life: Do Flexible Work Schedules Help?*, Philadelphia: Temple University Press.

BURKE, R. J. and GREENGLASS, E. R. (1987) 'Work and family', in C. L. Cooper and I. T. Robertson (eds), *International Review of Industrial and Organizational Psychology*, New York: Wiley, 273–320.

BURKE, R. J., WEIR, T. and DUWORS, R. E. (1980) 'Work demands on administrators and spouse well being', *Human Relations* **33**: 253–278.

CAPLAN, G. (1974) *Support Systems and Community Mental Health: Lectures on Concept Development*, New York: Behavioral Publications.

COBB, S. (1976) 'Social support as a moderator of life stress', *Psychosomatic Medicine* **38**: 300–314.

CROSBY, F. (1982) *Relative Deprivation and Working Women*, New York: Oxford University Press.

CROUTER, A. C. (1984) 'Spillover from family to work: the neglected side of the work-family interface', *Human Relations* **37**: 425–442.

CUMMINGS, T. G. and MOLLOY, E. S. (1977) *Improving Productivity and the Quality of Work Life*, New York: Praeger.

DENMARK, F. L., SHAW, J. S. and CIALI, S. D. (1985) 'The relationship among sex roles, living arrangements and the division of household responsibilities', *Sex Roles* **12**: 617–625.

DIGNAM, J. T., BARRERA, M., Jr. and WEST, S. G. (1986) 'Occupational stress, social support and burnout among correctional officers', *American Journal of Community Psychology* **14**: 177–193.

DUBIN, R. (1967) 'Industrial workers' worlds: a study of the central life interests of industrial workers', in E. Smigel (ed.), *Work and Leisure*, New Haven, CT: College and University Press. 143–174.

ENGLAND, P. and SWOBODA, D. (1988) 'The asymmetry of contemporary gender role change', *Free Inquiry in Creative Sociology* **61**: 157–161.

ETZION, P. and PINES, A. (1986) 'Sex and culture as factors explaining coping and burn out among human service professionals: a social psychological perspective', *Journal of Cross-Cultural Psychology* **17**: 191–209.

FAMILIES AND WORK INSTITUTE (1993) *An evaluation of Johnson and Johnson's work-family initiatives*, New York: Families and Work Institute.

FARRELL, D. and RUSBULT, C. E. (1981) 'Exchange variables as predictors of job satisfaction, job commitment and turnover: the impact of rewards, costs, alternatives and investments', *Organizational Behavior and Human Performance* **28**: 78–95.

FOLKMAN, S. and LAZARUS, R. S. (1980) 'An analysis of coping in a middle-aged community sample', *Journal of Health and Social Behavior* **21**: 219–239.

FRANKENHAEUSER, M. (1991) 'The psychophysiology of workload, stress and health: comparison between the sexes', *Annals of Behavioral Medicine* **13**: 197–204.

GILBERT, L. A. (1993) *Two Careers/One Family*, Newbury Park, CA: Sage.

GOODE, W. J. (1974) 'A theory of strain', *American Sociological Review* **25**: 483–496.

GRANROSE, C. S., PARASURAMAN, S. and GREENHAUS, J. H. (1992) 'A proposed model of support provided by two-earner couples', *Human Relations* **45**: 1367–1393.

GREENGLASS, E. (1973) 'Women: a new psychological view', *The Ontario Psychologist* **5**: 7–15.

GREENGLASS, E. (1982) *A World of Difference: Gender Roles in Perspective*, Toronto: Wiley.

GREENGLASS, E. (1988) 'Type A behaviour and coping strategies in female and male supervisors', *Applied Psychology: An International Review* **37**: 271–288.

GREENGLASS, E. R. (1991) 'Burnout and gender: theoretical and organizational implications', *Canadian Psychology* **32**: 562–572.

GREENGLASS, E. (1993) 'The contribution of social support to coping strategies', *Applied Psychology: An International Review* **42**: 323–340.

GREENGLASS, E. (1995) 'Gender, work-stress and coping: theoretical implications (Special issue), *The Journal of Social Behavior and Personality*.

GREENGLASS, E. R., BURKE, R. J. and ONDRACK, M. (1990) 'A gender-role perspective of coping and burnout', *Applied Psychology: An International Review* **39**: 5–27.

GREENGLASS, E. R., PANTONY, K. and BURKE, R. J. (1988) 'A gender-role perspective on role conflict, work stress and social support', *Journal of Social Behavior and Personality* **3**: 317–328.

GREENHAUS, J. H., BEDEIAN, A. G. and MOSSHOLDER, K. (1987) 'Work experiences, job performance and feelings of personal and family well-being', *Journal of Vocational Behavior* **31**: 200–215.

GREENHAUS, J. H. and BEUTELL, N. J. (1985) 'Sources of conflict between work and family roles', *The Academy of Management Review* **10**: 76–88.

GUTEK, B. (1993) 'Asymmetric changes in men's and women's roles', in B. C. Long and S. E. Kahn (eds) *Women, Work and Coping: A Multidisciplinary Approach to Workplace Stress*, Montreal and Kingston: McGill-Queen's University Press.

GUTEK, B. A., REPETTI, R. and SILVER, D. (1988) 'Nonwork roles and stress at work', in C. Cooper and R. Payne (eds) *Causes, Coping and Consequences at Work*, 2nd edn, Chichester, UK: Wiley, 141–174.

GUTEK, B. A., SERLE, S. and KLEPA, L. (1991) 'Rational versus gender role explanations for work-family conflict', *Journal of Applied Psychology* **76**: 560–568.

HALL, D. T. (1972) 'A model of coping with role conflict: the role behavior of college-educated women', *Administrative Science Quarterly* **17**: 471–485.

HALL, D. and RITCHER, J. (1988) 'Balancing life and home life: what can organizations do to help?', *Academy of Management Executive* **11**: 213–223.

HANNAH, J. A. and QUARTER, J. (1992) 'Could you do the bedtime story while I do the dishes?', *Canadian Journal of Community Mental Health* **11**: 147–162.

HAYGHE, H. V. (1990) 'Family members in the workforce', *Monthly Labor Review* **113**: 14–19.

HAYNES, S. G. and FEINLEIB, M. (1980) 'Women, work and coronary heart disease: prospective findings from the Framingham heart study', *American Journal of Public Health* **70**: 133–141.

HAYNES, S. G. and FEINLEIB, M. (1982) 'Women, work and coronary heart disease: results from the Framingham 10 year follow-up study', in P. Berman and

F. Ramey (eds) *Women: A Developmental Perspective*, NIH Publication No. 82-2298, Washington, DC: US Government Printing Office, 79–101.

HOBFOLL, S. E. (1988) *The Ecology of Stress*, Washington, DC: Hemisphere.

HOLAHAN, C. J. and GILBERT, L. A. (1979) 'Conflict between major life roles: women and men in dual career couples', *Human Relations* **32**: 451–467.

HOMANS, G. C. (1976) 'Fundamental processes of social exchange', in E. P. Hollander and R. G. Hunt (eds) *Current perspectives in social psychology*, 4th edn, New York: Oxford, 161–173.

HOSCHILD, A. (1989) *The Second Shift: Working Parents and the Revolution at Home*, New York: Viking.

HOUSE, J. S. (1981) *Work Stress and Social Support*, Reading, MA: Addison-Wesley.

HOUSE, J. S. and WELLS, J. A. (1978) 'Occupational stress, social support and health', in A. McLean, G. Black and M. Colligan (eds) *Reducing Occupational Stress*, DHEW Publication No. NIOSH 78-140, Washington, DC: US Government Printing Office.

HUGHES, D. L. and GALINSKY, E. (1994) 'Gender, job and family conditions and psychological symptoms', *Psychology of Women Quarterly* **18**: 251–270.

KAHN, R. C., WOLFE, D. M., QUINN, R. P., SNOEK, J. D. and ROSENTHAL, R. A. (1964) *Organizational Stress: Studies in Role Conflict and Ambiguity*, New York: Wiley.

KANDEL, D. B., DAVIES, M. and RAVEIS, V. H. (1985) 'The stressfulness of daily social roles for women: marital occupational and household roles', *Journal of Health and Social Behavior* **26**: 64–78.

KEITH, P. and SCHAFER, R. (1980) 'Role strain and depression in two-job families', *Family Relations* **29**: 483–488.

KELLER, E. F. (1992) *Secrets of Life, Secrets of Death: Essays on Language, Gender and Science*, New York: Routledge.

KESSLER, R. C., PRICE, R. H. and WORTMAN, C. B. (1985) 'Social factors in psychopathology: stress, social support and coping processes', *Annual Review of Psychology* **35**: 531–572.

KIRCHMEYER, C. (1992) 'Perceptions of nonwork-to-work spillover: challenging the common view of conflict-ridden domain relationships', *Basic and Applied Social Psychology* **13**: 231–249.

KIRCHMEYER, C. (1993) 'Non-work-to-work spillover: a more balanced view of the experiences and coping of professional women and men', *Sex Roles* **28**: 531–552.

KOPELMAN, R. E., GREENHAUS, J. H. and CONNOLY, T. F. (1983) 'A model of work, family and interrole conflict: a construct validation study', *Organizational Behavior and Human Performance* **32**: 198–215.

LAMBERT, S. J. (1990) 'Processes linking work and family: a critical review and research agenda', *Human Relations* **43**: 239–257.

LAZARUS, R. S. and FOLKMAN, S. (1984) *Stress, Appraisal and Coping*, New York: Springer.

LEE, M. D. (1994) 'Variations in career and family involvement over time: truth and consequences', in M. J. Davidson and R. J. Burke (eds) *Women in Management: Current Research Issues*, London: Paul Chapman, 242–258.

LUNDBERG, U., MARDBERG, B. and FRANKENHAEUSER, M. (1994) 'The total workload of male and female white collar workers as related to age, occupational level, and number of children', *Scandinavian Journal of Psychology* **35**: 315–327.

MAJOR, B. (1993) 'Gender, entitlement and the distribution of family labor', *Journal of Social Issues* **49**: 141–159.

MARKS, S. R. (1977) 'Multiple roles and role strain: some notes on human energy, time and commitment', *American Sociological Review* **42**: 921–936.

MATTIS, M .C. (1994) 'Organizational initiatives in the USA for advancing managerial women', in M. J. Davidson and R. J. Burke (eds) *Women in Management: Current Research Issues*, London: Paul Chapman, 261–276.

MERIKANGAS, K. (1985) 'Sex differences in depression', paper presented at the Murray Center Conferences on Mental Health in Social Context, Cambridge, MA.

PIETROMONOACO, P. R., MANIS, J. and FROHARDT-LANE, K. (1986) 'Psychological consequences of multiple social roles', *Psychology of Women Quarterly* **10**: 373–382.

PIOTRKOWSKI, C. (1979) *Work and the family system*, New York: The Free Press.

PIOTRKOWSKI, C. S., RAPOPORT, R. N. and RAPOPORT, R. (1987) 'Families and work', in M. B. Sussman and S. K. Steinmetz (eds) *Handbook of Marriage and the Families*, New York: Plenum.

PLECK, J. H. (1977) 'Work-family role system', *Social Problems* **63**: 81–88.

PLECK, J. H. (1985) *Working Wives/ Working Husbands*, Beverly Hills, CA: Sage.

REPETTI, R. L. (1987) 'Individual and common components of the social environment at work and psychological well-being', *Journal of Personality and Social Psychology* **52**: 710–720.

REPETTI, R. L. and CROSBY, F. (1984) 'Gender and depression: exploring the adult role explanation', *Journal of Social and Clinical Psychology* **2**: 57–70.

RUSSELL, D. W., ALTMEIER, E. and VAN VELZEN, D. (1987) 'Job related stress, social support and burnout among classroom teachers', *Journal of Applied Psychology* **72**: 269–274.

SANDERS, M. K. (1965) 'The new American female: demi-feminism takeover', *Harper's* **231** (1382): 37–34.

SARASON, I. G., JOHNSON, J. H. and SIEGEL, J. M. (1978) 'Assessing the impact of life changes: development of the life experiences survey', *Journal of Consulting and Clinical Psychology* **46**: 932–946.

SCHNEER, J. A. and REITMAN, F. (1993) 'Effects of alternate family structures on managerial career paths', *Academy of Management Journal* **36**: 830–843.

SHINN, M., WONG, N. W., SIMKO, P. A. and ORTIZ-TORRES, B. (1989) 'Promoting the well-being of working parents: coping, social support and flexible job schedules', *American Journal of Community Psychology* **17**: 31–55.

SIEBER, S. D. (1974) 'Toward a theory of role accumulation', *American Sociological Review* **39**: 567–578.

STAINES, G. L. and O'CONNOR, P. (1980) 'Conflicts among work, leisure and family roles', *Monthly Labor Review* **103**: 35–39.

STAINES, G. L., PLECK, J., SHEPPARD, L. and O'CONNOR, P. (1978) 'Wives' employment status and marital adjustment: yet another look', *Psychology of Women Quarterly* **3**: 90–120.

THOITS, P. (1983) 'Multiple identities and psychological well being', *American Sociological Review* **48**: 174–187.

US BUREAU OF LABOR STATISTICS (1992) *Labor Force Statistics derived from the Current Population Survey: A Databook*, Washington: DC.

VALDEZ, R. L. and GUTEK, B. A. (1987) 'Family roles: a help or a hindrance for working women?', In B. A. Gutek and L. Larwood (eds), *Women's Career Development*, Beverly Hills, CA: Sage, 157–169.

VERBRUGGE, L. M. (1982) 'Women's social roles and health', in P. Berman and E. Ramey (eds) *Women: A Developmental Perspective* (Publication No. 82-2298), Bethesuda, MD: National Institute of Health, 49–78.

WALDRON, I. and HEROLD, J. (1984) 'Employment, Attitudes toward Employment and Women's Health', Paper presented at the meeting of the Society of Behavioral Medicine, Philadelphia.

CHAPTER SIX

Women's ways of coping with employment stress: a feminist contextual analysis

BONITA C. LONG AND ROBIN S. COX

Research that considers the experience of women in the paid workforce is fairly recent and reflects the unprecedented influx, since the 1960s, of middle-class women into the labour market. Women's workforce participation has doubled in the last twenty-five to thirty years, and now six out of every ten women in North America are in the paid labour force (Spain and Bianchi, 1996; Statistics Canada, 1995). Despite inroads into traditionally male-dominated fields such as medicine and law, women are still predominantly confined to lower-level positions and to female sub-specialities (Acker, 1992). Moreover, three-quarters of all employed women are in teaching, health, clerical, or sales and service jobs and are disproportionately represented in low-status, low-paid, dead-end jobs (Gutek, 1993; Jacobsen, 1994).

In response to the increased globalization of economies, computerization and automation have facilitated a general deskilling of labour and an increased demand for cheap labour. Acker (1992) describes a 'feminization' of working conditions in industrialized countries such that the number of secure, adequately-paid jobs offering consistent lifetime employment continues to decrease in the face of a growing trend towards part-time, contract and temporary work. The growth in the number of these kinds of contingent jobs has occurred at the same time that women's paid work has become a necessity for the economic survival of most families. In addition, as women's participation in the paid workforce has increased, there has been a systematic undervaluation of their traditional roles as mothers and homemakers (Scarr, Phillips and McCartney, 1989). At the same time, society continues to expect and accept that women will take responsibility for the majority of childcare and work in the home. As a result, women are increasingly challenged to find ways to balance the demands of employment and family in a patriarchal society (Spain and Bianchi, 1996).

The purpose of this chapter is to provide direction for future research on the ways in which women cope with the stressors associated with being employed. We present a social, historical, and cultural analysis and argue that researchers need to acknowledge the socially constructed context of women's experience of employment. Moreover, we posit that women's coping takes place in a cultural context of patriarchy which reflects an unequal distribution of power and access to resources.

We begin with a discussion of the potential impact occupational stressors have on women's physical and mental health. Next we review research that focuses on women's ways of coping with stress associated with being employed and describe problems with current research and the conceptualization of stress and coping (see also, Banyard and Graham-Bermann, 1993; Handy, 1988; Wethington and Kessler, 1993). The term 'employed women' in this chapter refers to women who work outside the home for pay, who may be salaried or self-employed. This term is distinct from the term 'working women' which refers to women who work for pay outside the home and to women who receive no pay for work in the home.

The impact of occupational stressors on women's health

Although employed women are confronted with the same types of occupational stressor that male employees face (e.g. work overload, role ambiguity, lack of job security), women also encounter a number of stressors that are gender-specific and arise as a result of the male-dominated, patriarchal context in which women live and work (Nelson and Hitt, 1992). For example, Smith (1993) identified four intersecting sources of stress that may be associated with women's involvement in paid work: (a) a hostile work environment (e.g. discrimination, sexual harassment), (b) an unsupportive home (e.g. the 'double' shift of housework), (c) a disapproving social environment (e.g. conflict of values, especially between employment and parenting), and (d) the individual's own self-concept (e.g. conflict between 'real' and 'ought' selves, the 'super woman' syndrome). Thus, for women, the sources of stress associated with being employed are multiple and arise from the complexity and contradictions of their multiple locations within society. The term occupational stress, typically employed in this kind of research, may fail to encapsulate the diversity of stressors reflected in women's multiple locations as an employee, a mother, an intimate partner, and a woman within patriarchy. The effects these multiple positionings have on a woman's perception of stress and on her choice of coping strategies are interrelated; to separate out occupational stress as it is traditionally defined is to oversimplify the complex web of stressors associated with women's employment.

One means of understanding the impact of employment on women's health is to examine both the direct and indirect benefits and detriments of being employed. For example, Verbrugge (1986) identified three theoretical paths by which workforce participation might relate to health outcomes. The first is the health benefits model, which emphasizes the direct advantages of employment such as financial remuneration, self-esteem, greater feelings of control and social support. The second is the role accumulation model, which suggests that multiple roles (mother, wife, employee) enhance health through opportunities for reward and satisfaction. In the role accumulation model, satisfaction in the employment role may provide protection against strain in other roles, such as marital or parental roles. The third is the job stress model, which posits that the addition of the employment role will invariably lead to overload and harm to women's physical or psychological health. Although each of these models accounts for different but overlapping ways in which employment may enhance or impede women's health, most research emphasizes one model over another.

Occupational health research has until fairly recently neglected women as the focus of research (for reviews see, Repetti, Matthews and Waldron, 1989; Rodin and Ickovics, 1990). Although there has been a marked increase in the amount of research, there are still serious limitations in the scope of this body of work. With far too few exceptions (e.g. Bailey, D., Wolfe, D. and Wolfe, C. R., 1996; James, 1994; McGuire and Reskin, 1993; Melies, Norbeck, Laffrey, Solomon and Miller, 1989), research has focused on the employment experience of white, middle-class, professional women, and has viewed women through the lens of gender differences (Baum and Grunber, 1991). This latter approach oversimplifies women's lives and constructs gender as an individual attribute, rather than a product of the ongoing power relationships between men and women within a male-dominated social structure. Although the difference approach has, in part, countered some of the devaluing of so called 'female characteristics,' this tendency to dichotomize on the basis of a single trait results in a differential valuing of one side of the dichotomy. Hare-Mustin and Marecek (1990) point out that research results based on gender differences serve to provide an analysis of women as deviant in some way from a norm that historically has been male-defined. The by-product of the fact that so much of the occupational health research is based on the assumption of a male standard (Hall, 1991) is that women are often viewed from the standpoint of deficiency.

The focus on gender differences also tends to isolate gender as a fixed construct, obscuring the interaction of gender with other social categories, namely race, class, age, and sexual orientation and de-emphasizing the role that gender plays as an active process that structures social interactions. This oversimplification of gender as a variable is exacerbated by this almost exclusive focus of research on the experiences of white, middle-class, professional women. Despite the culturally and racially limited samples, generalizations continue to be made about women as though they were a homogeneous group, thereby minimizing or ignoring the within-group differences and between-group similarities that result from different social locations (Banyard and Graham-Bermann, 1993).

The Impact of Employment on Women's Mortality Rates

One of the ways in which the male standard of comparison is typically employed in occupational stress and coping research is evidenced in the popular expectation that women's exposure to occupational stress would reduce their more favourable mortality rates. However, findings from the Alameda County Study indicated that employment status, type of employment, and having more children did not predict mortality risk for women, although greater numbers of children in the home did elevate the risk in single working mothers (Kotler and Wingard, 1989). Based on data from the Framingham Heart Study, LaCroix and Haynes (1987) concluded that employed women appeared to be healthier than women who were not in the paid labour force. Their results, however, revealed that women at greatest risk for coronary heart disease were those employed in clerical occupations, and particularly those who also had major household responsibilities and a negative or limiting psychosocial work environment. These kinds of results demonstrate the need for research designs that do not isolate women's employment experiences but rather put them into context within the multiple locations they inhabit. Moreover, greater attention should also be paid to the specific nature of the stressors associated with the kinds of jobs the majority of women occupy; that is, jobs characterized by low pay, inflexibility, multiple supervisors, repetitive tasks, low status, limited benefits, and few opportunities for promotion.

The Impact of Multiple Roles on Women's Health

Considerable research has focused on the psychological impact of the increasing number of roles, particularly the paid worker role, that many women are now occupying. The multiple roles theory posits that multiple roles lead to role strain or overload (Barnett and Baruch, 1985; Verbrugge, 1986). Repetti et al. (1989) reviewed findings on multiple roles and concluded that there is mixed evidence for the association between multiple roles and distress. In fact, roles that are perceived to be meaningful and of high quality are consistently found to be associated with positive outcomes (e.g. Marshall and Barnett, 1992; Parker and Aldwin, 1994). Yet, role categories such as 'homemaker' again tend to oversimplify women's context, obscuring the many activities and demands within that superordinate category (e.g. child care, husband/partner care, housework, maintenance of extended family and social networks, elder care) and the ways in which these multiple roles intersect and interact with women's employment experiences (Facione, 1994). In addition, this focus on roles tends to obscure the differing social milieus in which roles are enacted and the situationally determined experience of each role (Banyard and Graham-Bermann, 1993; Lykes, 1983; Martikainen, 1995).

The Impact of Employment on Women's Physical and Mental Health

With regard to other research on the impact of women's employment on mental and physical health, Repetti *et al.* (1989) also found mixed evidence of benefits and costs to women. Employment had beneficial health effects on some women and detrimental effects on others depending on the character-istics of the woman (e.g. marital status and parental status, whether or not she wanted to work, or whether her husband helped with the housework and parental care), and the characteristics of her job (e.g. job classification, occupational hazards, and the social support at work). Repetti *et al.* (1989) concluded that employment was not universally detrimental to women's mental or physical health. These findings suggest that there can be no single understanding of women's experience of employment stress and argues against isolating gender as a variable. Instead, researchers should embrace an understanding and questioning of the complex and multiple ways women experience employment stress.

Women's ways of coping with employment stress

The fact that women face multiple sources of stress yet generally remain healthy is something of a paradox. Although healthier women may select employment and multiple roles, this 'hardiness' may also be due to the ways in which women cope with stress (coping being cognitive and behavioural efforts made by an individual when the demands of a given situation tax adaptive resources), or the personal or coping resources (i.e. relatively stable characteristics) that help maintain their well-being.

Gender Differences

Until recently the literature on women's coping strategies reflected stereo-typical beliefs about women. Hare-Mustin and Marecek (1990: 462) remind us that '[m]an is the hidden referent in our language and our culture,' and, in our theories of coping this referent is reflected in what are understood to be stressors, what is seen as valuable or effective coping, and what is recognized or revealed by researchers examining and theorizing about stress and coping in women's lives (Banyard and Graham-Bermann, 1993). For example, historically, women were believed to cope passively or emotionally, to be less rational than men, and to have poorer problem-solving skills. Early research tended to support these stereotypes; however, much of that research was badly flawed (Folkman and Lazarus, 1980; Pearlin and Schooler, 1978). For instance, men and women who were compared were not matched on important variables (i.e. the type of stressor, the participants' occupational level, their power and influence, or the resources available to them). The more recent studies that have sampled men and women employed in similar occupations and occupational levels report findings that contradict these earlier results (e.g. Long, 1990; Schwartz and Stone, 1993).

Recent research has found few gender differences in the use by managers or professional workers of problem-solving coping (for a review, see Korabik, McDonald and Rosin, 1993). Long (1990) found that both male and female managers used similar amounts of problem-solving coping strategies when dealing with stressors associated with being in the paid workforce. In addition, the female managers used avoidance coping and problem-reappraisal coping more frequently than the male managers, indicating that they had a broader repertoire of coping strategies. These results are consistent with those of other researchers who have found that men are more likely than women to persist with one type of coping strategy – problem-solving coping (Compas, Forsythe and Wagner, 1988). Folkman and Lazarus (1980), for instance, puzzled over their finding that men used more problem-focused coping than women in situations that had to be accepted; that is, that men tended to persist with a coping strategy even when it could be deemed to be situationally maladaptive.

Moreover, gender differences in coping strategies are often explained as being due to situational differences. For example, Hobfoll, Dunahoo, Ben-Porath and Monnier (1994) examined coping by identifying stressors from interpersonal versus professional domains. Yet there is evidence that women are more likely to define their experiences interpersonally, regardless of context (Gilligan, 1982; Gwartney-Gibbs and Lach, 1994). Although Hobfoll *et al.* (1994) noted that 'women are more influenced by social context than men' (*ibid.*, 50), they did not explore the underlying reasons for this tendency. Their analysis tends to frame women's behaviour as an innate quality obfuscating its socially constructed underpinning. A feminist analysis would attempt to explain women's tendency to emphasize social phenomena and relationships as the necessary by-product of surviving as subordinates in a male-dominated society (Miller, 1986). This kind of analysis is congruent with and furthers current reformulations of theories of coping that construe coping efforts as responses that are specific both to the individual and her/his context (Lazarus, 1991).

A fairly consistent gender difference articulated in the existing research is that women appear to be more likely than men to cope by turning to others and using distraction, whereas men are more likely to turn to sports or leisure activities (Korabik *et al.*, 1993; Schwartz and Stone, 1994; Statham, 1987). These differences can be attributed to socially sanctioned expectations of gendered behaviour as well as to differential opportunities that are open to men and women. Most women still assume or are expected to assume the major responsibility for housekeeping and parenting tasks (Scarr *et al.*, 1989). They may, therefore, have less opportunity than their male counterparts to employ coping strategies that focus on leisure activities. In addition, some observed differences may reflect the nature of the commonly researched job categories. Female managers, for instance, often have to cope with additional stressors related to their minority status and high visibility in male-dominated environments. Moreover, even when these women occupy roles of similar stature to males within an organization they often do not have the same access to power and resources (Kanter, 1977).

Coping Effectiveness

Much of the early research in this area tended to interpret women's coping strategies as being less effective than men's (e.g. Folkman and Lazarus, 1980; Pearlin and Schooler, 1978). Parasuraman and Cleek (1984), however, found that female managers used more 'adaptive' coping than male managers. Experts rated the effectiveness of self-reported coping strategies of 200 middle managers from a Midwest utility company. They found that women were more likely to use planning, organizing, prioritizing, enlisting support of others, and touching base with supervisors and co-workers. Male managers used more of what was considered by the experts as 'maladaptive' coping, i.e. they were more likely to work harder; they made more mistakes and promises they could not keep; they stuck to one solution to the problem; they left the workplace, and they avoided supervisors and peers. Thus, when specific coping strategies for a specific purpose in a particular context were examined for their effectiveness, gender differences emerged that did not fit the stereotypes, and in fact, may have contradicted them.

The over emphasis on instrumental or active coping reflects a Western value system that embraces the concept of the autonomous, individual self and is directly related to the privileged treatment of men and traditionally male behaviours within a patriarchal society. Based on these values, problem-solving has been regarded as the most effective coping strategy in many stressful situations. However, problem-solving efforts can be counter-productive, and problem-reappraisal coping (a form of coping that changes the meaning of what is happening) might offer the best coping choice and may be a very powerful device for regulating stress. These alternative coping strategies have often been overlooked or undervalued in the research as a result of the assumptions of the value of problem-solving. Although there are some notable exceptions (Hobfoll *et al.*, 1994; O'Brien and DeLongis, 1996), both interpersonal and collective coping strategies have also been largely ignored in this body of research. Not only does this have ramifications for those individuals from cultural and ethnic groups that emphasize such collective strategies, it has direct implications for women in the light of the research that supports the importance of the self-in-relation-to-others in women's experience (Gilligan, 1982). Because the scope of coping activities typically assessed by standardized coping inventories has been biased towards individual coping and masculine views of coping, it has tended to ignore crucial aspects of women's experiences and the potential strengths of using multiple strategies and the collective strategies of coping more common amongst women and members of various minority groups. Several conceptualizations of coping posit that a hallmark of effective adaptation is flexibility or variability in the way a person copes with different stressors (e.g. Cohen, 1984; Lazarus and Folkman, 1984; Meichenbaum, 1985; Moos and Billings, 1982). Thus, a broad repertoire of coping strategies could be construed as being particularly effective.

Generally, the effectiveness of coping strategies has not been adequately assessed so conclusions cannot be drawn as to which pattern of coping is most adaptive (Mattlin, Wethington and Kessler, 1990). For example, one can speculate that the use of avoidance coping can buy time to garner resources and thus facilitate problem-solving coping (e.g. Folkman, Lazarus, Gruen and DeLongis, 1986), or it may lead to ever-increasing feelings of helplessness and self-defeat. In order to determine 'good' or 'bad' coping it is important to know what is at stake for the individual and what the motivation is for using different coping strategies. For example, staying late at work to complete a project may be stressful for one woman because it is perceived as a threat to her beliefs that a mother should always be home at dinner time, or it could threaten another woman's sense of competence because she believes that she should be able to complete her tasks during the work day. In either case, which coping strategies are considered effective is relative to what is at stake in the person-environment transaction (Lazarus and Folkman, 1984). Due to the complexity of determining the efficacy of coping responses, much more research is needed that considers the appraisal of what is at stake, and thus what is being coped with.

Personal or coping resources

Occupational stress research that focuses on coping strategies tends to individualize problems which may be inherently structural, thereby minimizing the institutional barriers faced by women. This individualized approach implies that the solution to stress is to find more effective ways of coping. It is the status quo position wherein the goal is to facilitate women's adaptation to organizational structures and models of work that are essentially male. From this perspective, little attention has been paid to women's variable access to resources and the impact of social cultural values on the ways women cope with occupational stress (Marshall, 1993).

The importance of including an analysis of cultural themes and variations is apparent in Frye and D'Avanzo's (1994) study of Cambodian refugee women, even though this study does not focus on occupational stress. Their study highlights the role cultural mores and values play in dramatically shaping women's understanding of stress and their coping responses. Thus, if we extend our analysis to include a conceptualization of women as living within complex cultural categories that include gender, sexual orientation, race, ethnicity, and age, then it becomes clear that to ignore or factor out any of these variables will give us at best a very limited and incomplete understanding of employed women's experience of stress and coping.

Of particular concern is the paucity of research on working-class women and poor women, rendering them almost invisible in the stress and coping literature. The focus on women in professional and managerial positions presents a picture of women's work experiences that is in sharp contrast to the experiences of the majority of women who continue to work in low status, poorly paid, dead-end jobs in the clerical, sales and service sectors (Lowe, 1989) and who experience additional stressors related to poverty, classism, racism, and inadequate access to resources. Moreover they face additional stressors in the form of class-based criticism of their worth and fitness as women, mothers and members of society (Kollias, 1975). Little is known of their means of coping and whether their appraisal of stress or strategies of coping differ from those of middle-class, professional women. Yet, many of these women, particularly the working poor, experience a complex array of stressors and are the most vulnerable to this stress because of the very limited access they have to the resources that existing research indicates would help them to respond effectively (e.g. education, income, organizational support).

Integrative studies and the importance of context

The literature on employment and stress appears to be moving from the broad question of whether employment is harmful or beneficial to women, to an examination of the particular variables and processes that affect the complex relationships among employment, stress and coping and health. Research has identified a number of the personal or coping resources (i.e. relatively stable traits, beliefs, values or dispositional characteristics) that influence the experience of occupational stress, yet typically most studies consider only one or two of these factors in isolation (e.g. control beliefs, self-esteem, social support, gender-role orientation) and lack a theoretical framework for the selection of variables.

Richard Lazarus's (1991) stress and coping theory provides a theoretical framework that potentially identifies the impact of social and environmental forces on a woman's coping efforts. His model is one of the most extensive and cited theories of stress and may be particularly useful because it reflects the multidimensional, dynamic and complex nature of stress and coping, including the interactions between individuals and the structures or contexts in which they live. Lazarus's (1991: 1) cognitive phenomenological theory is 'transactional, process, contextual, and meaning-centred' and defines psychological stress as the transactional relationship 'between person and environment that is appraised by the person as taxing or exceeding his or her resources and endangering his or her well-being' (Lazarus and Folkman, 1984: 19).

Long, Kahn and Schutz (1992) provide an example of the potential usefulness of Lazarus's theory in the development of a more comprehensive and inclusive theory of women and coping. Drawing on Lazarus's theory, the researchers studied the process of coping with occupational stress among 249 managerial and professional women. The choice of constructs was based not only on the stress/coping literature, but also on career theory (Betz and Fitzgerald, 1987). The model was comprehensive in that both personal resources (e.g. income, marital and parental status, education, job level) and personal characteristics (e.g. gender-role attitudes, an optimistic sense of personal efficacy) were assessed, along with what was at stake in the stressor event and perceptions of the demands and supports in the work environment. Although this study begins to illustrate the ways in which social and environmental forces can be included in the study of stress and coping using Lazarus's theoretical framework, several limitations are still apparent. The study continued the tradition of focusing on professional women, and lacked an analysis of the intersection of class, race and sexual orientation with gender. It also did not include any examination of interpersonal or collective coping strategies.

As a further extension of Lazarus's theory of coping, Long (1998) examined the influence of institutionalized social roles on workplace stress and coping by cross-validating the model developed for women in managerial roles with a sample of clerical workers. The results revealed a pattern of relationships among the model's constructs that reflect differences in power and status between managers and clerical workers. For example, clerical workers had fewer coping resources, appraised the stress as less controllable, experienced more work demands and less support, and used relatively less engagement coping (i.e. active efforts aimed at managing both problem- and emotion-focused aspects of the stressful event). These results emphasize the need for researchers to consider contextual issues of power and status. Occupational stress reflects relationships between the employee and employer that constrain the availability of coping resources and is not just about the individual woman and her coping pattern.

Important research directions

The perspective that underlies much of the previously described research has tended to exclude any focus on the complexity of the interrelationship between social conditions and subjective experience. Handy (1988) points out in her review of occupational stress and burnout research, the vast majority of researchers in the field have neglected to examine the inter-relationship between an individual's subjective experience and her/his social-political context. Handy (1988) calls for an analysis of the reciprocal relationships between social and organizational structures and individual subjectivity and action, and the relative power imbalances between the individual and the organization. Overall the tendency has been to eschew any in-depth analysis of power as it relates to the stress and coping process. Despite research that supports the importance of access to resources in the use of effective coping strategies (Hall, 1991; Mainiero, 1986; Rosenfield, 1989), little attention has been paid to women's differential abilities to access resources based on colour, age, sexual orientation and economics; or to the nature of the roles that women inhabit as a subordinate in a male-dominated society. It seems incumbent upon researchers now to address this lack and to begin to integrate more fully an analysis of the power structures that shape employed women's experiences and relationships and to counter the assumptions of equality inherent in the individualized approach (Colwill, 1993; Kanter, 1977).

Although researchers often include self-reflexive criticisms about the limitations of using self-report or phenomenological designs, they rarely offer any analysis of how dominant social beliefs and value systems, and internalized sexism, racism, classism, agism, and homophobia might be reflected in self-report responses and influence both the participants' and the researchers' understanding and articulation of their experiences. Most of the research makes invisible the multiple locations of the researcher and the dynamic interaction between the researcher's position and that of the participants. Such work tends to use what Haraway (1988: 584) describes as the 'God trick;' that is, the narrative speaks from 'nowhere and everywhere, equally and fully.'

Historically, there has been a tendency for research to simplify the complex by controlling for variables rather than examining the dynamic relationships of one to another. From this vantage point, the within-group differences among women are made invisible; women are presented as a homogeneous group and the theories that arise from the experience of white, middle-class women are generalized to apply to all women regardless of class, race, sexual orientation and other self- or other-identified categories of oppression. Such research tells us little about how these variables interact with gender to generate stress in an employed woman's life; nor does it tell us how they may affect her appraisal of that stress or her access to resources for coping with that stress.

Women's experience of employment takes place in the multi-layered social and cultural context of patriarchy in which there is the real potential for conflicting values and differential access to power and resources. Their lives include membership in many complex relationships from which they garner both costs and benefits. Thus, to study the ways women manage the demands of employment requires movement away from linear conceptions of cause and effect towards explanatory models that embrace complexity and view power relations as the basis of patriarchal social arrangements (Banyard and Graham-Bermann, 1993). Reformulated theories of coping need to be developed that challenge overly simplistic dichotomies (male versus female, problem-focused versus emotion-focused coping); it is no longer sufficient to add women to theories developed for, by, and about men (Harding, 1991). Researchers need to acknowledge their own and their participants' specific locations in a complex and shifting socio-political and cultural context and recognize that women's ways of coping are a reflection of an unequal distribution of power based on gender, colour, economic sufficiency, age, and sexual orientation.

Acknowledgements

This research was supported by a grant to the first author from the Social Sciences and Research Council of Canada.

References

ACKER, J. (1992) 'The future of women and work: ending the twentieth century', *Sociological Perspectives* **35**: 53–68.

BAILEY, D., WOLFE, D. and WOLFE, C. R. (1996) 'The contextual impact of social support across race and gender: implications for African American women in the workplace', *Journal of Black Studies* **26**: 287–307.

BANYARD, V. L. and GRAHAM-BERMANN, S. A. (1993) 'Can women cope? A gender analysis of theories of coping with stress', *Psychology of Women Quarterly* **17**: 303–318.

BARNETT, R. C. and BARUCH, G. K. (1985) 'Women's involvement in multiple roles and psychological distress', *Journal of Personality and Social Psychology* **49**: 135–145.

BAUM, A. and GRUNBER, N.E. (1991) 'Gender, stress and health', *Health Psychology* **10**: 80–85.

BETZ, N. E. and FITZGERALD, L. F. (1987) *The Career Psychology of Women*, San Diego, CA: Academic Press.

COHEN, F. (1984) 'Coping', in J. D. Matarazzo, S. M. Weiss, A. J. Herd, N. Miller and S. M. Weiss (eds), *Behavioral Health: A Handbook of Health Enhancement and Disease Prevention*, New York: Wiley, 261–274.

COLWILL, N. L. (1993) 'Women in management: power and powerlessness', in B. C. Long and S. E. Kahn (eds) *Women, Work, and Coping: A Multidisciplinary Approach to Workplace Stress*, Montreal and Kingston: McGill-Queen's University Press, 73–89.

COMPAS, B. E., FORSYTHE, C. J. and WAGNER, B. M. (1988) 'Consistency and variability in causal attributions and coping with stress', *Cognitive Therapy and Research* **12**: 305–320.

FACIONE, N.C. (1994) 'Role overload and health: the married mother in the waged labor force', *Health Care for Women International* **15**: 157–167.

FOLKMAN, S. and LAZARUS, R. S. (1980) 'An analysis of coping in a middle-aged community sample', *Journal of Health and Social Behavior* **21**: 219–239.

FOLKMAN, S., LAZARUS, R. S., GRUEN, R. J. and DELONGIS, A. (1986) 'Appraisal, coping, health status and psychological symptoms', *Journal of Personality and Social Psychology* **50**: 571–579.

FRYE, B. A. and D'AVANZO, C. D. (1994) 'Cultural themes in family stress and violence among Cambodian refugee women in the inner city', *Advances in Nursing Science* **16**: 64–77.

GILLIGAN, C. (1982) *In a Difference Voice: Psychological Theory and Women's Development*, Cambridge, MA: Harvard University Press.

GUTEK, B. (1993) 'Asymmetric changes in men's and women's roles', in B. C. Long and S. Kahn (eds) *Women, Work, and Coping: A Multidisciplinary Approach to Workplace Stress*, Montreal and Kingston, ONT: McGill-Queen's University Press.

GWARTNEY-GIBBS, P. A. and LACH, D. H. (1994) 'Gender differences in clerical workers' disputes over tasks, interpersonal treatment, and emotion', *Human Relations* **47**: 611–639.

HALL, E. (1991) 'Gender, work control, and stress: a theoretical discussion and an empirical test', in J. V. Johnson and G. Johansson (eds) *The Psychosocial Work Environment: Work Organization, Democratization, and Health*, Amityville, NY: Baywood, 89–108.

HANDY, J. A. (1988) 'Theoretical and methodological problems within occupational stress and burnout research', *Human Relations* **41**: 351–369.

HARAWAY, D. (1988) 'Situated knowledges: the science question in feminism and the privilege of partial perspective', *Feminist Studies* **14**: 575–599.

HARDING, S. (1991) *Whose Science? Whose Knowledge? Thinking From Women's Lives*, Ithaca, NY: Cornell University Press.

HARE-MUSTIN, R. T. and MARECEK, J. (1990) 'Gender and meaning of difference: postmodernism and psychology', in Hare-Mustin, R. and Marecek, J. (eds) *Making a Difference: Psychology and the Construction of Gender*, London: Yale University Press, 22–54.

HOBFOLL, S. E., DUNAHOO, C. L., BEN-PORATH, Y. and MONNIER, J. (1994) 'Gender and coping: the dual-axis model of coping', *Journal of Community Psychology* **22**: 49–82.

JACOBSEN, J. P. (1994) 'Trends in work force sex segregation, 1960–1990', *Social Science Quarterly* **75**: 204–211.

JAMES, K. (1994) 'Social identity, work-stress, and minority workers' health', in G. P. Keita and J. J. Hurrell, Jr. (eds) *Job Stress in a Changing Workforce: Investigating Gender, Diversity, and Family*, Washington, DC: American Psychological Association, 127–145.

KANTER, R. M. (1977) *Men and Women of the Corporation*, New York: Basic Books.

KOLLIAS, K. (1975) 'Class realities: create a new power base', *Quest* **1**: 28–43.

KORABIK, K., MCDONALD, L. M. and ROSIN, H. M. (1993) 'Stress, coping, and social support among women managers', in B. C. Long and S. E. Kahn (eds) *Women, Work, and Coping: A Multidisciplinary Approach to Workplace Stress*, Montreal and Kingston: McGill-Queen's University Press, 133–153.

KOTLER, P. and WINGARD, D. L. (1989) 'The effect of occupational, marital and parental roles on mortality: the Alameda County study', *American Journal of Public Health* **79**: 607–611.

LACROIX, A. and HAYNES, S. G. (1987) 'Gender differences in the health effects of workplace roles', in R. C. Barnett, L. Biener and G. K. Baruch (eds) *Gender and Stress*, New York: Free Press, 96–121.

LAZARUS, R. S. (1991) 'Psychological stress in the workplace', *Journal of Social Behavior and Personality* **6**: 1–13.

LAZARUS, R. S. and FOLKMAN, S. (1984) *Stress, Appraisal, and Coping*, New York: Springer.

LONG, B. C. (1990) 'Relation between coping strategies, sex-typed traits and environmental characteristics: a comparison of male and female managers', *Journal of Counseling Psychology* **37**: 185–194.

LONG, B. C. (1998) 'Coping with workplace stress: a multiple-group comparison of female managers and clerical workers', *Journal of Counseling Psychology* **45**: 65–78.

LONG, B. C., KAHN, S. E. and SCHUTZ, R. W. (1992) 'A causal model of stress and coping: women in management', *Journal of Counseling Psychology* **39**: 227–239.

LOWE, G. (1989) *Women, Paid/Unpaid Work and Stress: New Directions for Research*, Ottawa: Canadian Advisory Council on the Status of Women.

LYKES, M. (1983) 'Discrimination and coping in the lives of black women: analyses of oral history data', *Journal of Social Issues* **39**: 79–100.

McGUIRE, G. and RESKIN, B. (1993) 'Authority hierarchies at work: the impacts of race and sex', *Gender and Society* **7**: 487–506.

MAINIERO, L.A. (1986) 'Coping with powerlessness: the relationship of gender and job dependency to empowerment strategy usage', *Administrative Science Quarterly* **31**: 633–653.

MARSHALL, J. (1993) 'Patterns of cultural awareness: coping strategies for women managers', in B. C. Long and S. E. Kahn (eds) *Women, Work, and Coping: A Multidisciplinary Approach to Workplace Stress*, Montreal and Kingston: McGill-Queen's University Press, 90–110.

MARSHALL, N. L. and BARNETT, R. C. (1992) 'Work-related support among women in caregiving occupations', *Journal of Community Psychology* **20**: 36–42.

MARTIKAINEN, P. (1995) 'Women's employment, marriage, motherhood and mortality: a test of the multiple role and role accumulation hypotheses', *Social Sciences and Medicine* **40**: 199–212.

MATTLIN, J. A., WETHINGTON, E. and KESSLER, R. C. (1990) 'Situational determinants of coping and coping effectiveness', *Journal of Health and Social Behavior* **31**: 103–122.

MEICHENBAUM, D. (1985) *Stress Inoculation Training*, New York: Pergamon.

MELIES, A. I., NORBECK, J. S., LAFFREY, S., SOLOMON, M. and MILLER L. (1989) 'Stress, satisfaction, and coping: a study of women clerical workers', *Health Care for Women International* **10**: 319–334.

MILLER, J. B. (1986) *Toward a New Psychology of Women*, 2nd edn, Boston: Beacon Press.

MOOS, R. H. and BILLINGS, A. G. (1982) 'Conceptualizing and measuring coping resources and processes', in L. Goldberger and S. Breznitz (eds) *Handbook of Stress: Theoretical and Clinical Aspects*, New York: Free Press, 212–230.

NELSON, D. L. and HITT, M. A. (1992) 'Employed women and stress: implications for enhancing women's mental health in the workplace', in J. C. Campbell, L. R. Murphy and J. J. Hurrell, Jr. (eds) *Stress and Well-being at Work: Assessments and Interventions for Occupational Mental Health*, Washington, DC: American Psychological Association, 164–177.

O'BRIEN, T. and DELONGIS, A. (1996) 'The interactional context of problem-, emotion-, and relationship-focused coping: the role of the big five personality factors', *Journal of Personality* **64**: 1277–1292.

PARASURAMAN, S. and CLEEK, M. A. (1984) 'Coping behaviors and manager's affective reactions to role stressors', *Journal of Vocational Behavior* **24**: 170–193.

PARKER, R. A. and ALDWIN, C. M. (1994) 'Desiring careers but loving families: period, cohort, and gender effects in career and family orientations', in G. P. Keita and J. J. Hurrell, Jr. (eds) *Job Stress in a Changing Workforce: Investigating Gender, Diversity, and Family Issues*, Washington, DC: American Psychological Association, 23–38.

PEARLIN, L. I. and SCHOOLER, C. (1978) 'The structure of coping', *Journal of Health and Social Behavior* **19**: 2–21.

REPETTI, R. L., MATTHEWS, K. A. and WALDRON, I. (1989) 'Employment and women's health', *American Psychologist* **44**: 1394–1401.

RODIN, J. and ICKOVICS, J. R. (1990) 'Women's health: review and research agenda as we approach the 21st century', *American Psychologist* **45**: 1018–1034.

ROSENFIELD, S. (1989) 'The effects of women's employment: personal control and sex differences in mental health', *Journal of Health and Social Behavior* **30**: 77–91.

SCARR, S., PHILLIPS, D. and MCCARTNEY, K. (1989) 'Working mothers and their families', *American Psychologist* **44**: 1402–1409.

SCHWARTZ, J. E. and STONE, A. A. (1993) 'Coping with daily work problems: contributions of problem content, appraisals, and person factors', *Work and Stress* **7**: 47–62.

SMITH, C. A. (1993) 'Evaluations of what's at stake and what I can do', in B. C. Long and S. E. Kahn (eds) *Women, Work, and Coping: A Multidisciplinary Approach to Workplace Stress*, Montreal and Kingston: McGill-Queen's University Press, 238–265.

SPAIN, D. and BIANCHI, S. M. (1996) *Balancing Act: Motherhood, Marriage, and Employment Among American Women*, New York: Russell Sage Foundation.

STATHAM, A. (1987) 'The gender model revisited: differences in management styles of men and women', *Sex Roles* **16**: 198–429.

STATISTICS CANADA (1995) *Women in Canada*, 3rd edn, catalogue no. 89-503E, Ottawa, Canada: Ministry of Supply and Services.

VERBRUGGE, L. M. (1986) 'Role burdens and physical health of women and men', *Women and Health* **11**: 47–77.

VERBRUGGE, L. M. (1987) 'Role responsibilities, role burdens and physical health', in F. J. Crosby (ed.) *Spouse, Parent, Worker: On Gender and Multiple Roles*, New Haven, CT: Yale University Press, 154–166.

WETHINGTON, E. and KESSLER, R. C. (1993) 'Neglected methodological issues in employment stress', in B. C. Long and S. Kahn (eds) *Women, Work, and Coping: A Multidisciplinary Approach to Workplace Stress*, Montreal and Kingston: McGill-Queen's University Press, 269–295.

CHAPTER SEVEN

Organizational change: adaptive coping and the need for training

MICHAEL P. LEITER

Training and coping

By considering stress as an appraised imbalance of demands over resources (Cox, 1978; Lazarus and Folkman, 1984), the transactional model generates diverse perspectives on coping strategies. In the most general sense, coping involves reducing demand, increasing resources or some combination of the two. While strategies that enhance resources may encompass the acquisition of material resources, they may primarily involve increasing skills or gaining confidence in existing abilities. Bandura (1977) emphasized this latter aspect of coping with regard to the concept of self-efficacy: people who are confident of their capacity to apply effective coping responses are less likely to experience stress when encountering demands. Skill enhancement is an aspect of resource development which promotes self-efficacy in the process of executing occupational tasks. Skills are held by individual employees, who determine their application to occupational demands, to the extent that professional autonomy is exercised. Professional skills empower human service professionals as experts (French and Raven, 1960); to the extent that their employers value these skills, they recognize employees' individual autonomy regarding their use. Within health service institutions, the enhancement of professional skills is one of the most fundamental coping responses within the occupational domain.

The Social Context of Mid-Career Training

The enhancement of occupational skills extends the coping capacity of the individual while enhancing the effectiveness with which he or she performs their job. Despite the importance of training to the individual coping capacity, decisions to undertake training often exceed the discretion of individual employees. People make these decisions within a complex social context. The enhancement of skills requires the investment of financial and personal resources: learning takes time and money. Employees often look to their employers to provide training through the provision of in-house programmes during working hours, or through the funding of employees' participation in extramural training in salaried release time. The process of enlisting substantial financial or time support from employers may have a considerable impact on employees' training plans. It may result in training taking a form, content or extent quite different from that which the employees originally perceived to be appropriate to their needs. Aspects of funding, location and timing may change as well. As an investor in occupational training, an employer becomes an important stakeholder in training decisions.

The employer is not by any means the only stakeholder with whom an employee must interact in pursuit of his or her training aspirations. Training providers – be they intramural training departments or extramural agencies – determine the availability and form of training opportunities. Although there are obvious reasons for training providers to be responsive to the wishes of prospective students, they may not complement each other perfectly. For example, Leiter, Dorward and Cox (1994) described a persistent mismatch between the training aspirations of occupational health nurses and the offerings of training providers. Available training opportunities were largely restricted to clinical skills, despite the nurses' expressed interest in pursuing the full range of professional issues relevant to a broadly defined role as an occupational health nurse.

At times the format of training deliberately restricts access. An individual may seek a specific skill which is only available as part of a wider training programme. For example, training to administer intelligence scales may not be available outside the context of broader professional training programmes in psychology or education. The process of developing training opportunities involves interactions between employees and training providers. Each of these parties pursues somewhat distinct goals in regard to training, while maintaining ongoing dialogues with employers, who provide yet another perspective.

The interactions of employees, employers and training providers occur within a broader institutional environment of values and regulations. Professional associations develop training policies that define requirements for entry into the profession as well as maintenance of that status. They directly influence training providers through accreditation procedures, and monitor to varying degrees employers' acknowledgement of professional privileges. Government agencies have a parallel influence, particularly through health and safety regulations which require training that may not be a personal priority of the other parties.

Individuals make decisions about their own training plans in the context of these varied, overlapping and often conflicting interests. This context places constraints upon the extent to which they may pursue their own goals – whether they are the enhancement of professional efficacy or of personal marketability. On the one hand service professionals are expected to function as autonomous individuals, yet on the other their autonomy is only legitimate within a social context, which often imposes constraints. The resolution of these dilemmas involves a process of negotiation.

Individual discretion

In an organizational context individual discretion – defined as 'freedom or power to make one's own judgment' (Funk and Wagnalls, 1980: 182) – is the capacity of an individual member of an organization to make decisions about organizational issues on the basis of professional judgement. Organizational systems that explicitly use individual discretion are in contrast both to systems in which individuals must refer decisions up a hierarchy and to systems in which detailed policies and procedures predominate in decision making. A primary advantage of systems that promote individual discretion is responsiveness to unique problem situations. The primary disadvantage for the organization is a vulnerability to fragmentation, as the power to make decisions with consequential implications is dispersed. The challenge for organizations in managing discretion is to provide for the requisite decision-making capacity, which Butler (1991) defined as 'one sufficiently crisp to minimize decision-making costs and sufficiently fuzzy to achieve adaptability' (*ibid.*: 57).

The Role of Individual Discretion in Human Services

Organizational support of individual discretion is not unique to human service institutions, but it is intrinsic to the culture of these organizations. Human service organizations depend on their employees for sophisticated expertise. Effective counselling, medical care and teaching proceed from skills developed through extensive training and refined in practical experience. In the broadest sense, human service work encompasses the assessment of unique situations, creative problem solving, implementation of interventions, and assessment of their impact. Utilizing that expertise requires flexibility. Policies and procedures that permit workers latitude to choose approaches pertinent to the situation facilitate the adaptation of services to specific client requirements.

For example, health-care professionals employed by a hospital provide services to patients, each of whom presents a unique cluster of symptoms and life conditions. The assessment information gathered through interactions between health-care professionals and their patients creates the conditions under which providers may apply medical knowledge and technology to the problem. The problem situation evolves through the interactions of health-care professionals with patients as the intervention progresses, creating new information and demanding more decisions. Health-care professionals use hospital policies and established medical practice to guide their work in a general sense. However, they decide on the relevance of the policies to the situation, adapting them to the unique and changing aspects. Front-line health-care professionals require wide discretion to implement this form of treatment.

In contrast, systems that require the involvement of parties external to treatment situations introduce demands on the information processing capacities of systems. The individual overseeing decisions would require access to specific information about each treatment situation in order to make a viable contribution appropriate to that situation. While such interactions occur in regard to unusual or problematic cases requiring expert consultation, they would overwhelm an organization's communication capacity were they to occur with every encounter of health-care professionals with patients. Systems which acknowledge individual discretion lighten the information processing load, because a large part of the relevant information remains close to the problem situation. Decentralization of decision making reduces some of the demands on management.

Limitations of Individual Discretion

Despite the advantages or even the necessity to have systems that support professional judgement, the exercise of individual discretion brings with it problems of integrating this into the social context of organizational life. One aspect of this is the assignment of responsibility. The organization is liable for the consequences of the decisions of individual employees working within their mandate of discretion. By not overseeing every decision, the employer permits clients to suffer the consequences of employees' mistakes.

Another problem arises from the cooperative nature of most tasks within a service organization. One individual's professional discretion may conflict with that of another when they both are interacting with the same patient, using the same space, drawing upon the same support staff, or depending on the same administrative budget. The need for ongoing resolution of conflicts constrains the boundaries of individual discretion, while contributing to the overall demands on service providers. In addressing this conflict, human service professionals make a fundamental change in focus from concentrating on the requirements of clients to that of coordinating the functioning of an organization. The resolution of conflicts among them regarding their professional judgement is a fundamental management function. Interdisciplinary rivalries are signs of inadequate resolution of conflicts.

The mutual coordination of autonomous professionals differs fundamentally from the application of professional judgement to problems of the recipients of their services. First, mutual coordination requires communication among colleagues. These interactions among peers make demands that are distinct from those with service recipients. Second, coordination requires cooperative decision making. Sharing authority in this manner requires a distinct perspective on the process of solving problems. Individual decision making requires much less emphasis on the process; it is solely concerned with solutions. Third, mutual coordination requires the capacity to resolve conflicts among peers, further increasing the demands of communication. These communication tasks require skills, temperament, time and energy that are to some extent distinct from those demanded of service provision. These skills and resources are often scarce.

The Role of Training in an Expanding Mandate

This rudimentary management function – the coordination of autonomy – serves as a foundation for expanding the administrative mandate of clinical managers. The need for decentralized interventions for resolving conflicts among service providers is intrinsic to the exercising of professional judgement in an organizational environment. The demands of organizational life often far exceed such tasks. The devolution of authority to organizational units closer to service provision increases the scope of those tasks. The process of devolution, as is occurring in many health-care systems throughout North America and Europe, highlights the demands on service providers. Their additional responsibilities often arrive without a commensurate increase in resources; in fact, devolution is often part of a process of resource reduction. A successful resolution of this apparent contradiction requires a fundamental change in the way in which work is done.

Devolution of managerial responsibilities does increase an organizational unit's degree of self-determination, but at a cost. To some extent additional responsibilities are a further tax on the time and energy of members of the unit, and this burden differs significantly from that associated with service provision. When providing services, professionals draw upon training that they have augmented through their experience. This knowledge base is much less extensive in regard to management issues. While service providers do amass a body of tacit knowledge through intrinsic learning in organizational settings (Sternberg, 1985; Wagner and Sternberg, 1987), this knowledge base is generally more limited than that of their professional skills. Training could address some aspects of this shortfall, but the appropriate training may not be accessible.

Deterrents to involvement in training during organizational transitions arise from all four stakeholders noted earlier. The first of these, the *employees* are pressed for time and energy as their responsibilities expand. They often fail to recognize the extent to which their new responsibilities are discontinuous with the skills they had developed through their previous work experience. Often they are simultaneously confronting demands to enhance or maintain their clinical credentials. The financial constraints that prompted devolution further limit the willingness of *employers* to devote resources to training costs. *Training providers* are reluctant to commit significant resources to the development of new programmes in an uncertain and changing situation. *Professional bodies and government agencies* are often in a reactive position: they are reluctant to develop policies regarding management functions that are somewhat outside their range of expertise until they confront concrete problems in practice. As a result, employees who may have the time and commitment to acquire training rarely encounter programmes that they consider to be appropriate to their situation.

Shortfall In Organizational Resources

A crisis in management skills is part of a widespread shortfall in organizational resources encountered during transition periods. Hackman (1986) and others (Manz, 1986; Manz and Sims, 1984) have examined the exercise of individual discretion in management, under the label of self-management, in organizations that do not necessarily utilize service professionals. Hackman (1986) outlined five enabling conditions which were necessary for effective self-management, with special emphasis on clarity of goals. A clear understanding of the organization's goals provides a general context within which self-managers pursue objectives pertinent to their organizational unit. Without shared goals they have no basis for selecting objectives or for assessing their success. To work effectively towards goals, managers require an enabling task structure, access to appropriate training and information, and expert coaching. Within this model self-managers are not merely given a task and left to their own devices. A supportive organization treats the judgement and initiative of managers as valued organizational resources. The foundation of successful self-management is adequate material resources, which make it possible to attain objectives and signal that the organization values the unit.

Self-Management and Stress

In his discussion, Hackman (1986) did not consider the problems that individuals encounter when these enabling conditions are lacking, although the potential for personal problems is implicit in his discussion. The strong personal investment of someone in a self-management position brings opportunities both for exhilaration and for burnout, in that the working context of self-managers encourages them to take responsibility for outcomes. A context which assigns responsibilities to a self-manager without providing the enabling conditions, puts incumbents into a condition of ever-depleting resources. Accepting the self-management role, they invest personal time and energy to compensate for organizational shortcomings. Failures, which are inevitable without sufficient support, undermine their sense of personal well-being, which Hobfoll (1989) underscored as the fundamental personal coping resource.

Support Shortfalls in an Expanding Mandate

Health care organizations have difficulty in maintaining sufficient support for employees when expanding their mandate. A mandate can be expanded much more quickly than an organization can develop the enabling conditions to support it. During these transitions, clinical managers are – at least temporarily – in a position of working beyond their resources, setting the stage for them to experience stress. If the organization can develop support fairly rapidly, the event is experienced as an acutely stressful period, during which the exhilaration of new challenges balances increased anxiety and confusion. If the organization's response is particularly slow, managers are vulnerable to burnout (Cherniss, 1989; Leiter, 1991), in which individuals become discouraged and cynical as the novelty wears off.

Interview study

The applicability of this framework was explored in a series of interviews with health-care workers regarding their perspectives on enabling conditions for self-management during a period of transition. Their experience of supportive conditions during the transition is illustrated in this chapter by quotes from the interviews. The interviews demonstrate the vital role played by training in the expansion of occupational responsibilities.

Setting

The people interviewed in this study were health-care professionals working at an 800-bed adult acute care hospital employing 4,000 people in a wide range of health-care services and specialities. The hospital provides a general hospital function for the surrounding community. It is the place to which smaller community hospitals across the province refer difficult cases, and the regional treatment centre for complex procedures (e.g. organ transplants, oncology treatment) for Eastern Canada. It is also a teaching hospital in which the heads of its clinical departments are chairs of their corresponding departments at the local university medical school.

Participants

From May to July 1990, 100 employees of the hospital participated in an interview about health-care management, conducted by two researchers.

After receiving permission for the study from the hospital's research review board and the management committee, the researchers recruited participants by attending departmental meetings. Table 7.1 indicates the areas of the hospital in which the participants worked. All were health-service professionals.

Procedure

The rationale for the study was summarized before each individual interview. The interviewers stated that participation was entirely voluntary, and that the interview was being tape recorded, and would be transcribed by a typist. None of the participants exercised their right to withdraw participation at any time; in fact many stated that they enjoyed the process, and actively encouraged their colleagues to participate.

We conducted interviews individually at the hospital. The tapes were transcribed by typists at the university into a computer word-processing programme. Themes were recorded using the index and footnote functions of the word processor. The vignettes presented here were selected from the index generated in the initial analysis.

Background

The hospital was confronting three major management issues during the time of this study. First, it was beginning a multi-year process of moving from the status of a government agency to that of a free-standing hospital. During this transition, employees would cease to be civil servants: government offices

Table 7.1. Departments participating in the study.

Department	Number of interviews
Special Services	2
Social Work	10
Physiotherapy	12
Occupational Therapy	11
Volunteer Services	3
Pastoral Care	6
Nursing	35
Senior Administration	4
Clinical Department Heads	5
Medical Administration	12
Total	100

would cease to make and implement major administrative decisions. Second, the hospital was introducing a vice-presidential level into its management structure, thereby cutting down substantially on the number of people reporting directly to the Chief Executive Officer. Part of this management restructuring included a move to devolve authority for many budgetary and management decisions, as explained below. Third, the hospital was charged by its Board of Directors with the responsibility of reducing its operating budget by 5 per cent in response to constraints on the provincial health budget. The senior administration decided to effect this reduction by requiring each budget unit in the hospital to cut 5 per cent from its budget. As a result many middle-level managers were receiving their initial experience of budgetary responsibility at the time of a considerable reduction in their operating resources and in the context of a move from civil service which caused many employees anxiety about the security of their jobs or their benefits.

The management ethic espoused by the hospital emphasized the devolution of authority down to middle management, and the promotion of participatory decision making among direct care units. Whereas first-order clinical managers (e.g. head nurses, support service department heads) had been accustomed to considerable autonomy regarding clinical decisions, responsibility for the financial state of their units was a new thing. Previously, the hospital had been bailed out of budgetary deficits by the provincial government; the hospital in turn bailed out the clinical units. In response to many forces (broad-based concerns about the cost of health care, pressures to reduce government budgetary deficits), the senior administration of the hospital had embraced a value of stricter financial responsibility.

The position of the hospital as an organization providing high quality professional services in a government bureaucracy had resulted in chronic tension between the professional and bureaucratic models. The tension between these perspectives persisted in a system in which an outdated administrative bureaucracy formed a backdrop against which individuals or units operated nearly autonomously in their pursuit of clinical missions. The professional model dominated within these units; the bureaucratic model dominated across the overall structure. Each model carried with it an organizational culture. An insular perspective from which one focused exclusively on one's own clinical department had been a functional survival strategy.

Results of the study

Building upon the Clinical Professional Model

The introduction of change was built upon the existing knowledge, skills, orientations, and abilities of clinical managers. The clinical background of managers in these areas was often the basis of their credibility with their staff and the hospital management. The abilities to process large amounts of information and to make decisions under pressure were particularly important. Overall, the application of professional autonomy to management roles drew upon existing individual attributes and the organizational culture of a teaching hospital. While this approach was intended to extend the clinical managers' effectiveness, it also had its limitations.

Reliance on Professional Judgement

In the interviews, many of the participants' remarks indicated that they were accustomed to the exercise of professional judgement in the course of their clinical responsibilities. Their remarks conveyed an implicit awareness that the discretion that permitted them to be responsive to changing conditions of patient was essential for effective service. For example, one nurse stated in reference to patients recovering from a major surgical intervention:

> We have about five medications going, a little bit of this, a little bit of that, then when they start to warm up, they dilate the vessels so you may have a blood pressure of good 140 or something. All of a sudden they bottom out because they have dilated their vessels so they only have that much circulating so you have to clamp off this and open up this and give some blood, so you can't wait for a doctor to say do it, you have to do it.

Communication

The coordination of clinical treatment, including the resolution of conflicts, relies of necessity on informal communication networks. Participants shape the content and extent of the networks to the requirements that emerge in pursuit of clinical goals. Communication networks represent the strength of existing professional systems, as they are the mechanism through which autonomous members of the organization coordinate their activities and resolve conflict.

The interviews suggested that the hospital's unacknowledged reliance on informal professional networks to support management initiatives introduced a level of uncertainty. For example, the hospital lacked a proactive means for repairing the break in communication when a person departed from a key position within a network. Departure had an even greater impact when an individual was key in linking otherwise separate networks:

> I was fortunate to come through an established network of people all through the institution so that if I want to get something done in some area, I usually take the direct approach to somebody that I know in that department that will be able to do the task. I think that kind of networking is very useful . . . Moving on to another position, a lot of people are starting to realize that when I am gone, there is going to be a hole in the network and they are uncertain as to who is going to fill that one up.

or

> So you sort of build your own networks, and you have to open the doors to those networks to some people, to new people, and that's difficult sometimes.

or

> The most important is establishing and maintaining communication networks. That has just come into vogue in the last ten years, but it is a thing that makes a modern hospital a modern hospital – keeping everybody informed of things around them.

People whose networks are relatively restricted in space or in the organizational hierarchy can feel quite isolated.

> I think everyone works well together with their own team. You don't sort of see a lot of interaction with the higher levels and I think that's one of the things that isn't conducive to a lot of trust.

Intrinsic Education

Continuous learning had been an integral part of the existing professionally based system. The focus of this learning was primarily on clinical issues.

> There's virtually no end to what you could learn in this job in terms of infectious diseases and infection control.

or, from a head nurse on an intensive care unit:

> So, I trained them all by myself. They were RNs (Registered Nurses) but then they had to be trained for a specialty, and I would be here [from 7: 00 am] until 7: 00 at night but it was a challenge, it didn't matter. As time progressed we have the ICU intensive care unit course now which helps us a bit, but we still have to continue with on-the-job training.

The Expansion of the Self-Management Mandate

Devolution as an Organizational Objective

The senior administration decided to build on the professionally based model existing in the clinical units to encompass a wider range of management issues. They undertook this change while aware of its potential impact on employees, intending to provide the necessary support systems. The following statement by a senior administrator at the hospital sums up the approach which the hospital management instituted.

> So if you take it to its logical conclusion, and this is not a unique thought, it seemed to me the way to manage a hospital is from the patient care area. As a sort of triumvirate in essence, in which there is a physician manager, a nurse manager and an administrator, who look at the problem as it affects their patients, their relations with the ancillary services, their demands on the system and said okay, this is your task, this is the amount of bucks you've got to deal with, tell me how best you can do this.

The senior management recognized that the move to expand the financial mandate of its clinical managers required the development of substantial organizational supports:

You can't devolve management all the way down the line, unless you devolve it from yourself, and we needed a strengthening of management in order to have that devolution take place.

Administrative efforts to support the change were considerable. They included training sessions, hospital-wide meetings to discuss the transitions, and open meetings of budget unit heads. Further, the self-management in clinical responsibilities had clearly been accepted as an emerging part of the organizational culture.

Demands of an Expanding Mandate

An expanding mandate by necessity requires new tasks. People who are already working a full day must decide what old tasks to relinquish. This can be a difficult decision, particularly when senior management has not provided clear signals. As one respondent said of her fellow nurses:

I think sometimes they may feel even put upon by the fact that we're able to carry out these added responsibilities without getting anything taken away that doesn't fall under the realm of nursing responsibilities.

Some of those affected were surprised by the decision to expand their mandate. They were suspicious of the contradiction implicit in a top-down decision to increase the level of participatory management.

[The idea of business plans for units] came from a hospital-wide committee that had a mandate to develop, an orientation to develop, business plans. It was one of these things that was, in some opinions, inflicted upon us.

While devolution brought new responsibilities, clinical managers had difficulty relinquishing existing tasks:

People are feeling that they are at the max already. We've been short-staffed for a couple of months, so it was a bad time to be [taking on new responsibilities], because staff are already feeling that they are pushed. But there were things like, cut out our research review committee, which now meets quarterly to once a year. People were just feeling, 'now what can we do?' [The staff in] outpatients have said 'we're not going to do any more back education sessions for hospital staff', which again – we do a fair number of those, maybe one every three weeks. But they weren't able to pinpoint something and say that this is an activity that we don't need to be doing.

Clinical managers frequently cited a fundamental problem in the devolution of budgetary responsibility: the person responsible for the budget was rarely the person controlling the expenditures. The use of materials on a hospital ward depended on patient census, which was controlled primarily by physicians. The head nurse, who was responsible for the budget, did not have the authority to refuse admissions. Senior management recognized this problem, as the statement about the management triumvirate above indicated, and was intending to develop a solution. While waiting for this solution to be put in place, the managers experienced uncertainty and resentment.

So, [the 5 per cent budget reduction programme] is a good opportunity for us to examine the function and the role that we have on a day-to-day basis. But also the majority of nurses, . . . and I tend to agree with them . . . felt 'why is it back to nursing again?' We have no control! For instance, I can save the nickels and dimes on equipment, supplies and stuff, but if the physicians keep bringing patients in, who perhaps do not require ICU, then it's a vicious cycle, I mean you never win.

or

Now the role of the head nurse is really managing the unit . . . When I talk to the [medical] staff and I say, 'look we do have to do the 5 per cent cuts, we need to do this and this', they say, . . . 'Hey, the budget is yours. This nursing budget is not my problem.' But when they want to bring patients in, then it's their predicament. They want to bring ten patients in, whether you have six nurses or not. That makes it difficult.

The Limitations of Participatory Decision Making

As noted above, a pervasive challenge within professionally based management systems is the resolution of disputes regarding the limits of individual discretion. In the social context of health care, professionals from various disciplines provide services to the same patients. Having a diversity of perspectives introduces conflict, which ideally prompts a thorough consideration of alternative approaches. Often, conflict is resolved through a forceful presentation by a person in authority.

In many ways the systems for resolving such conflicts are more relevant to expanding management roles than to the clinical aspects of individual discretion. Various participants made comments that reflected shortcomings in aspects of cooperative management within organizational units:

I think there is room for improvement in terms of participatory management, even in nursing. I think we talk about it a lot, but I don't think we're practising it much.

or

I mean one simple thing. Even changing a chart: it might seem foolish to you, but when our patients went up to the floor, or vice versa when a patient got transferred down to here, the old way we used to do it is we'd take the chart apart and put all the paper in an envelope for transfer to and from the ICU. A memo came out with the new chart plan that we must bring the whole chart up to the floor and take an empty one down. I have a major problem with that. But that's the word that came down and I talked to a lot of head nurses – they have a problem with that, too. To me, that's a simple thing. Why not let us have a say in it? That's very simple. It's just a chart.

These comments indicate dissatisfaction with decision making in clinical units. Although there is a professed intention to move towards more inclusive decision-making formats, inadequacies of communication networks and a lack of experience with participatory decision making prevent the realization of these values in day-to-day work.

Training and Information

Senior management were aware that training was needed to accompany expanding roles. Most clinicians had at best a superficial acquaintance with financial management from their initial or mid-career training. Not all of the training programmes to support the increased mandate found their mark; some who were charged with increased responsibility for financial management found the programmes wanting:

> No, I wouldn't say that we have been provided with the training. It was something that evolved into the head nurse's role over time, by bits and pieces . . . When something comes up, sure there are people that you can phone, but sometimes you have to make two or three phone calls to try to straighten something out. We do have sessions to introduce us to budget processes and also to this transitional change. We have had many opportunities to go to meetings about the transition and all that we've gotten into was corporate structure. I'm very uncomfortable with that because I still, like everybody else in my group and around, don't really know what this means to us in our jobs.

or

> I guess, I really feel the need for more information on the financial side of it. I really do not understand. The first time I did a ledger, I didn't know which end was up, literally. So I just asked a phenomenal number of questions and checked it out with our accounting department and finance department and kind of got through it. But I think it would have probably helped me if it had have been a little more prepared.

Managers need up-to-date financial information in order to apply their training to a mandate that has been expanded to include budget management. The system at the hospital was seen to be less than adequate by everyone: the information provided was late, inaccessible, indecipherable, and wrong. Senior management put a high priority on its improvement. Nonetheless, the front line managers received the mandate to control their budgets before the system had been improved. That improvement may take a few years, which could strain the coping capacities of individuals.

> The financial information system is less than adequate. That seems to be a constant source of frustration. It does come out at our meetings, definitely. Also, there's a concern that people are not really in charge of their own judgements. The system hasn't been decentralized, so when you have a department with 100 or so people in it, the information is less than adequate.

or

> They're working on [the financial information system], and they will be for a while. I think part of my job is to try to impart to head nurses that we can't allow ourselves to become overly frustrated with that particular point because that will be one of the strong problems with reducing by 5 per cent. How do we know we're reducing any percent? . . . [The head nurses] keep track of all that, now, by hand. Mainly because we can't trust the system.

The weakness of the management information systems was markedly at variance with the central role of information processing in the hospital's clinical mandate:

> Everything that happens in a hospital is a result of information management. When patients come in we examine them, we test them, we do all kinds of information gathering on every patient that comes in here. After we have done that we have to share all that information with a mass of people who provide professional opinions as to what treatment or what diagnosis or what prognosis exists for each patient. Then you take that mass of information, and you translate it into a care plan – an individualized care plan for this one person – and you get that implemented. And that creates more information in terms of the monitoring process. We have to monitor how the treatment plan is going, and if it works well then that patient is discharged and ceases to be part of the information management of the hospital. It is just one huge information management system that we have going.

Managing the Transition

Coaching as a Coping Resource

Individual coaching supplements formal training as a means of enhancing occupational skills. It is a resource-intensive form of training that consumes expertise for a small audience. Yet, it is one with special benefits for individual coping capacity. It complements the flexibility implicit in the exercise of individual discretion. Through one-to-one exchanges mentors may adapt teaching to the specific problems that individuals encounter while attempting to apply developing skills to a novel situation. This form of instruction is particularly appropriate to the specific application of a complex skill. For example, most people utilize a small range of the special features of sophisticated computers (e.g. word processors, spreadsheets). By asking specific questions of a mentor, they avoid having to learn to use an extensive system that exceeds their requirements, or eliminates the need for training to decipher written documentation on the program. This time-saving is even more pertinent to the complex demands of management.

Such mentoring relationships further enhance coping by providing social support. One person in a training programme for the development of business plans required a substantial degree of support to address the conflict between managerial and clinical responsibilities. Her immediate supervisor described their discussions:

> She hooked up with me after each session to bring me up to date with what they were doing. I think it was to keep me updated and to also just chat about her feelings about the business plan. Did I think they were legitimate feelings? Was she not knowledgeable about certain things in relation to how the business runs? And whatever. So we had good chats about how she felt about doing those plans, what it meant to her in relation to the time frame on the floor. It took a long time to develop that first business plan, and hers evolved around her nursing

coverage . . . I suppose our chats gave her probably the comfort of knowing that there was another person who thought the same way she did, or we sort of came to grips together with her thought that there's no way that we could make business plans for everything we do on [her unit].

While this individual's needs were being met by a concerned and capable supervisor, it did not appear that this level of support was available to all who needed it. Experienced people who serve as mentors draw upon their propensity as health-care professionals to provide supportive relationships. What is important to them in mentoring are the skills and perspectives they have developed outside formal training; they are continuing that aspect of implicit learning through their coaching.

I have a lot of skills now that I didn't have then, so I'm trying to put myself in the place of a new staff nurse coming in, and I think it's really important how the environment that you're in is going to help make or break a situation. But I think when you've had more skills behind you, as you get older and you get more developed in your roles and responsibilities – as whatever, mother, wife, person – that you have more capabilities of handling those kinds of difficult situations.

Value Conflicts Within a Self-Management Mandate

Clinical vs Managerial Values

An additional source of stress during transition is the potential for conflicts between management and clinical values. To some extent a new mandate simply represents more work and new opportunities. For many health-care professionals the requirements of a new role conflict with their established clinical values:

The hospital as a whole seems to be gearing us more to business person/paperwork and not to the patient who is out there. So for me it is to learn to balance the two of them because I can't be someone who sits in the office all day, because I have to know what is happening.

or

I'm not convinced that the 5 per cent [budget reduction] will be easily achieved, especially when you hear some of the clinicians being very honest about their concern for quality patient care and they are not going to go beyond the point where patient care is jeopardized.

or

I suppose it's called priority setting. I guess part of the professional thing is being able to set those priorities and knowing what you can leave out. One of the things that we're preparing ourselves for is the fact that we're going to have more complaints from patients. That we're not doing enough. So and so didn't get shaved today. So we've developed the approach that it's probably best to deal with it at the time, with the patient. So the staff nurse will have to develop skills in putting a caring and compassionate communication along with 'I can't shave you today'.

Confronting Self-Management from Subordinates

While middle level managers are developing their own self-management initiatives, they have to be prepared to address initiatives from people whom they supervise. Without confidence in their individual discretion and access to resources, they are unable to respond in a definitive fashion. An expanding administrative mandate increases the demands on their mutual coordination.

> A situation occurred just a couple of weeks ago that was really unusual for me and I had some difficulty dealing with it. We had an in-service [training] which we often have, and after the in-service the girls on the floor asked me if they could use my office for a few minutes and have sort of a general meeting of the staff. They talked about their concerns and I wasn't invited to that, which is really unusual because I'm used to them being very open with me. When we met and discussed . . . if there were concerns I couldn't really make them feel better about. Some of them I just had to say, 'I hear what you're saying and understand what you're saying but I can't assure you that it won't be as your fearing it might be.' They were concerns that we may lose staff, and I could assure them that we wouldn't function at unsafe levels. But they would still feel very pressured even before they got to the unsafe level of staffing . . . They would be suffering burnout in order to give the standard of care that they were comfortable with. I mean if it would just be too much for them to continue to give quality patient care.

Conclusions

The interview material presented here has illustrated the manner in which health-care professionals in a Canadian hospital articulated their experience of shortfalls in training and other coping resources when confronting an expansion of their management roles. While this material has demonstrated that the framework developed here is relevant to their experiences as health-care professionals, assessing the pervasiveness of these experiences requires more extensive research. Future studies that supplement interview material with quantitative data on supportive conditions for expanding self-management mandates may assess the impact of training programmes on the capacity to cope with major organizational change.

The interviews, however, do serve to identify the difficulties encountered in a hospital setting by clinical managers who were confronting an expansion of their budgetary responsibility at a time when budget management was becoming more challenging. These people reported problems in finding support for their increased mandate. The responses also pointed out additional problems in terms of value conflicts and the maintenance of communication networks within the organization.

An expansion of the responsibilities of clinical managers in this hospital was built upon an existing system of self-management within the clinical domain. Systems in place to support the mutual coordination of professional autonomy among health-care professionals included extensive communication networks, experience at making decisions, coordinating action and resolving conflicts. However, participants in the interview study encountered shortcomings in the transference of these skills, experiences and orientations to a more comprehensive management role. The planned expansion of the organization's management support systems lagged behind the expansion of this mandate. While the principal stakeholders in this transition recognized the relevance of training to addressing this shortfall, they also encountered deterrents to providing appropriate training programmes during this period of uncertainty.

The use of mentors illustrated an adaptive approach to skill enhancement as a means of coping with problems presented by the transition. While individual employees had difficulty in accessing formal training programmes appropriate to their needs, many used their formal or informal communication networks to access knowledgeable people within the organization. Through these interactions, they not only increased their relevant skills, but also obtained emotional support and gained practical information about organizational developments. This suggests that the capacity of employees to cope with organizational transitions of this sort depends upon the extensiveness of their communication networks throughout the organization.

Individual adaptive responses within an organization result in an uneven reaction to change, often at a considerable cost in terms of individual distress. Skill enhancement is of particular importance in health-care organizations within which sophisticated clinical skills are the basis of major interventions and of status within the broader institutional environment. The absence of clear frameworks for articulating the institutional context of mid-career training programmes hinders the development of proactive adaptation to major organizational changes. The shortfall in administrative skills and support systems underscores the importance of training as a means of coping with major organizational transitions, for which organization-wide enhancement of skills is imperative.

Note

This research was conducted with support from a research grant from Strategic Grants Program, Work and Education, of the Social Sciences and Humanities Research Council of Canada. The author wishes to thank Mr. Neil Pilkington for his assistance in conducting these interviews. He also wishes to thank many people at the participating hospital for their time and insights which made this research possible.

References

BANDURA, A. (1977) 'Self-efficacy: towards a unifying theory of behavioural change', *Psychological Review* **84**: 191–215.

BUTLER, R. (1991) *Designing organizations: a decision-making perspective*, London: Routledge.

CHERNISS, C. (1989) 'Career stability in public service professionals: a longitudinal investigation based on biographical interviews', *American Journal of Community Psychology* **17**: 399–422.

COX, T. (1978) *Stress*, London: Macmillan.

FRENCH, J. R. P., Jr. and RAVEN, G. H. (1960) 'The bases of social power', in D. Cartwright and A. Zander (eds) *Group dynamics: Research and Theory*, 2nd edn, New York: Row, Peterson, 607–623.

FUNK AND WAGNALLS STANDARD DESK DICTIONARY (1980) New York: Funk and Wagnalls.

HACKMAN, J. R. (1986) 'The psychology of self-management in organizations', in M. S. Pallak and R. Perloff (eds) *Psychology and Work: Productivity, Change, and Employment*, Washington, DC: American Psychological Association.

HOBFOLL, S. (1989) 'Conservation of resources: a new attempt at conceptualizing stress', *American Psychologist* **44**. 513–524.

LAZARUS, R. and FOLKMAN, S. (1984) *Stress, Appraisal, and Coping*, New York: Springer Publishing.

LEITER, M. P. (1991) 'The dream denied: professional burnout and the constraints of service organizations', *Canadian Psychology* **32**: 547–555.

LEITER, M. P., DORWARD, A. L. and COX, T. (1994) 'The social context of skill enhancement: training decisions of occupational health nurses', *Human Relations* **47** 1233–1249.

MANZ, C. C. (1986) 'Self-leadership: toward an expanded theory of self-influence processes in organizations', *Academy of Management Review* **11**: 585-600.

MANZ, C. C. and SIMS, H. P. (1984) 'Searching for the "unleader": organizational member views on leading self-managed groups', *Human Relations* **37**: 409–424.

STERNBERG, R. J. (1985) *Beyond IQ*, Cambridge: Cambridge University Press.

WAGNER, R. K. and STERNBERG, R. J. (1987) 'Tacit knowledge in managerial success', *Journal of Business and Psychology* **1**: 301–312.

CHAPTER EIGHT

Coping with subjective health problems in organizations

HEGE R. ERIKSEN, MIRANDA OLFF AND HOLGER URSIN

Subjective health – a problem?

In recent years several industrialized countries have invested a major effort in reducing sickness absence in the workforce. The emphasis typically has been on reducing somatic disease. However, data on Norwegian sickness certificates has shown that subjective health complaints without objective signs or symptoms play a prominent role in short-term and long-term sickness absence (Tellnes *et al.*, 1989; Ursin *et al.*, 1993). Furthermore, even with somatic diagnoses, the subjective evaluation of the health state is an important factor in reaching decisions – by both patient and doctor – on sickness absence. Since about half of long-term sickness absence and permanent disabilities derive from subjective states, it is time for preventive work medicine to direct its efforts towards the main cause of absence behaviour, which is no longer the traditional somatic conditions.

Factors contributing to sickness certification are the assessment of disease by the medical doctor, the subjective feelings of the patient, the sickness role that the patient assumes and the patient's own feeling of being unfit for work. The term *disease* is used for physiological and psychological dysfunction, and is classifiable according to the existing systems of diagnosis. *Illness* is used for the patients' own symptoms, health problems or feeling of disorder. Illness, therefore, is a subjective state of a person who is aware of not being well. *Sickness* is used to describe the sickness role assumed by a person who believes himself or herself to have a disease or an illness. Sickness is therefore a state of social dysfunction (Last, 1983; Meads, 1983). These states may or may not be the reason for absence from work. *Absenteeism* is used by most workers simply for the state of not being present at work. *Sickness absence* may be used when the absence is caused by disease, injuries or illness (Tellnes,

1989). It is up to the individual himself or herself to decide whether the subjective state of health exists or not. It is even possible to have a feeling of positive health in spite of somatic handicaps or a chronic somatic disease. The issue is not disease versus no disease, but a feeling of illness versus no illness.

Absence of health may be as important for society as presence of disease. A way to operationalize positive health might be to define this state as an absence of subjective complaints rather than an absence of disease. The World Health Organization's (WHO) (1985) definition of health is that it is a state of complete physical, mental, and social well-being, and not merely the absence of disease or infirmity. This state is difficult to obtain, and even more difficult to measure. Some problems in the Scandinavian and Western European welfare states derive from a lack of definitions and operationalizations of the absence of the positive health state. A lack of definitions in this area leads to a lack of clear policy, and may be the root of one of our most difficult problems: the large amount of sickness absence and sickness compensation granted for subjective states.

What are the subjective health complaints?

In a large self-report investigation of 4,000 Norwegians the most frequently reported subjective health complaints were exhaustion, fatigue, muscle pain, and gastro-intestinal complaints (Norwegian Bureau of Statistics, 1993). The most common subjective complaints in a sample of 400 Norwegian males, and seventy-four females, were anxiety, sleep disturbance, muscle pain, and problems associated with the gastro-intestinal tract (Ursin et al., 1988). In an unpublished study of a stratified sample of 4,000 subjects, 1,000 from each Nordic country (Denmark, Finland, Norway, and Sweden), a similar panorama of subjective complaints was found (Eriksen et al., 1998). Muscle pain and headache were common in all three samples. These epidemiological data correspond well with other results from Scandinavian, English and American investigations. In a postal survey of a stratified random sample of 2,034 adults from two general practices in Cheshire, England, Croft et al. (1993) found that the prevalence of 'chronic widespread pain' was 11.2 per cent. Therefore, even if we only include 'serious' pain, still about one in ten of the normal population has this problem at any given time. In one study, more than half the respondents reported pain that lasted for more than twenty-four hours (Croft et al., 1993). In a postal study in Sweden, Brattberg et al. (1989) found that 66 per cent of the respondents had 'some pain or discomfort.' Low back pain was the most common musculoskeletal pain reported (35 per cent). Low back pain was also the most frequently mentioned 'disease' in a Danish survey of musculoskeletal diseases (12 per cent) (Bredkjær, 1991).

Subjective complaints rank high in the reasons for seeing a general practitioner (Meads, 1983). The complaints may be common, even banal or everyday experiences for a large proportion of the population, but they may reach levels that are deemed intolerable or problematic for a certain percentage of the population.

In a study by Tellnes *et al.* (1989) on the sickness certification practice of Norwegian general practitioners, the most frequent causes of sickness certifications were diseases of the respiratory system, musculoskeletal/connective tissue diseases, mental disorders, and injuries. Among the single diagnoses, non-articular rheumatism and myalgia were more frequent among females than males, while back pain (without radiating symptoms) was most commonly found in males. As many as 59 per cent of the days lost because of sickness absence were attributable to diagnostic groups that solely or mainly depended on subjective statements from the patient. These concerned, in particular, musculoskeletal and connective tissue diseases, ill-defined conditions and mental disorders. Taken together, no other single diagnosis was used more frequently than musculoskeletal pain. Patients with musculo skeletal pain were not only frequently absent from work, they also showed a high risk of long-term sickness certification, ranking third after diseases of the circulatory system and neoplasms (Tellnes *et al.*, 1989).

The high number of sickness absences for muscle pain reported by Tellnes *et al.* (1989) should be supplemented by the high number of people within the normal, working population complaining of such pain. In a sample of healthy working men and women, Ursin *et al.* (1988) found a very high incidence of muscle pain, particularly in women. Fifty-four per cent of the women and 40 per cent of the men reported some form of muscle pain in the neck, back and/or shoulders during the preceding month. Serious complaints were only reported by 8 per cent of women and 3 per cent of men, corresponding with the numbers found in the Canadian health service (Lee *et al.*, 1985). If healthy means at work, the 'healthy population' includes a high number of people who report muscle pain without seeking medical advice, and without asking for or expecting sickness compensation. However, these people do not fall within the WHO health definition, the positive state of well-being without any complaints at all.

In addition to the above, we should also add subjective complaints in conditions having a clear somatic basis. Even under such conditions, seeking sickness certification depends on individual factors. For many of the patients their condition is acceptable, they are able to continue a normal life, and will often refer to themselves as healthy albeit having a chronic disease. Differences in sickness certification depending on social class suggest that this difference in tolerance may be related to the total resource situation for the individual.

Psychological factors contributing to lack of health

Traditionally psychosomatic medicine has concentrated on the psychological contribution to somatic conditions, for instance psychological factors contributing to cardiac infarctions or duodenal ulcers. These contributions are significant, but generally explain a small part of the variance. Therefore, psychologists and psychiatrists have not acquired a main role in the medical care and prevention of these diseases.

A very clear contribution from psychology and the behavioural sciences has been focused on programmes aimed at bringing about changes in lifestyle that may affect long-term risk factors for somatic health. These programmes, concentrating on for instance smoking behaviour or diet, will not be addressed in this chapter. The most interesting aspects from our point of view are the psychosomatic conditions associated with somatic complaints but without any demonstrable somatic condition or with very slight somatic changes. These conditions are, per definition, within the subjective complaints we address in this chapter. In the more extreme form, required to meet the criteria for 'somatisation' within the Diagnostic and Statistical Manual, DSM IV (Task force on DSM IV, 1994), they only account for a small number of the total subjective health complaints.

The traditional view in this field is that somatic complaints are secondary to a psychological disturbance. This condition has had many names through the history of medicine, within neurology, psychiatry and psychology. One of the first terms used seemed to be 'asthenia', or 'Beards disease' (Chatel and Peele, 1970). In 1869 Beard described a syndrome consisting of tiredness and multiple complaints, which he suggested derived from emotions causing an exhaustion of the nervous system. Medicine has reacted negatively and even antagonistically to the concept, expressing concern that psychotherapy and psychotherapists might become a problem for the general practice of medicine (Nolan, 1959).

Terms related to asthenia have also been used for similar conditions, like neurasthenia and psychasthenia. Pavlov used such terms for his dogs. At present these terms have acquired precise and rather restricted definitions within the DSM and International Classification of Diseases (ICD) diagnostic systems. The complaints seem to occur in clusters, and, when the level of complaints is high enough, the patients may be given diagnoses like epidemic fatigue, chronic fatigue, burnout, stress, a variety of intoxications, functional dyspepsia, fibromyalgia, irritable colon, food intolerance, myalgic encephalitis, postviral syndrome, 'yuppie flu', or 'vital exhaustion'. The most reasonable diagnosis may be *somatization* (Malt, 1991; Stewart, 1990) or neurasthenia (Wessely, 1990). However, if the present international diagnostic criteria are followed, most people having such 'normal' complaints will not reach the threshold for inclusion and do not qualify for the diagnosis.

Malt (1986) has repeatedly pointed to the lack of rationality in the way contemporary medicine reacts to conditions with these clusters of symptoms. We are spending a lot of resources in a futile search for single-factor attributions for the conditions. Given that the conditions are most probably multifactorial and multicausal, a proper understanding requires an integration of cultural, social, psychological and somatic factors (Malt, 1986).

Lack of coping and lack of health

Lack of coping, which can be defined as lack of positive response outcome expectancies, may contribute to health complaints through *sustained activation*. The sustained activation hypothesis states that individuals with a positive expectancy of the outcome of their actions in dangerous situations show low or only moderate physiological activation. Activation is defined as an increased activity in the central nervous system as judged from the electroencephalogram, activation of the vegetative nervous system, changes in the hormonal system indicative of activation or arousal, changes in the immune system, and, finally, increases in muscle tension (Levine and Ursin, 1991). When the individual is unable to develop positive response outcome expectancies, as in situations of helplessness (no relationships between acts and outcome) or hopelessness (negative response outcome expectancies) (Ursin, 1980; Levine and Ursin, 1991) sustained activation may occur. It is the heightened psychological factors in interaction with work factors that determine whether an individual enters into helplessness, hopelessness and a sickness role, or is able to fight his/her way back to and through the challenges of working life.

In a study on coping styles and health Olff *et al.* (1993) identified four main coping styles: instrumental mastery-oriented coping, emotion-focused coping, cognitive defence and defensive hostility. In particular a lack of instrumental mastery-oriented behaviour was found to be related to poor subjective health, depression, burnout and to a higher number of days on sick leave. Instrumental mastery-oriented coping reflects general positive expectancies about the ability to handle the environment and a tendency to use more instrumental, problem-focused coping. This particular way of dealing with stressors apparently has a beneficial effect on the subject's health experience.

Reducing subjective complaints

Sickness absence for somatic disease has traditionally been reduced through work medicine, hygiene, safety, and improvements in the working environment. Most of these initiatives are directed towards well-known risk factors for somatic disease. On a long-term basis, absence may be reduced by initiatives designed to improve the lifestyle and eliminate well-established risk factors for somatic disease, such as smoking and alcohol abuse. However, such factors as the individual's coping styles, motivation, work climate and work morale have a crucial impact on the decision to seek sickness absence. Short term sickness absence was reduced by 15.5 per cent from 1990 to 1993 (Andersen, 1994) in a joint initiative sponsored by the trade unions and professional organizations in Norway, directed at work moral and team spirits providing peer pressure to reduce absenteeism.

It is often assumed that there are two basic approaches to interventions in working life. Interventions may take place at the specific individual level, strengthening the individual, such as stress management training or physical exercise. Alternatively, they may emphasize changes in the organization such as, for instance, job redesign (Ursin et al., 1993).

In the field of coping and health most of the interventions focus on the individual. Individual-level coping strategies may be grouped into four main categories: those aimed at psychological conditions, those aimed at physical conditions, those aimed at changing the individual's behaviour, and those aimed at changing their work environment (Newman and Beehr, 1979). There is an extensive literature on such interventions, but relatively few well-designed and controlled studies. This is mainly due to the difficulties in running such studies in a systematic fashion.

The literature on organizational-level intervention programmes is even larger, but with very few controlled studies, and even less focus on health consequences. The main idea behind interventions directed at the organizational level is to attack the sources of stress, under the assumption that it is the work situation that causes stress. Most of the interventions discussed below are controlled studies, or reviews of such studies, from a literature search covering a five-year period up to 1994.

General health-promoting programmes

Health promotion programmes are primarily concerned with reducing risk factors for somatic disease and represent a typical 'primary prevention' strategy (Roman and Blum, 1988). The list of possible health promotion programmes is very long, but few have been evaluated thoroughly. The benefits of particular programmes are often asserted rather than identified (Maynard, 1991). Reviews of the effects of health promotion programmes are found in Glasgow et al. (1993), Pelletier (1991), and Roman and Blum (1988). It is well established that programmes directed at specific risk behaviours (smoking, overeating) may show beneficial health effects in the long term perspective.

Health promotion includes health risk appraisal and feedback, achieving and maintaining physical and mental fitness, controlling alcohol use, avoiding or quitting tobacco and other drugs, and providing opportunities to control high blood pressure, and reduce elevated blood cholesterol, obesity and other health hazards (Breslow et al., 1990). Interventions aiming at controlling alcohol use, avoiding or quitting smoking and other drugs, will not be examined further here. For reviews see for example Matson et al. (1993), Fisher et al. (1990), and Berridge and Cooper (1993).

Despite a burgeoning interest in, and acceptance of corporate health promotion, the overall economic effects of these programmes are not clear (Kaman and Patton, 1994). Many believe that health and fitness promotion at the worksite is firmly accepted by all concerned. Others believe that only health-care cost containment interests allow these programmes to survive. Of concern is the low participation rate in programmes when the entire employee population is considered (Wanzel, 1994).

Success is often measured through the achievement of corporate objectives, including employee participation, increased morale, improved productivity, and/or reduced medical costs (Pelletier, 1991). Success is achieved when clear objectives are established at the start of a programme, investment is made in achieving the stated goals, and accurate measurement is built into the programme. Pelletier (1991) concludes that comprehensive health promotion programmes are both health- and cost-effective. In a literature review up to 1988 Gebhardt and Crump (1990) concluded that in general, fitness and wellness programmes resulted in increased levels of fitness and a reduction in the risk factors for coronary heart disease. Research using control groups found relationships between reduction in health-care costs, absenteeism and turnover, and the implementation of comprehensive health promotion programmes. Issues related to participation rates, programme implementation, and evaluation were also addressed.

In the literature from 1990 to 1994 we identified fifteen randomized controlled studies, some published in several papers, on health promotion programmes (see Table 8.1), including wellness programmes, cholesterol screening, and so on. The majority of randomized health promotion studies encompassed interventions aiming at lower cholesterol levels and blood pressure levels. However, health promotion programmes are also often designed to produce changes at the organizational and employee level (Glasgow et al., 1994).

Sickness absence may also be affected by general health promotion interventions. Shi (1993), for instance, examined the relative effectiveness of different levels of health promotion interventions on changes in the number of sick days and in medical care use. Four groups were compared: assessment only (control), high intensity health promotion intervention (high-risk targeting), medium intensity (group support), and low intensity (self care) health promotion interventions. Only the high intensity group exhibited a consistent decline in visits to the doctor, hospitalization, injury and illness days in both men and women.

Comprehensive programmes seem to have more effect than health appraisal alone. Spilman et al. (1986) evaluated the effects of a health promotion programme comparing the experimental group that was given an initial health-risk appraisal and offered health education modules with a control group that was given the health-risk appraisal with no modules. A second control group was neither given the health-risk appraisal nor offered modules. The health promotion programme was found to lower health risks and improve health-related and job-related attitudes.

Organizational interventions

Job redesign, employee assistance programmes, job rotation, flexible scheduling and supervisory role clarification are all examples of organizational

Table 8.1. Randomized controlled studies in health promotion.

Reference	Intervention	Main variables examined
Jeffery *et al.* (1993)	Weight control	Absenteeism
Cockcroft *et al.* (1994)	Receive advice Set targets for lifestyle change Given results	Weight, exercise, perceptions of health, compliance
Pruitt (1992)	Stress management as part of overall fitness programme	Stress-related physical symptoms
Glasgow *et al.* (1994)	Early vs. delayed intervention Tobacco use Dietary fat intake Serum cholesterol	Organizational and employee data
Gomel *et al.* (1993)	Health risk assessment Risk factor education Behavioural counselling Behavioural counselling + incentives	Cardiovascular disease risk factors
Elton *et al.* (1994)	Knowing serum cholesterol concentration Control group	Cholesterol level
Fitzgerald *et al.* (1991)	Follow-up mailing after cholesterol screening Control group	Cholesterol level
Francisco *et al.* (1994)	Health fair and incentive Control group	Cholesterol level
Barratt *et al.* (1994)	Screening + self-help package or nutrition course Control group	Cholesterol level
Fielding *et al.* (1994)	Physician referral only (BP) Referral, counselling, and mailings (BP)	Blood pressure
Frommer *et al.* (1990)	Counselling programme Control group	Blood pressure
Ellis *et al.* (1994)	One-stage screening Two-stage screening	Blood pressure
Mayer *et al.* (1991)	Co-worker-delivered reminders Control group	Breast self-examination (cancer)
Girgis *et al.* (1994)	Skin screening + education Control group	Solar protection behaviour (cancer)
Hebert *et al.* (1993a) Hebert *et al.* (1993b) Sorensen *et al.* (1992)	Nutrition intervention Control sites	Consumption of cancer- related nutritients; dietary habits

interventions that may improve subjective health and job satisfaction, and reduce sick leave (Karasek and Theorell, 1990). Providing supervisor support, for instance, has been shown to be an effective preventive strategy (Amick and Celentano, 1991). Educational programmes for supervisors are viable and may lead to changes in attitude and behaviour related to assumed risks for musculoskeletal injuries, and increased support for individuals returning to work following low-back pain episodes (Linton, 1991).

Several studies have also shown that flexible scheduling and shift work have effects on subjective health (e.g. Knauth and Kiesswetter (1987); Totterdell and Smith (1992)). Dalton and Mesch (1990), for instance, assessed the effects of a flexible-scheduling programme on absenteeism and turnover in a division implementing such a programme, and a comparable control group. They reported gross reductions in employee absenteeism after the flexible-scheduling intervention for the experimental group, while no such changes were found in the control group. Two years after the programme ended, absenteeism had returned to base-rate levels.

Burke (1993) reviewed ten organization-level interventions to reduce stress at work. It was concluded that these interventions generally had positive effects. Given the limited success of individual-level interventions in addressing occupational sources of stress, Burke claimed that organizational interventions should be encouraged. However, it is even more difficult to find controlled studies in this area, and difficult to identify the causal factors for any improvement. A few organization-level interventions will be highlighted below.

Job redesign has been a key concept in the field for a long time. A review of this tradition is available from the two researchers who have a central position in this field, Karasek and Theorell (1990). They review the theoretical background and the empirical literature up to 1989. The critical dimensions are job demand and decision latitude. They believe lack of decision latitude to be the source of psychosocial contributions to somatic disease, and to what they characterize as psychosomatic illness. This is very similar to what we refer to as subjective complaints.

Karasek and Theorell (1990) review a considerable empirical base for the beneficial effects of workplace reforms and legal regulations of the workplace (see also Petty et al., 1992; Mesch and Dalton, 1992; Hanlon and Taylor, 1991; Turner, 1991). However, many of these reforms have not been evaluated empirically, at least not if a strict scientific design is required. Subjective complaints were not identified as an entity, and most of their arguments are based on traditional epidemiology of somatic diseases, in particular cardiovascular disease.

One of the few studies that explicitly focused on the effect of organizational interventions on sickness absence was done by Werner (1992). The effectiveness of an employee recognition programme aimed at improving work attendance, was evaluated. The programme was conducted for one year as part of a three-year study using the awarding and posting of attendance certificates as the primary method of employee recognition. The employees in the recognition group decreased their use of sick leave by 28 per cent as compared with the levels in the year prior to intervention. In the year following the end of the recognition programme, the average sick leave returned to a level higher than that prior to the intervention. The sick leave of the control group showed an increase each year from the year prior to implementation of the recognition programme to the year following the end of the recognition programme.

Karasek (1990) investigated associations between increased job control and health status. Subjects in a job reorganization group who had influenced the reorganization process, and obtained increased task control as a result, had lower levels of illness symptoms. Coronary heart disease was significantly lower in circumstances of increased job control for males. For females absenteeism and depression were lower, however, smoking was significantly higher for women. The process of job reorganization was associated with significantly higher levels of illness. However, for males, when job reorganization was accompanied by increased control, symptom levels were often as low or lower than when no job reorganization had taken place.

Landau (1993) examined the effectiveness of a change in an attendance control policy in reducing absenteeism and tiredness. After establishing a stricter disciplinary system, there was a decrease in absenteeism once the incentive system was added. However, there were no control groups.

Individual interventions

In contrast with organizational programmes, individual programmes are directed at how each person may develop coping strategies to handle or reinterpret the job stress they experience. The literature on the following individual interventions will be discussed below: physical training, stress management training, relaxation, biofeedback, and mental training. Low back schools and other educational programmes will not be described further here. For reviews see Linton and Kamwendo (1987), and King (1993).

Physical Exercise

It has been reasonably well established that regular physical exercise results in physiological benefits in normal and clinical populations, in particular with respect to cardiovascular disease and mortality (Dannenberg et al., 1989; Rauramaa and Salonen, 1994). Adequately vigorous and continuing physical activity is conducive to maximal good health, high quality of life, and longevity (Paffenberger et al., 1994; Caspersen et al., 1994). Promoting physical activity is a high-priority area because of the repeated demonstrations in epidemiological studies of an association of the 'sedentariness' fostered by modern lifestyles with risk of coronary disease and other prevalent contemporary medical conditions (Ainsworth *et al.*, 1994).

Exercise has been proposed to produce increases in academic performance, assertiveness, confidence, emotional stability, independence, intellectual functioning, internal locus of control, mood, and perception (Thomas *et al.*, 1994). In agreement with this there are several reports of specific and general effects of physical training in ordinary working populations. Early studies of regular physical exercise have reported decreased absenteeism, reduced health problems, reduced health-care costs, and increased productivity (Cox *et al.*, 1981; Falkenberg, 1987). Physical activity has been used as a treatment for patients with mental problems, in particular depression and anxiety (Martinsen, 1990). Psychological benefits of physical exercise have also been reported in mentally normal subjects (Dienstbier, 1984; Moses *et al.*, 1989; Grønningsæter *et al.*, 1992). Exercise has also been shown to reduce the seizure frequency in patients with epilepsy (Eriksen *et al.*, 1994), possibly due to the psychological effect of regaining control and mastery.

There is still controversy over the psychological benefits of company-sponsored fitness programmes. In a review Jex (1991) concluded that evidence of the psychological impact of exercise has been inconclusive. For more detailed reviews of the relationship between physical activity, fitness and psychological effects see Landers and Petruzzello (1994) and Morgan (1994).

The correlations between exercise and beneficial health effects are generally statistically significant, but are at relatively low predictive levels, leaving the decision makers with a good deal of uncertainty regarding the determination of the cost benefit of employee fitness programmes (Kaman and Patton, 1994). In a critical analysis of work site fitness programmes and their postulated benefits Shephard (1992) stated that work-site fitness programmes have often been initiated through the enthusiasm of an occupational health department or a corporate executive who has personally experienced the benefits of an improved lifestyle. The stated intent has been altruistic: to demonstrate good corporate citizenship, to boost the morale of employees or to improve health, rather than to engender fiscal savings. However, sometimes there has also been an underlying belief that programme investment might have instrumental value, boosting productivity or containing rising expenditure on medical care.

For corporate decision making, the interesting data concerns the effect of physical exercise interventions on absenteeism, and productivity. The Shephard (1992) review identified twenty-six reports from 1960–1990 (ten of which were without proper controls). Twenty-three of these reported increases in productivity, but the gain varied from 2 per cent to 52 per cent, with most reporting effects of between 2 per cent to 10 per cent. The effects were on productivity itself, reduction of fatigue, reduced perceived workload, and, in one report, greater creativity.

Concerning absenteeism, the data is backed by more reports. Shephard (1992) found thirty-nine published reports, three without controls. The absenteeism decreased in thirty of these studies, and increased in one. The range of the decrease varied: most studies found a reduction in a couple of days per year. With regard to the changes in medical costs for American and Canadian companies (sixteen references, one Dutch reference) the amounts range from $60 to $450 per year. However, there are many problems with interpretation of these data.

Changes in lifestyle factors were reported in twenty-one of twenty-two studies involving physical exercise (Shephard, 1992). Stress reduction was reported in four of these studies, and improvement of subjective or perceived health in two. Compliance and motivation to continue training are important factors for the effects of exercise programmes. The intention to exercise is influenced by habit, perceived barriers and attitude (Godin and Gionet, 1991). Employees with high health risks perceive more and higher barriers than do employees with low risks (Harrison and Liska, 1994). Management should take this into account when establishing such programmes.

In an analysis of the economic implications of worksite health promotion programmes, Warner et al. (1988) attempted to subject literature reports to a systematic analysis of outcomes. In fact, due to the lack of methodological rigour found throughout the 400 articles chosen for analysis, the authors ultimately relied on subjective judgements for their conclusions. Twenty-eight of the articles that they reviewed focused on exercise. The authors acknowledged the positive relationship between exercising and improvements in physiological functioning. They could not, however, find sufficient research knowledge satisfying the criteria for conclusions to be drawn regarding causality to economically justify such worksite programmes. Specifically, flaws in the economic analyses themselves, non-standardisation of definitions, limited participation and self-selection were listed as problems with the research outcomes claimed.

In our own literature search from 1990 to 1994 several well-designed studies with randomized controls (Ruskin et al., 1991; Gamble et al., 1993; Grønningsæter et al., 1992; Norvell and Belles, 1993; Hilyer et al., 1990; Takala et al., 1994; Gundewall et al., 1993) were identified (see Table 8.2).

The physical flexibility of participants was found to have improved along with other physiological indicators in two of these studies (Hilyer et al., 1990; Norvell and Belles, 1993). Physiological improvements were also reported by Gamble et al. (1993), Grønningsæter et al. (1992), and Gundewall et al. (1993). Ruskin et al. (1991) found that randomly dividing workers into physical activity versus sedentary social activity showed the effect of the physical activity on job satisfaction, willingness to participate in physical activity and reduced health problems. In a study of circuit weight training Norvell and Belles (1993) found not only significant increases in cardio-vascular fitness, but also significant improvements in mood and decreased somatization (improved subjective health), anxiety, depression and hostility. They found improvements in job satisfaction. Participants who dropped out of the exercise training programme had significantly greater anxiety, depression, and hostility at pre-treatment than subjects who completed the

Table 8.2. Randomized studies on physical exercise.

Reference	Intervention	Main variables examined
Ruskin *et al.* (1991)	Physical activity Social activity control group	Job satisfaction, need and willingness for physical activity, health problems, psychological measurements
Gamble *et al.* (1993)	Soccer and circuit-training Control group	Physical fitness
Grønningsæter *et al.* (1992)	Aerobic dancing (endurance, strength, flexibility, coordination) Stress management training Control group	Health, job stress, job satisfaction, psychological measurements
Norvell and Belles (1993)	Circuit-training Control group	Mood, perceived stress, job satisfaction,
Hilyer *et al.* (1990)	Flexibility training Control group	Joint injuries
Gundewall *et al.* (1993)	Exercise (endurance, strength, coordination) Control group	Strength, absenteeism, pain
Takala *et al.* (1994)	Gymnastics Control group	Neck pain, disability

programme. In a study by Grønningsæter *et al.* (1992) employees were randomly assigned to participation in a three-month occupational exercise programme, to stress management training, or to a no treatment group. Both interventions had significant effects, which were specific to the type of intervention. The exercise group showed some improved physiological indicators, especially in women, but paradoxically reported significantly reduced job satisfaction. Compared to controls, the exercise group reported fewer health complaints post test. Gundewall *et al.* (1993) randomized personnel at a geriatric hospital to either exercise during working hours or a control group. The aim of the study was to prevent back symptoms and absence from work. After thirteen months, the training group had increased back muscle strength. There were also significant effects on absenteeism and pain complaints in the exercise group compared to the control group. In another study Takala *et al.* (1994) did not find any significant effect of gymnastics on neck pain among forty-four women in a printing company in a randomized cross-over design. Seasonal variations seemed to override any effect of the intervention.

Some of the negative or inconclusive findings on reduced health-care costs may derive from the fact that the high expenses to individuals and society of musculo-skeletal pain had not been considered. Instead, many of the health cost-benefit studies have been directed to the gains related to cardiovascular disease, which accounts for a relatively low proportion (14 per cent) of the total health care costs (Yen *et al.*, 1991). Difficulties in obtaining precise measures of cost, sustaining randomly selected experimental and control populations and obtaining reliable and standardized outcome data, all contribute to the uncertainties of these analyses. Generally, lack of control groups, low compliance rates and lack of follow-up evaluations make it difficult to reach final conclusions (Murphy, 1987).

Despite the difficulties, there appear to be correlative if not causal links between improved fitness and reduced cost. Participants in worksite fitness programmes do seem to be at lower risk for ill health, utilize fewer health-care services, be absent from work less often, may be more productive, and may stay with their companies longer than non-participants. The precise quantification of these apparent benefits and a causality link awaits continued progress in research design and data analysis (Der-Karabetian and Gebharbp, 1986; Gamble *et al.*, 1993; Harma *et al.*, 1988).

Stress Management Training

Stress Management Training (SMT), or Worksite Stress Management, has become an increasingly popular type of intervention (Murphy and Sorensen, 1988; Bernier and Gaston, 1989; Auerbach, 1989). However, the evaluation of this type of intervention is complicated, since the terms are used to encompass several types of intervention, for instance biofeedback, muscle relaxation, meditation, and assorted cognitive techniques (Murphy and Sorenson, 1988). By Stress Management Training we will include psychological techniques for reducing 'stress', and information on somatic 'stress'. This includes time management, goal planning and problem-solving techniques including cognitive modification techniques such as cognitive self-instruction. Within the emotional domain, such training involves developing or improving the support network, improving interpersonal relationships (often through development of communication skills) and the expression of anger. Biofeedback muscle relaxation, and meditation will be treated as separate interventions.

Stress management training is directed at the experience of 'job stress', at feelings of helplessness and hopelessness, and at how the individual may cope with stress. Terms such as control and feedback, and the positive feelings of being able to cope with the environment, are explained and appropriate training given. 'Control' is a common element in many types of stress management training, which may differ widely in theoretical background (Newman and Beehr, 1979; Ganster *et al.*, 1982; Bruning and Frew, 1987).

There are many books that cover this field. Bernier and Gaston (1989) counted 100 books published between 1978 and 1986. Remarkably, this number is much higher than the number of controlled studies. There are, however, controlled studies in the literature before 1990, reviewed by Auerbach (1989), demonstrating that various types of stress management techniques reduce anxiety, stress-related symptoms and self-reported negative emotions, and result in lower EMG levels and improved performance. Stress management programmes have been found to produce small to moderate changes in subjective well-being and psychophysiological arousal (Murphy, 1987). In a controlled study, Frayn and Latham (1987) found that a self-management training programme increased attendance at the worksite. The intervention was directed at managing personal and social obstacles to job attendance. Perceived self-efficacy was increased as participants learnt that they could exercise influence over their behaviour. The higher the perceived self-efficacy, the better the subsequent job attendance.

In a review, Bernier and Gaston (1989) concluded that in five controlled and randomized studies of stress management programmes compared with no treatment groups, there were effects in four of the five studies. They also reviewed three controlled studies comparing combined somatic and cognitive programmes with either of the treatments on its own. The results were mixed: there was either no effect or better than no treatment on some, but not all, dependent variables. Comparing the efficiency of different stress reduction programmes demonstrated the usefulness of a great variety of programmes and their relative equivalence as measured at the end of the treatment. These interventions covered simple relaxation techniques, cognitive techniques, self-instruction techniques, educational stress-reduction programmes, and more somatic methods such as aerobics and biofeedback. In all these studies there were proper and randomized controls. When compared with no-treatment groups, most stress management programmes were found to lower all or some somatic measures of stress and cognitive measures of anxiety, as well as to improve academic or work performance. It should be noted, however, that no differential effectiveness of stress reduction programmes was found in this analysis. This is in accordance with results obtained in research into other fields of psychological treatment such as relaxation techniques (Shapiro, 1982) and psychotherapy (Smith et al., 1980).

In our own literature search from 1990 to 1994 we found six randomized and controlled studies (Pruitt, 1992; Stanton, 1991; Grønningsæter et al., 1992; Russler, 1991; Bru et al., 1994; Cecil and Forman, 1990) (see Table 8.3).

Three studies on stress related complaints showed significant effects. In a fourth study, carried out on nursing students, there was no significant effect of SMT as compared to a placebo group and a control group (Russler, 1991). Grønningsæter et al. (1992) found improved coping ability in employees receiving SMT, but no significant changes in somatic or psychological health. Bru et al. (1994). used an SMT programme which was a further development from the one used by Grønningsæter et al. (1992), and compared this with relaxation training and a combined group. The stress management group ('cognitive group') showed a reduction in pain in the neck and shoulders.

Table 8.3. Randomized studies on stress management.

Reference	Intervention	Main variables examined
Pruitt (1992)	Stress management as part of overall fitness programme	Stress-related physical symptoms
Stanton (1991)	Stress management Control group	Stress level
Grønningsæter et al. (1992)	Aerobic Stress management training Control group	Health, job stress, job satisfaction, psychological measurements
Russler (1991)	Stress management Control group	Psychological measurements
Bru et al. (1994)	Cognitive group Relaxation group Combination group Control group	Neck, shoulder and low back pain
Cecil and Forman (1990)	Coworker support Stress inoculation training Control group	Job stress Coping skills

The combined groups had the same effects, and the relaxation group showed reduced pain in low back and shoulders.

Stress inoculation training (Meichenbaum and Deffenbacher, 1988), a form of stress management, may reduce self-reported stress in teachers, with a change in the motoric manifestations of anxiety in the classroom (Cecil and Forman, 1990). Reduced stress levels in secretarial workers have also been reported after as little as two treatment sessions (Stanton, 1991). In a controlled study of stress counselling (Cooper and Sadri, 1991) there were significant reductions in sickness absence, anxiety and depression, and an increase in self-esteem. There were no effects with regard to job satisfaction or organizational commitment.

However, there have also been negative reports, which should be given attention considering the difficulties involved in getting such results published. Nicholson et al. (1989) found a stress management programme to be ineffective in improving psychological well-being, anxiety proneness or situation-specific anxiety compared to a no-treatment control group.

Relaxation, Mental Training and Biofeedback

Muscle relaxation training as a technique for coping with subjective health problems was evaluated by Murphy and Sorenson (1988) in a quasi-experimental design. Those authors reported that employee absenteeism, performance ratings, equipment accidents, and work injuries may be influenced by muscle relaxation training supported by biofeedback. A group receiving muscle relaxation training had significant lower absenteeism and higher attendance ratings the year after training. In another study Tunnecliffe *et al.* (1986) evaluated the effects of relaxation training compared with collaborative behavioural consultation training and a waiting list control group. At the end of the intervention period the use of collaborative behavioural consultation had proved superior to relaxation training, and this effect had been maintained at follow up three months later. These results can be compared with reports from Larsson (1987), who did not find any effect of mental training (relaxation, meditation and imagery rehearsal – alone or combined) on physical and mental well-being, as compared with a control group. However, Alexander *et al.* (1993) examined the effect of transcendental meditation on stress reduction, health and employee development in two settings in the automotive industry. The experimental group was trained and then instructed to practise transcendental meditation twice daily for three months. The experimental group had greater decreases in trait anxiety and state anxiety, greater increase in job satisfaction and improved more in general health, sleep/fatigue, employee effectiveness, job worry and tension, and personal relationships at work.

The use of biofeedback as treatment for musculoskeletal pain has been evaluated by Flor and Birbaumer (1993). They compared the effects of EMG biofeedback, cognitive-behavioural therapy and conservative medical treatment as treatments for chronic musculoskeletal pain. Improvements were noted in all treatment groups, with the biofeedback group displaying the most substantial change. At six- and twelve-months follow up, only the biofeedback group maintained significant reductions in pain severity, interference, affective distress, pain-related use of the health-care system, stress related reactivity of the affected muscles, and an increase in active coping self-statements. However, in a study by Thomas *et al.* (1993) no significant effects were found in an attempt to treat carpal-tunnel syndrome with biofeedback.

Discussion and conclusions

Subjective complaints, as used in this report, cover conditions that are frequent sources of absenteeism, consultations with a doctor and the granting of sickness certificates. The most common subjective health complaints are muscle pain states, exhaustion, fatigue, gastro-intestinal complaints, anxiety, and sleep disturbances. Since about half of long-term sickness absence and permanent disabilities probably derive from subjective states, priority should be given to these. However, the conditions are still ill-defined, and will remain so if we do not accept that they really are subjective states. It is no longer the traditional somatic status, but the patient's own subjective feeling of illness that matters for treatment and for granting sickness compensation. Therefore, it is also futile to search for single-factor causations (Malt, 1986).

Absenteeism – being away from work – may have many causes. The effect may be related to changes in the interpretation of complaints, or changes in threshold for acceptance of complaints. Psychological factors, therefore, may play a role, along with social and motivational factors for the patient. The doctor is also subject to these psychological factors. Practice and opinions about what conditions should lead to sickness certificates and absence from work appear to vary, particularly for subjective complaint states. General practitioners do not seem to have any consistent policies for these decisions, at least in Norway (Håland Haldorsen et al., 1995).

There seems to be one general theme throughout the literature on interventions concerning reduction of sick leave in organizations. Increased control, improved coping skills and increased trust in one's own coping abilities are important ingredients in the many very different types of interventions that may reduce subjective health complaints. These factors seem to be important both when the intervention is directed towards the individual, such as in physical or psychological training, and towards the organization and design of the worksite. Experience of increased control may be obtained through a variety of techniques, some of which have been reviewed in this chapter. Physical exercise, for instance, may increase the sense of physical well-being through the feeling of regaining control and mastery of one's own body. But mental well-being also seems to increase. The mechanism behind this may be that the individual feels he or she has better coping abilities. Therefore, it is not surprising that a wide variety of psychological effects have been reported following physical training, including reductions of days on sick leave. Many reports have suggested that fitness and lifestyle programmes can decrease absenteeism, particularly short and disruptive uncertified absences (see Shephard, 1992). However, definitions of absenteeism have been inconsistent, and relatively few studies have used controls. It has also been difficult to avoid a Hawthorne effect in which individuals respond to the fact that they are under investigation or the influence of self-selection.

Even for interventions currently in use, and demanding high costs from society, the data available is far from sufficient. There is also a bias in the literature, due to an unfortunate mixture of low adherence and low compliance for many of the interventions, and a significant reluctance by both scientists and the editors of scientific journals to publish negative results. In this field low compliance may be one of the really important dimensions for the evaluation of the cost-effectiveness of a programme.

There is really no good scientific or economic reason why issues concerning subjective complaints should not be addressed with the same seriousness and vigour as the serious somatic conditions. It is true that people seldom die from subjective health complaints, although suicides do occur, but in terms of days of absence, and in economic costs for the welfare society, these conditions rank as number one. Even if we do not have full data sets permitting proper analyses, according to our own calculations based on published results, we believe it is still possible to estimate that treatment effects both from single-method approaches, and even more from multiple approaches, should be significant and predictable when groups consist of 100–200 subjects.

Therefore we argue for the same scope and the same rigour in experimental design as for the large worksite cancer control trials. The Working Well Trial was a five-year study with fifty-seven matched pairs of worksites, for a total of 114 worksites. Each worksite was randomized into intervention or control. The comprehensive health promotion intervention addressed dietary change and smoking cessation, delivered by a participatory strategy that targeted individuals and the worksite environment. Experimental intervention and control intervention were standardized across all study centres by specifying a common set of core requirements directed at the smoking and dietary change endpoints (Abrams *et al.*, 1994). This type of well organized initiative is definitely needed for the evaluation of the many types of current intervention – many of them are expensive, few of them are evaluated.

Rough estimates from cost benefit analyses suggest to us that there may be a considerable profit for society in increased efforts in this field. Even more important, increased efforts are necessary, since these conditions affect so many of our fellow human beings, and cause so much needless reduction in the quality of life.

Note

This chapter is based on a review of literature up to and including 1994. A first draft of the chapter was submitted in 1995. The review was sponsored by the Norwegian Research Council.

References

ABRAMS, D. B., BOUTWELL, W. B., GRIZZLE, J., HEIMENDINGER, J., *et al.* (1994) 'Cancer control at the workplace: the working well trial', *Preventive Medicine. An International Journal Devoted to Practice and Theory* **23(1)**: 15–27.

AINSWORTH, B. E., MONTOYE, H. J. and LEON, A. S. (1994) 'Methods of assessing physical activity during leisure and work', in Bouchard, C., Shephard, R. J., and Stephens, T. (eds) *Physical Activity, Fitness, and Health: International Proceedings and Consensus Statement*, Champaign, IL: Human Kinetics Publishers, 146–159.

ALEXANDER, C. N., SWANSON, G. C., RAINFORTH, M. V., CARLISLE, T. W., *et al.* (1993) 'Effects of the transcendental meditation program on stress reduction, health, and employee development: a prospective study in two occupational settings', Special Issue, Stress and stress management at the workplace, *Anxiety, Stress and Coping: An International Journal* **6(3)**: 245–262.

AMICK, B. C. and CELENTANO, D. D. (1991) 'Structural determinants of the psychosocial work environment: introducing technology in the work-stress framework', *Ergonomics* **34(5)**: 625–646.

ANDERSEN, L. (1994) 'Sykefraværsprosjektet 1991–1993. Erfaringer fra bransjer i LO/NHO-området, Trondheim: SINTEF IFIM.

AUERBACH, S. M. (1989) 'Stress management and coping research in the health care setting: an overview and methodological commentary', *Journal of Consulting and Clinical Psychology* **57(3)**: 388–395.

BARRATT, A., REZNIK, R., IRWIG, L., CUFF, A., SIMPSON, J. M., OLDENBURG, B., HORVATH, J. and SULLIVAN, D. (1994) 'Work-site cholesterol screening and dietary intervention: the staff healthy heart project, steering committee', *American Journal of Public Health* **84(5)**: 779–782.

BERNIER, D. and GASTON, L. (1989) 'Stress management: a review', *Canada's Mental Health* **37(3)**: 15–19.

BERRIDGE, J. and COOPER, C. L. (1993) 'Stress and coping in US organizations: the role of the employee assistance program', Special Issue: Coping with stress at work, *Work and Stress* **7(1)**: 89–102.

BRATTBERG, G., THORSLUND, M. and WIKMAN, A. (1989) 'The prevalence of pain in a general population. The results of a postal survey in a county of Sweden', *Pain* **37**: 215–222.

BREDKJÆR, S. R. (1991) 'Musculoskeletal disease in Denmark. The Danish Health and Morbidity Survey 1986-87', *Acta Orthopaedica Scandinavica* **62 (suppl 241)**: 10–12.

BRESLOW, L., FIELDING, J., HERRMAN, A. A. and WILBUR, C. S. (1990) 'Worksite health promotion: its evolution and the Johnson & Johnson experience', *Preventive Medicine* **19(1)**: 13–21.

BRU, E., MYKLETUN, R. J., BERGE, T. and SVEBAK, S. (1994) 'Effects of different psychological interventions on neck shoulder and low backpain in female hospital staff', *Psychology and Health* **9**: 371–382.

BRUNING, N. S. and FREW, D. R. (1987) 'Effects of exercise, relaxation, and management skills training on physiological stress indicators: a field experiment', *Journal of Applied Psychology* **72**: 515–521.

BURKE, R. J. (1993) 'Organizational-level interventions to reduce occupational stressors', Special Issue: Coping with stress at work, *Work and Stress* **7(1)**: 77–87.

CASPERSEN, C. J., POWELL, K. E. and MERRITT, R. K. (1994) 'Measurement of health status and well-being', in Bouchard, C., Shephard, R.J. and Stephens, T. (eds), *Physical activity, Fitness, and Health: International Proceedings and Consensus Statement*, Champaign, IL: Human Kinetics Publishers, 180–202.

CECIL, M. A. and FORMAN, S. G. (1990) 'Effects of stress inoculation training and coworker support groups on teachers' stress', *Journal of School Psychology* **28(2)**: 105–118.

CHATEL, J. C. and PEELE, R. (1970) 'The concept of neurasthenia', *International Journal of Psychiatry* **9**: 36–49.

COCKCROFT, A., GOOCH, C., ELLINGHOUSE, C., JOHNSTON, M. and MICHIE, S. (1994) 'Evaluation of a programme of health measurements and advice among hospital staff', *Occupational Medicine (Oxford)* **44(2)**: 70–76.

COOPER, C. L. and SADRI, G. (1991) 'The impact of stress counselling at work', Special Issue: Handbook on job stress, *Journal of Social Behavior and Personality* **6(7)**: 411–423.

COX, M., SHEPHARD, R. J. and COREY, P. (1981) 'Influence of an employee fitness programme upon fitness, productivity and absenteeism', *Ergonomics* **24**: 795–806.

CROFT, P., RIGBY, A. S., BOSWELL, R., SCHOLLUM, J. and SILMAN, A. (1993) 'The prevalence of chronic widespread pain in the general population', *The Journal of Rheumatology* **20**: 710–713.

DALTON, D. R. and MESCH, D. J. (1990) 'The impact of flexible scheduling on employee attendance and turnover', *Administrative Science Quarterly* **35(2)**: 370–387.

DANNENBERG, A. L., KELLER, J. B., WILSON, P. W. F. and CASTELLI, W. P. (1989) 'Leisure time physical activity in the Framingham offspring study', *American Journal of Epidemiology* **129**: 76–87.

DER-KARABETIAN, A. and GEBHARBP, N. (1986) 'Effect of physical fitness program in the workplace', 93rd Annual Convention of the American Psychological Association, 1985, Los Angeles, California. *Journal of Business and Psychology* **1(1)**: 51–58.

DIENTSBIER, R. A. (1984) 'The effects of exercise on personality', in Sacs, M. H. and Buffone, G. W. (eds) *Running as Therapy*, London: University of Nebraska Press, 253–272.

ELLIS, E., KOBLIN, W., IRVINE, M. J., LEGARE, J. and LOGAN, A. G. (1994) 'Small, blue collar work site hypertension screening: a cost-effectiveness study', *Journal Occupational Medicine* **36(3)**: 346–355.

ELTON, P. J., RYMAN, A., HAMMER, M. and PAGE, F. (1994) 'Randomised controlled trial in northern England of the effect of a person knowing their own serum cholesterol concentration', *Journal of Epidemiology and Community Health* **48(1)**: 22–25.

ERIKSEN, H. R., ELLERTSEN, B., GRØNNINGSÆTER, H., LØYNING, Y., NAKKEN, K. O. and URSIN, H. (1994) 'Physical exercise in women with intractable epilepsy', *Epilepsia* **35**: 1256–1264.

ERIKSEN, H. R., SVENSRØD, R., URSIN, S., and URSIN, H. (1998) 'Prevalence of subjective health complaints in the Nordic European countries in 1993', *European Journal of Public Health* **8**: 294–298.

FALKENBERG, L. E. (1987) 'Employee fitness programs: their impact on the employee and the organization', *Academy of Management Review* **12(3)**: 511–522.

FIELDING, J. E., KNIGHT, K., MASON, T., KLESGES, R. C. and PELLETIER, K. R. (1994) 'Evaluation of the IMPACT blood pressure program', *Journal of Occupational Medicine* **36(7)**: 743–746.

FISHER, K. J., GLASGOW, R. E. and TERBORG, J. R. (1990) 'Work site smoking cessation: a meta-analysis of long-term quit rates from controlled studies', *Journal of Occupational Medicine* **32(5)**: 429–439.

FITZGERALD, S. T., GIBBENS, S. and AGNEW, J. (1991) 'Evaluation of referral completion after a workplace cholesterol screening program', *American Journal of Preventive Medicine* **7(6)**: 335–340.

FLOR, H. and BIRBAUMER, N. (1993) 'Comparison of the efficacy of electromyographic biofeedback, cognitive-behavioural therapy, and conservative medical interventions in the treatment of chronic musculoskeletal pain', *Journal of Consulting and Clinical Psychology* **61(4)**: 653–658.

FRANCISCO, V. T., PAINE, A. L., FAWCETT, S. B., JOHNSTON, J. and BANKS, D. (1994) 'An experimental evaluation of an incentive program to reduce serum cholesterol levels among health fair participants', *Archives of Family Medicine* **3(3)**: 246–251.

FRAYNE, C. A. and LATHAM, G. P. (1987) 'Application of social learning theory to employee self-management of attendance', *Journal of Applied Psychology* **72(3)**; 387–392.

FROMMER, M. S., MANDRYK, J. A., EDYE, B. V., HEALEY, S., BERRY, G. and FERGUSON, D. A. (1990) 'A randomised controlled trial of counselling in a workplace setting for coronary heart disease risk factor modification: effects on blood pressure', *Asia-Pacific Journal of Public Health* **4(1)**: 25–33.

GAMBLE, R. P., BOREHAM, C. A. and STEVENS, A. B. (1993) 'Effects of a 10-week exercise intervention programme on exercise and work capacities in Belfast's ambulance men', *Occupational Medicine (Oxford)* **43(2)**: 85–89.

GANSTER, D. C., MAYES, B. T., SIME, W. E. and THARP, G. D. (1982) 'Managing organizational stress: a field experiment', *Journal of Applied Psychology* **67**: 533–542.

GEBHARDT, D. L. and CRUMP, C. E. (1990) 'Employee fitness and wellness programs in the workplace', Special Issue: Organizational psychology, *American Psychologist* **45(2)**: 262–272.

GIRGIS, A., SANSON FISHER, R. W. and WATSON, A. (1994) 'A workplace intervention for increasing outdoor workers' use of solar protection', *American Journal of Public Health* **84(1)**: 77–81.

GLASGOW, R. E., McCAUL, K. D. and FISHER, K. J. (1993) 'Participation in worksite health promotion: a critique of the literature and recommendations for future practice', *Health Education Quarterly* **20(3)**: 391–408.

GLASGOW, R. E., TERBORG, J. R., HOLLIS, J. F., SEVERSON, H. H., FISHER, K. J., BOLES, S. M., PETTIGREW, E. L., FOSTER, L. S., STRYCKER, L. A. and BISCHOFF, S. (1994) 'Modifying dietary and tobacco use patterns in the worksite: the take heart project', *Health Education Quarterly* **21(1)**: 69–82.

GODIN, G. and GIONET, N. J. (1991) 'Determinants of an intention to exercise of an electric power commission's employees', *Ergonomics* **34(9)**: 1221–1230.

GOMEL, M., OLDENBURG, B., SIMPSON, J. M. and OWEN, N. (1993) 'Work-site cardio-vascular risk reduction: a randomized trial of health risk assessment, education, counseling, and incentives', *American Journal of Public Health* **83(9)**: 1231–1238.

GRØNNINGSÆTER, H., HYTTEN, K., SKAULI, G., CHRISTENSEN, C. C. *et al.* (1992) 'Improved health and coping by physical exercise or cognitive behavioral stress management training in a work environment', *Psychology and Health* **7(2)**: 147–163.

GUNDEWALL, B., LILJEQVIST, M. and HANSSON, T. (1993) 'Primary prevention of back symptoms and absence from work. A prospective randomized study among hospital employees', *Spine* **18(5)**: 587–594.

HÅLAND HALDORSEN, E., BRAGE, S., JOHANNESSEN, T.S., TELLNES, G. and URSIN, H. (1996) 'Musculoskeletal pain: concepts of disease, illness and sickness certification in health professionals in Norway', *Scandinavian Journal of Rheumatology* **25**: 224–232.

HANLON, S. C. and TAYLOR, R. R. (1991) 'An examination of changes in work group communication behaviors following installation of a gainsharing plan', *Group and Organization Studies* **16(3)**: 238–267.

HARMA, M. I., ILMARINEN, J., KNAUTH, P., RUTENFRANZ, J. and HANNINEN, O. (1988) 'Physical training intervention in female shift workers: II. The effects of intervention on the circadian rhythms of alertness, short-term memory, and body temperature', *Ergonomics* **31(1)**: 51–63.

HARRISON, D. A. and LISKA, L. Z. (1994) 'Promoting regular exercise in organizational fitness programs: health-related differences in motivational building blocks', *Personnel Psychology* **47(1)**: 47–71.

HEBERT, J. R., STODDARD, A. M., HARRIS, D. R., SORENSEN, G., HUNT, M. K., MORRIS, D. H. and OCKENE, J. K. (1993a) 'Measuring the effect of a worksite-based nutrition intervention on food consumption', *Annals of Epidemiology* **3(6)**: 629–635.

HEBERT, J. R., HARRIS, D. R., SORENSEN, G., STODDARD, A. M., HUNT, M. K. and MORRIS, D. H. (1993b) 'A work-site nutrition intervention: its effects on the consumption of cancer-related nutrients', *American Journal of Public Health* **83(3)**: 391–394.

HILYER, J. C., BROWN, K. C., SIRLES, A. T. and PEOPLES, L. (1990) 'A flexibility intervention to reduce the incidence and severity of joint injuries among municipal firefighters', *Journal of Occupatonal Medicine* **32(7)**: 631–637.

JEFFERY, R. W., FORSTER, J. L., DUNN, B. V., FRENCH, S. A., McGOVERN, P. G. and LANDO, H. A. (1993) 'Effects of work-site health promotion on illness-related absenteeism', *Journal of Occupational Medicine* **35(11)**: 1142–1146.

JEX, S. M. (1991) 'The psychological benefits of exercise in work settings: a review, critique, and dispositional model', *Work and Stress* **5(2)**: 133–147.

KAMAN, R. L. and PATTON, R. W. (1994) 'Costs and benefits of an active versus an inactive society', in Bouchard, C., Shephard, R.J. and Stephens, T. (eds) *Physical Activity, Fitness, and Health: International Proceedings and Consensus Statement*, Champaign, IL: Human Kinetics Publishers, 134–144.

KARASEK, R. (1990) 'Lower health risk with increased job control among white collar workers', *Journal of Organizational Behavior* **11(3)**: 171–185.

KARASEK, R. A. and THEORELL, T. (1990) *Healthy Work. Stress, Productivity, and the Reconstruction of Working Life*, New York: Basic Books.

KELDER, S. H., JACOBS, D. R., JEFFERY, R. W., MCGOVERN, P. G. *et al.* (1993) 'The worksite component of variance: design effects and the healthy worker project', *Health Education Research* **8(4)**: 555–566.

KING, P. M. (1993) 'Back injury prevention programs: a critical review of the literature', *Journal of Occupational Rehabilitation* **3(3)**: 145–158.

KNAUTH, P. and KIESSWETTER, E. (1987) 'A change from weekly to quicker shift rotations: a field study of discontinuous three-shift workers', 2nd CEC Workshop: irregular and abnormal hours of work. *Ergonomics* **30(9)**: 1311–1321.

LANDAU, J. C. (1993) 'The impact of a change in an attendance control system on absenteeism and tardiness', *Journal of Organizational Behavior Management* **13(2)**: 51–70.

LANDERS, D. M. and PETRUZZELLO, S. J. (1994) 'Physical activity, fitness, and anxiety', in Bouchard, C., Shephard, R.J. and Stephens, T. (eds) *Physical Activity, Fitness, and Health: International Proceedings and Consensus Statement*. Champaign, IL: Human Kinetics Publishers, 868–882.

LARSSON, G. (1987) 'Routinization of mental training in organizations: effects on performance and well-being', *Journal of Applied Psychology* **72(1)**: 88–96.

LAST, J. M. (ed.) (1983) *A Dictionary of Epidemiology*, New York: Oxford University Press.

LEE, P., HELEWA, A., SMYTHE, H. A., BOMBARDIER, C. and GOLDSMITH, C. H. (1985) 'Epidemiology of musculoskeletal disorders (complaints) and related disability in Canada', *Journal of Rheumatology* **12**: 1169–1173.

LEVINE, S. and URSIN, H. (1991) 'What is stress?', in: Brown, M. R., Rivier, C. and Koob, G. (eds) *Stress, Neurobiology and Neuroendocrinology*, New York: Marcel Decker, 3–21.

LINTON, S. J. (1991) 'A behavioral workshop for training immediate supervisors: the key to neck and back injuries?' *Perceptual and Motor Skills* **73(3 Pt 2)**: 1159–1170.

LINTON, S. J. and KAMWENDO, K. (1987) 'Low back schools. a critical review', *Physical Therapy* **67(9)**: 1375–1383.

MALT, U. F. (1986) 'Philosophy of science and DSM-III. Philosophical, idea-historical and sociological perspectives on diagnoses', *Acta Psychiatrica Scandinavica* **73**: 10–17.

MALT, U. F. (1991) 'Somatization: an old disorder in new bottles?', *Psychiatria Fennica* **22**: 79–91.

MARTINSEN, E. W. (1990) 'Benefits of exercise for the treatment of depression', *Sports Medicine* **9**: 380–389.

MATSON, D. M., LEE, J. W. and HOPP, J. W. (1993) 'The impact of incentives and competitions on participation and quit rates in worksite smoking cessation programs', *American Journal of Health Promotion* **7(4)**: 270–280, 295.

MAYER, J. A., BEACH, D. L., HILLMAN, E., KELLOGG, M. C. and CARTER, M. (1991) 'The effects of co-worker-delivered prompts on breast self-examination frequency', *American Journal of Preventive Medicine* **7(1)**: 9–11.

MAYNARD, A. (1991) 'The relevance of health economics to health promotion', in

Badura, B. and Kickbusch, I. (eds) *Health Promotion Research*. Copenhagen: WHO, 29–54.

MEADS, S. (1983) 'The WHO reason-for-encounter classification', *WHO Chronicle* **37**: 159–162.

MEICHENBAUM, D. H. and DEFFENBACHER, J. L. (1988) 'Stress inoculation training', *Counseling Psychologist* **16(1)**: 69–90.

MESCH, D. J. and DALTON, D. R. (1992) 'Unexpected consequences of improving workplace justice: a six-year time series assessment', *Academy of Management Journal* **35(5)**: 1099–1114.

MORGAN, W. P. (1994) 'Physical activity, fitness, and depression', in Bouchard, C., Shephard, R. J. and Stephens, T. (eds) *Physical Activity, Fitness, and Health: International Proceedings and Consensus Statement*. Champaign, IL: Human Kinetics Publishers, 851–867.

MOSES, J., STEPTOE, A., MATHEWS, A. and EDWARDS, S. (1989) 'The effects of exercise training on mental well-being in the normal population: a controlled trial', *Journal of Psychosomatic Research* **1**: 47–61.

MURPHY, L. R. (1987) 'A review of organizational stress management research: methodological considerations', *Journal of Organizational Behavior and Management* **8**: 215–227.

MURPHY, L. R. and SORENSON, S. (1988) 'Employee behaviors before and after stress management', *Journal of Organizational Behavior* **9(2)**: 173–182.

NEWMAN, J. E. and BEEHR, T. A. (1979) 'Personal and organizational strategies for handling job stress: a review of research and opinion', *Personnel Psychology* **32(1)**: 1–43.

NICHOLSON, T., BELCASTRO, P. A. and DUNCAN, D. F. (1989) 'An evaluation of a university stress management program', *College Student Journal* **23(1)**: 76–81.

NOLAN, D. C. (1959) 'American psychiatry', in Arieti, S. (ed.) *American Handbook of Psychiatry*, New York: Basic Books.

NORVELL, N. and BELLES, D. (1993) 'Psychological and physical benefits of circuit weight training in law enforcement personnel', *Journal of Consulting and Clinical Psychology* **61(3)**: 520–527.

NORWEGIAN BUREAU OF STATISTICS (1993) *Arbeidslivsundersøkelsen* 1993.

OLFF, M., BROSSCHOT, J. F. and GODAERT, G. (1993) 'Coping styles and health', *Personality and Individual Differences* **15(1)**: 81–90.

PAFFENBARGER, R. S. Jr., HYDE, R. T., WING, A. L., LEE, I. M. and KAMPERT, J. B. (1994) 'Some interrelations of physical activity, physiological fitness, health, and longevity', in Bouchard, C., Shephard, R. J. and Stephens, T. (eds) *Physical Activity, Fitness, and Health: International Proceedings and Consensus Statement*. Champaign, IL: Human Kinetics Publishers, 119–133.

PATTON, J. P. (1991) 'Work-site health promotion: an economic model', *Journal of Occupational Medicine* **33(8)**: 868–873.

PELLETIER, K. R. (1991) 'A review and analysis of the health and cost-effective outcome studies of comprehensive health promotion and disease prevention programs', *American Journal of Health Promotion* **5(4)**: 311–315.

PETTY, M. M., SINGLETON, B. and CONNELL, D. W. (1992) 'An experimental evaluation of an organizational incentive plan in the electric utility industry', *Journal of Applied Psychology* **77(4)**: 427–436.

PRUITT, R. H. (1992) 'Effectiveness and cost efficiency of interventions in health promotion', *Journal of Advanced Nursing* **17(8)**: 926–932.

RAURAMAA, R. and SALONEN, J. T. (1994) 'Physical activity, fibrinolysis, and platelet aggregability', in Bouchard, C., Shephard, R.J. and Stephens, T. (eds) *Physical Activity, Fitness, and Health: International Proceedings and Consensus Statement*. Champaign, IL: Human Kinetics Publishers, 471–479.

ROMAN, P. M. and BLUM, T. C. (1988) 'Formal intervention in employee health:

comparisons of the nature and structure of employee assistance programs and health promotion programs', *Social Science and Medicine* **26(5)**: 503–514.

RUSKIN, H., HALFON, S. T., ROSENFELD, O. and TANNENBAUM, G. (1991) 'Effects of a daily physical activity program on health status and function of worker industry', *Journal of the International Council for Health, Physical Education and Recreation (Reston, VA)* **27(3)**: 20–25.

RUSSLER, M. F. (1991) 'Multidimensional stress management in nursing education', *Journal of Nursing Education* **30(8)**: 341–346.

SHAPIRO, D. H. (1982) 'Overview: clinical and physiological comparison of meditation with other self-control strategies', *American Journal of Psychiatry* **139**: 267–274.

SHEPHARD, R. J. (1992) 'A critical analysis of work-site fitness programs and their postulated economic benefits', *Medicine and Science in Sports and Exercise* **24(3)**: 354–370.

SHI, L. (1993) 'Worksite health promotion and changes in medical care use and sick days', *Health Values, the Journal of Health Behavior Education and Promotion* **17(5)**: 9–17.

SMITH, M. L., GLASS, G. U. and MILLER, T. I. (1980) *The Benefits of Psychotherapy*, Baltimore: The Johns Hopkins University Press.

SORENSEN, G., HUNT, M. K., MORRIS, D. H., DONNELLY, G. *et al.* (1990) 'Promoting healthy eating patterns in the worksite: the Treatwell intervention model', Special Issue: Nutrition education, *Health Education Research* **5(4)**: 505–515.

SORENSEN, G., MORRIS, D. M., HUNT, M. K., HEBERT, J. R., HARRIS, D. R., STODDARD, A. and OCKENE, J. K. (1992) 'Work-site nutrition intervention and employees' dietary habits: the Treatwell program', *American Journal of Public Health* **82(6)**: 877–880.

SPILMAN, M. A., GOETZ, A., SCHULTZ, J., BELLINGHAM, R. and JOHNSON, D. (1986) 'Effects of a corporate health promotion program', *Journal of Occupational Medicine* **28(4)**: 285–289.

STANTON, H. E. (1991) 'The reduction in secretarial stress', *Contemporary Hypnosis* **8(1)**: 45–50.

STEWART, D. E. (1990) 'The changing faces of somatization', *Psychosomatics* **31**: 153–158.

TAKALA, E. P., VIIKARI JUNTURA, E. and TYNKKYNEN, E. M. (1994) 'Does group gymnastics at the workplace help in neck pain? a controlled study', *Scandinavian Journal of Rehabilitation Medicine* **26(1)**: 17–20.

TASK FORCE ON DSM IV (1994) *Diagnostic and Statistical Manual of Mental Disorders*, 4th ed, Washington, DC: American Psychiatric Association.

TELLNES, G. (1989) 'Sickness certification in general practice: a review', *Family Practice*, **6(1)**: 58–65.

TELLNES, G., SVENDSEN, K. O., BRUUSGAARD, D. and BJERKEDAL, T. (1989) 'Incidence of sickness certification. Proposal for use as a health status indicator', *Scandinavian Journal of Primary Health Care* **7(2)**: 111–117.

THOMAS, J. R., LANDERS, D. M., SALAZAR, W. and ETNIER, J. (1994) 'Exercise and cognitive function', in Bouchard, C., Shephard, R. J. and Stephens, T. (eds) *Physical Activity, Fitness, and Health: International Proceedings and Consensus Statement*. Champaign, IL: Human Kinetics Publishers, 521–529.

THOMAS, R. E., VAIDYA, S. C., HERRICK, R. T. and CONGLETON, J. J. (1993) 'The effects of biofeedback on carpal tunnel syndrome', *Ergonomics* **36(4)**: 353–361.

TOTTERDELL, P. and SMITH, L. (1992) 'Ten-hour days and eight-hour nights: can the Ottawa shift system reduce the problems of shiftwork?', *Work and Stress* **6(2)**: 139–152.

TUNNECLIFFE, M. R., LEACH, D. J. and TUNNECLIFFE, L. P. (1986) 'Relative efficacy of using behavioral consultation as an approach to teacher stress management', *Journal of School Psychology* **24(2)**: 123–131.

TURNER, J. T. (1991) 'Participative management: determining employee readiness', *Administration and Policy in Mental Health* **18(5)**: 333–341.

URSIN, H. (1980) 'Personality, activation, and somatic health. A new psychosomatic theory', in Levine, S. and H. Ursin (eds) *Coping and health*, New York: Plenum Press, 259–279.

URSIN, H., ENDRESEN, I. M. and URSIN, G. (1988) 'Psychological factors and self-reports of muscle pain', *European Journal of Applied Physiology* **57**: 282–290.

URSIN, H., ENDRESEN, I. M., SVEBAK, S., TELLNES, G. and MYKLETUN, R. (1993) 'Muscle pain and coping with working life in Norway: a review', *Work and Stress* **7**: 247–258.

WANZEL, R. S. (1994) 'Decades of worksite fitness programmes, Progress or rhetoric?', *Sports Medicine* **17(5)**: 324–337.

WARNER, K. E., WICKIZER, T. M., WOLFE, R. A., SCHILDROTH, J. E. and SAMUELSON, M. H. (1988) 'Economic implications of workplace health promotion programs: review of the literature', *Journal of Occupational Medicine* **30(2)**: 106–112.

WERNER, G. A. (1992) 'Employee recognition: a procedure to reinforce work attendance', *Behavioral Residential Treatment* **7(3)**: 199–204.

WESSLEY, S. (1990) 'Old wine in new bottles: neurasthenia and "ME"', *Psychological Medicine* **20**: 35–53.

WORLD HEALTH ORGANIZATION (1985) *Targets for Health for All*, Copenhagen: WHO.

YEN, L. T., EDINGTON, D. W. and WITTING, P. (1991) 'Associations between health risk appraisal scores and employee medical claims costs in a manufacturing company', *American Journal of Health Promotion* **6(1)**: 46–54.

Organizational interventions

CHAPTER NINE

Organizational healthiness, work-related stress and employee health

TOM COX AND LOUISE THOMSON

Most of the employed people in the United Kingdom work in organizations and think of themselves as working for organizations. For them, their employer is an organization: that organization provides them with not only their job but also the context within which that job exists. Not surprisingly, many of the problems that they report, which are known to affect their health and behaviour at work, and which they consequently have to cope with, appear to originate in organizational and related management issues (Cox, Leather and Cox, 1990). More widely, the organization, and the context that it offers for work, shape the behaviour of its employees and can determine not only the quality of their working life but also their general health. It is, therefore, important to examine the nature of organizations, and the relationship between the characteristics of those organizations, and the experiences, behaviour and health of their employees. It is suggested that an important mediating factor may be the perceived 'quality' of the organization and its healthiness. Adopting this framework has obvious implications for the management of work-related stress and improving employee health, emphasizing the importance of organizational-level interventions and coping responses. This chapter provides a framework for those that follow, and explores the organizational context to employee health, stress and coping in terms of the concept of organizational healthiness.

The nature of organizations

Perhaps the first question to answer is 'What is an organization?' and, more particularly, 'What is a work organization?' Contrasting these two questions immediately implies the important point that not all organizations are work organizations, and that whatever definition of an organization is adopted, it must be capable of modification to clearly distinguish between those organizations which are work organizations and those which are not, such as social clubs or voluntary organizations (Cox and Cox, 1996). In answering these questions, researchers have used a number of levels of analysis (see Blau, 1957). Three are clearly identifiable in the relevant literature; they focus on either (1) individual behaviour within organizations, (2) the characteristics and function of particular aspects of organizational structure, or (3) the characteristics and behaviour of the organization itself as a collective identity (Scott, 1981). Here the authors are primarily concerned with the first level of explanation, and exploring individual behaviour within the context of the organization. The characteristics and behaviour of the organization itself provide the context, or environment within which individual behaviour is enacted. What the authors are attempting is an explanation of how such a context determines and, at the same time, reflects the attitudes, behaviour and health of employees. There are several different types of explanation available at this level. These have been classified in terms of their historical development and the types of theory that they reflect: first classical organization theory, then the human relations school, and most recently, systems theory (see Bryans and Cronin, 1983). Arguably, the latter theories – the human relations school and systems theory – have proved the most influential in the present context. Theories within the human relations school have often been labelled social psychological and are exemplified, over the years, by the writings of March and Simon (1958), Porter, Lawler and Hackman (1975), and Duncan (1981). Systems theory has been integrated with the social psychological definitions of organizations to produce *sociotechnical systems theory*, developed largely by Emery and Trist (1960, 1981). The arguments put forward in this chapter derive from both social psychological and systems theory. These are dealt with in turn.

Social Psychological Theory

In their book, *Organizational Behaviour*, Buchanan and Huczynski (1985) suggest that 'organizations are social arrangements for the controlled performance of collective goals'. There are three important elements to such a definition. The first is that organizations are social arrangements, and are thus represented in the existence and behaviour of a group, or groups, of people. Organizations are thus about the interactions both between individuals in those groups and between such groups. People and groups are key components in the organization as a social system. The second important element is that organizations are concerned with achieving collective goals. This implies that organizations, like all systems, exist for a purpose and that purpose is represented in its collective goals. It is interesting to ask whether or not most, if not all, members of organizations know and understand, if not share, those collective goals. This question might provide one dimension whereby the healthiness of the organization could be assessed. The third important element of the definition is that organizations function through controlled performance in pursuit of those collective goals. This introduces the notion of management control and with it that of a division of labour.

A somewhat similar social psychological definition was offered, earlier in 1987, by Duncan (1981): organizations are 'a collection of interacting and interdependent individuals who work towards common goals and whose relationships are determined according to a certain structure'. Duncan's (1981) definition emphasizes the interaction and interdependence of the individuals who make up the social groups, which in turn form the organization, and introduces the idea that organization structure, in addition to its management function, plays a part in controlling individual behaviour.

Buchanan and Huczynski's (1985) and Duncan's (1981) approaches to defining organizations draw attention to their reality as social constructions, and this may be contrasted with a more physical or technical view of organizations as embodied in their site, plant, machinery and specified work systems, procedures and hierarchies. Thus, we might be able to contrast the social organization – the organization as a people system – with the technical organization, the organization as hardware and software systems – and ask to what extent these different aspects of the organization are consistent. Does the technical organization support and facilitate the social organization and vice versa? Questions such as this may again offer dimensions whereby the healthiness of organizations might be assessed (Cox and Howarth, 1990).

Social psychological definitions of organizations treat them as social constructions, systems in which the main components are people and groups of people. Such organizations exist for the controlled achievement of shared goals. The control exercised within these organizations is partly through structure, and partly through management and management systems. The latter is said to be both adaptive and purposeful; the former is control by design (see Cox and Cox, 1996). To treat organizations as if they were only social systems, however, would be to deny their reality. It is recognized here that the social organization has to be considered in relation to the technical organization because these are complementary and interdependent aspects of the total organizational system. Any adequate theory of the organization as a total system has to embrace both aspects. Such an approach to organizations has given rise to sociotechnical systems theory.

Sociotechnical Systems Theory

Some early systems theorists tended to view the organization as a closed system (see Emery and Trist, 1981) despite the inadequacies of such an approach. For example, there is an inherent tendency of closed systems to grow towards maximum homogeneity of their parts, with steady states eventually achieved by the cessation of all activities. This is clearly at odds with the life history of many organizations. Such theorists also believed that organizations could be sufficiently independent to allow most of their problems to be analysed with reference to their internal structures and without reference to the external environment in which they exist. Such a belief was clearly misplaced. Again, this is clearly at odds with the experience of effective practice. The alternative approach considers organizations as open systems.

Open systems depend on exchanges with their wider environments in order to survive; they import things (materials, information, etc.) across their boundary with that environment, transform those imports and then export things back out to that environment. Open systems are self-regulating and adaptable. They grow through a process of internal elaboration (Herbst, 1954) and may spontaneously reorganize towards states of greater complexity and heterogeneity (Waldrop, 1989). They manage to achieve a steady state while working – a dynamic equilibrium – in which the organization as a whole remains constant, with a continuous throughput and despite a considerable range of external changes (Lewin, 1951). Their survival and success is dependent on the extent to which they can both adapt to external change and manage it and its consequences. This approach to organizations would seem more appropriate than the closed systems model, and, indeed, current systems thinking applied to organizations treats them as if they are, to a large extent, open systems (Buchanan and Huczynski, 1985; Emery and Trist, 1981).

The open systems view of organizations has also been termed the 'organic analogy' (see, for example, Flood and Jackson, 1991; Miller and Rice, 1967; Rice, 1963) because it implies that organizations have properties in common with living organisms, including the capacity for adaptation, growth and contraction. The concept of organizational healthiness draws directly on this analogy in terms of a model derived from the physiological mechanisms which underpin individual health (Cox and Howarth, 1990).

While the notion of the organization as an open system is appealing because the introduction of systems theory allows the development of a total systems view, in itself it was not fully compatible with social psychological definitions of organizations. Some further reconciliation of these approaches was always required, and the integration of these two sets of ideas produced sociotechnical systems theory. The development of sociotechnical thinking is largely attributed to the work of Emery and Trist and their colleagues (for example, Emery and Trist, 1960; Emery and Trist, 1981; Trist and Bamforth, 1951; Trist et al., 1963; see also Pasmore and Khalso, 1993).

The concept of the organization as a sociotechnical system derives from the fact that all work systems require both a technology (hardware and software: equipment, work systems and tasks, and environments) and a social organization of the people who use that technology and work those systems. While these different components of the organization are or should be, to some degree, interdependent, they are, at the same time, independent. Their independence was demonstrated in the studies of Trist and his colleagues (1951; 1963) in the coal mines around Durham, UK, and in a textile mill in Ahmedabad, North West India. These studies showed that the social organization can have social and psychological properties that are not dependent on the demands of the technology used. Findings such as this make clear that sociotechnical systems theory is not simply a re-dressed technical determinist approach as those espoused by Woodward (1965) and Perrow (1967).

Sociotechnical theory seeks to describe the nature of the technical and social components of organizations, analyse their interdependence and design the 'best fit' between them. Theorists have argued that organizations must be designed in such a way that the needs of each component are met to some extent, and that the components are consistent with each other. As suggested earlier, the extent to which these two aspects of the organization are consistent and mutually supportive can be treated as a measure of the health of that organization. An unhealthy organization is one, for example, where the technical system has been designed or is managed without taking into account the needs of the social system: under such circumstances the system as a whole will not work effectively. Interestingly, however, Trist and his colleagues (1951, 1963) have argued that an effective sociotechnical system design cannot completely satisfy the needs of either the technical or the social systems: this sub-optimization appears to be a necessary feature of sociotechnical design. Such a strategy has been referred to as 'satisficing' (Simon, 1981; see also Cox and Cox, 1996).

In their early studies in Durham, Trist and his colleagues (Trist and Murray, 1948; Trist and Bamforth, 1951) compared what was then a conventional form of work organization for mining, the longwall system, with a composite system effectively based on group working. The conventional system combined a complex formal structure with simple work roles: miners worked a single part-task, had only limited interactions with others in their particular task group, and were sharply divided from those outside that group. The composite system, in contrast, combined a simple formal structure with complex work roles: miners performed multiple tasks with a commitment to the overall group task, had much contact with others in the group and were involved in the self regulation of the group. The two different systems worked with the same technology and the same coal seam; however, their social systems and the miners' task profiles were very different. The composite group system, in principle, represented a good sociotechnical design.

The available data suggested that, in terms of safety and health, the composite 'group' system was the superior. The employee absence data (expressed as a percentage of possible shifts actually worked) clearly indicated this superiority, with 20 per cent absence recorded for the conventional system, and only 8.2 per cent recorded for the composite group system. The figures included those for absence due to accidents, the data for which also reflected the superiority of the composite group system over the conventional system: 6.8 per cent for the conventional system vs 3.2 per cent for the composite group system.

The work of Trist and his colleagues (for example, Trist and Bamforth, 1951; Emery and Trist, 1960; Trist *et al.*, 1963), and other researchers based in the Tavistock Institute, London (for example, Herbst, 1962), fuelled the conclusion that work in groups is more likely to provide meaningful work, develop responsibility and satisfy and support human needs than work that is allocated to separately supervised individuals. However, Emery and Trist (1981) emphasized that their findings did not suggest that work-group autonomy should be maximized in all organizational settings. They argued that there is an optimal level of grouping which needs to be determined in relation to the requirements of the technical system but that the relationship between level of technology and level of grouping is not a simple one. They also suggested that the psychological needs that are met by grouping are not employees' needs for friendship while working (as the Human Relations school of thought might suggest):

> Grouping produces its main psychological effects when it leads to a system of work roles such that the workers are primarily related to each other by way of the requirements of task performance and task interdependence. When this task orientation is established the worker should find that he (or she) has an adequate range of mutually supportive roles. As the role system becomes more mature and integrated, it becomes easier for a worker to understand and appreciate his (or her) relation to the group.
>
> (Emery and Trist, 1981: 174)

Despite these caveats, the work of sociotechnical systems theorists led to the general promotion of autonomous work groups as a preferred form of work organization. The implementation of such groups has been viewed as a form of job enrichment. Autonomous group working can be effectively applied to groups of people whose work is related or interdependent. Buchanan (1979) has attempted to set out what such work group organization should involve.

1 job rotation or physical proximity especially when individual tasks are: interdependent, stressful and lack perceivable contribution to the end product;

2 the grouping of interdependent jobs to give: whole tasks which contribute to the end product, control over work standards and feedback of results, and control over boundary tasks; and

3 communication and promotion channels.

Systems theory approaches to organizations have been criticized for treating the organization as if it were an entity separate from the individuals who are part of it, and for ascribing goals and actions to it rather than to those individuals (see Beazley, 1983). Sociotechnical systems theory overcomes this criticism to some extent, making explicit the role and importance of the social component of the organizational system. However, the emphasis is still not strictly on the individual, as the notions of work group and work roles are arguably more important for such theorists than that of the individual employee. Whatever the exact nature of the emphasis in systems theory approaches to organizations, the very nature of the theory lends itself to the analysis of organizational-level problems and actions, and can lead to a variety of organizational development strategies for change. Therefore, its emphasis for the reduction of work-related stress and employee health is on the reduction of organizational-level problems through organizational-level actions and development, rather than on increasing individuals' coping resources.

Systems thinking has been variously and successfully applied to a variety of organizational issues over the last thirty years (see, for example, Miller and Rice, 1967; Scott and Mitchell, 1972; Open Systems Group, 1981; Beer, 1985; Waring, 1989; Happ, 1993; Jones, 1995). Furthermore, recent developments in systems thinking, such as the interest in complexity theory (see Cox and Cox, 1996), are now having an increasing influence on organizational thinking (for example, Browning, Beyer and Shelter, 1995; Howarth, 1995; Brown and Eisenhardt, 1997; Chen, 1997; White, Marin, Brazeal and Friedman, 1997; Murray, 1998; Miller, Crabtree, McDaniel and Stange, 1998). Among a plethora of relatively new concepts being introduced is that of organizational healthiness. This has been offered, operationalized and used in several different forms (e.g. DudekShriber, 1997; Hoy and Feldman, 1987; Hoy and Hannum, 1997; Hoy, Tarter, and Bliss 1990; Hoy and Woolfolk, 1993; Miles, 1965; Peterson and Wilson, 1998); several of them rather unimaginative and not novel. What is discussed here is the concept as developed at the Centre for Organizational Health & Development, now incorporated into the Institute of Work, Health, & Organizations, at the University of Nottingham, by the authors and their colleagues.

Organizational healthiness

The concept of organizational healthiness is based on an analogy with individual health and is a derivation of sociotechnical systems thinking. It is about the nature and viability of organizations as systems, and includes measures of the perceived quality of the social organization and its relationships with the technical organization.

The term 'the health of the organization' can be thought of as referring to its condition, in the same sense that the parallel term 'the health of the individual' refers to the general condition of the person. In itself, introducing the notion of the 'condition' of the organization is intellectually insufficient, and two further refinements need to be made. First, the health of the individual is often defined in terms of their condition of body, mind and spirit (Longman's Dictionary of the English Language, 1992). In parallel terms we can talk about the health of the organization as being the condition of its structure and function, management systems and culture. Second, a distinction needs to be made between what is healthy and what is not, in terms of 'general condition'. Healthy individuals, and healthy organizations, are those which are seemingly sound, that is fit-for-purpose, thriving and able to adapt in the longer term. The concept of 'quality' may be appended to this notion.

It has been suggested by Smewing and Cox (1996) that the health of the organization is:

> the general condition of its structure and function, management systems and culture.

This may be rephrased as 'the quality' of its structure and function, management systems and culture. Expanding on this, a healthy organization is:

> an organization in which the different components, which define its general condition, sum to it being 'fit-for-purpose', thriving and adaptable, and which is perceived positively by its employees.

Cox and Howarth (1990) presented a model of the health of an organization built upon an analogy with an individual's physical health. The model placed particular emphasis on the sub-systems[1] which together describe the internal functioning of the organization and underpin its ability to adapt to a changing environment. These sub-systems relate to the total organization in much the same way as physiological systems, such as the cardiovascular or respiratory systems, relate to the whole person and their behaviour.

In a study of education provision in Britain, the first author and his colleagues (see Cox, Kuk and Leiter, 1992) identified, through teachers' descriptions of their schools, three possible sub-systems: these represented the school in terms of task completion, problem-solving and staff development. They found that the quality of the task and problem-solving sub-systems could be related to teachers' experience of work-related stress and also to their report of general symptoms of ill-health, to their absence behaviour and to their declared intentions to leave. The study demonstrated that the health of employees and (some aspects) of their work behaviour were related to the health of the organization (school) in which they worked. Such employees, it was generally argued by Cox and Howarth (1990), functioned more effectively when (1) they had a coherent and positive perception of the organization, and when (2) this perception was largely congruent with organizational reality. It has been variously suggested that the coherence which exists within employees' perceptions of their organization, between its various sub-systems, reflects at least two factors: the effectiveness of its communication systems, and the strength of its social structure.

Organizational Sub-Systems

Cox and Leiter (1992) offered a preliminary commentary on the nature of the three sub-systems of organizations: task completion, problem solving and staff development. This commentary focused on health-care organizations, and has been elaborated, both empirically and theoretically, by Smewing and Cox (1996).

Task Completion

The absence of support at the level of primary task completion for strongly espoused organizational goals and objectives may reflect unresolved conflicts regarding policy throughout the organization. Argyris (1982) proposed that such contradictions of an organization's espoused goals and objectives and its actual practice were signs of fundamental flaws in organizational functioning. The strain of such contradictions may undermine organizational effectiveness by increasing unproductive demands on staff while simultaneously undermining their resource base, thereby increasing their potential for experiencing work-related stress (see Cox and Griffiths, 1995). In addition to contradictions between aims and actual practice, other potential organizational-level barriers to task completion include lack of knowledge of aims, lack of actual aims, and lack of agreement over aims.

Problem Solving

The nature and effectiveness of problem-solving processes are sensitive indicators of organizational healthiness. Problem solving is the occupational activity that offers the widest scope to the creative competence and professional judgement of employees, both as a group and as individuals. It also points up the inadequacies and strengths of group work. Problem solving conducted solely as an individual responsibility or under pressure or hampered by arbitrary constraints denies the value to the organization of staff competence and group cohesion. This loss will be exacerbated when the target of ineffective problem solving is the organization itself. By undermining efforts to improve organizational functioning, poor problem solving diminishes hope for future improvement. It encourages the perception of the organization as stagnant with little potential for growth; that is, a closed system. The thoroughness of problem-solving procedures, formal and informal, and the broadness of staff participation in those processes arc defining characteristics of an organization's culture (Janis and Mann, 1977).

Staff Development

Employees often judge an organization's potential for growth through their own aspirations for personal and professional staff development. The thwarting of these aspirations can undermine staff confidence in the organization's future (Hendry, 1991). Therefore, a key aim is to develop systems in which individual development and enhancement are consistent with a drive for organizational success and growth.

In addition to the three sub-systems themselves, the strength of the social context of the organization is considered to be important to enable them to work together (Kuk, 1993). Harmonious social relationships may reflect consensus regarding major goals and objectives, coherence of work and integrated work systems, as well as a balance between those goals and objectives and the availability of appropriate resources. An organization characterized by harmonious social relationships may generate less stress for its members. The associated consistency in goals and values may facilitate the setting of priorities, thereby reducing conflict between organizational members regarding the provision of resources (Hackman, 1986). Similarly, consistency may make it easier for individuals to effectively allocate their time and energies. In contrast, organizations lacking such congruence place increasing pressures on their members (Holland, 1985).

Organizational healthiness, employee stress and health

The health of organizations, it is suggested, is the reflection of several different levels of organizational analysis. At the highest levels, the health of the organization is a reflection of how well matched are the technical and social aspects of organizations and how positive and realistic are the employees' perceptions of it. At the lowest level, it is a reflection of the perceived quality[2] of one or all of its sub-systems. At the intermediate level of analysis, health is reflected in the quality and degree of integration of the three sub-systems. This type of thinking about organizational health is based on a 'level of description' model and implies the possible existence of a hierarchy of measures varying in 'granularity'.

Recent research has suggested that measures of health at the finest level of granularity – the quality of the system's components – are relatively strong predictors of outcome measures such as employees' experience of stress, their self-reported well-being, and absence behaviour. There is, therefore, evidence that the health of the organization is related to that of its employees. The question is how – by what mechanisms are organizational healthiness, employee stress and health interrelated?

It has been argued by Cox and Cox (1996) that there are at least three ways in which threats to employee safety can arise within organizations: this argument may be extended to answer the issue of employee health. According to this model, the main threats to employee health may be a breakdown in systems management and adaptive control; unpredictable events arising from the complexity of the system; and failures of system design. In addition to these aspects of systems design, behaviour and management, threats to health may occur because of unhealthy or 'risky' behaviour at all levels of employment in the system; including errors and violations. Such behaviour may be secondary to a failure of systems design or to a breakdown in adaptive control. or it may be a reflection of the organizational culture or the interface between the system and its wider environment. Dissatisfaction, disaffection and distraction, which may partly drive unhealthy behaviour, may themselves be the result of an unhealthy organization. Therefore, the lack of quality relating to the task, problem-solving and development environments may give rise to dissatisfaction, disaffection and distraction and, in turn, drive unhealthy behaviour. The experience of stress by employees may be the mechanism by which their perceptions and cognitions about the organization are translated into unhealthy behaviour.

This argument is represented in Figure 9.1. The model outlined in this figure suggests that the health of an organization can affect employee health both (1) through the design and management of its work systems and procedures and (2) through the experience of stress and the organization's impact on employee behaviour at work. It is argued here that the latter can moderate the effects of the former on health. The interplay between the two is represented in a true interaction. Healthy organizations are those which, among other things, not only design and effectively manage healthy systems of work, but also seek explicitly to enhance the health of their employees, encouraging healthy work behaviour. The health of employees in unhealthy organizations is expected to be poor, with employees under-performing and showing reduced commitment as those organizations attempt to implement inadequately designed and managed work systems.

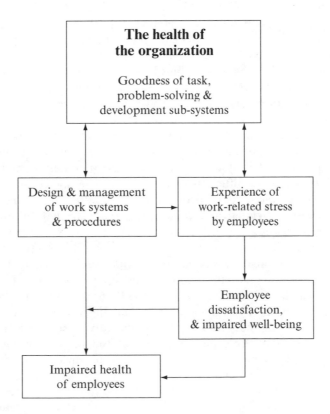

Figure 9.1. Organizational healthiness and employee health.

Smewing and Cox (1995) considered the possible impact of Employee Assistance Programmes (EAPs) on both individual and organizational health. If positive perceptions of the organization lead to better well-being, then the manner in which an EAP is introduced, marketed and managed will be an important factor in determining its effectiveness. The study found that access to an EAP affected two aspects of organizational healthiness: the quality of the development and the problem-solving environments. Furthermore, it demonstrated the importance of perceptions of organizational health with regard to the way EAPs affect individual health, with both problem-solving and EAP presence predicting aspects of employee well-being.

Organizational healthiness and culture

The concept of organizational health is very close to and, in some ways, overlaps with that of organizational culture. It is possible that they are two different views of the same set of phenomena, the conceptual difference being generated largely by the differences in methodology associated with their study. The issue of their relationship is not resolved, although the closeness of the two concepts is obvious. For example, organizational culture has been defined by Schein (1985) in terms of employees', shared values and perceptions of the organization, beliefs about it, and common ways of solving problems within the organization. As such, it must have relative stability and not change on an hourly, daily or even weekly basis. Such a definition is very close to the concept of organizational healthiness. However, a distinction between the two can be made in terms of subjective and objective levels of the organization as defined by Cox and Howarth (1990). The *objective* level of an organization is defined by its written policies, rules of operation and procedures, its communication channels and its physical interchanges of materials and products within and between its different environments. The *subjective* level of an organization is represented in people's understanding of it and attitudes towards it, both individual and collective. Organizational culture may be defined as representing the subjective aspects of the environment, whilst organizational health requires consideration of both aspects, such that a healthy organization requires its culture (subjective aspects) to be in some way consistent with its structure, policies and procedures (objective aspects) (Cox and Howarth, 1990). Furthermore, Ho (1996) describes the distinctions between organizational culture and climate and suggests that the concept of organizational health incorporates the quantitative aspect of the climate concept, used for modelling purposes, and at the same time encompasses the qualitative component of the culture concept in, for example, its reliance on the perceptual data elicited from individuals.

In terms of systems theory, organizational culture can be treated as an emergent property of the organization as a system. That is, culture is a property of the whole system. It is a reflection of the state and function of those individual components and processes, and their interactions. It influences them, but it is not located in any single or particular component, process or interaction. It is a gestalt: it resides in (and is greater than) the sum of the system's parts and not in any one of them. This observation may offer a further differentiation of the concepts of organizational culture and healthiness. The former is an emergent property of the organization as a system, while the latter is a measure of the perceived quality of that system and of the compatibility of its components.

Any system can be deconstructed into its component sub-systems and many, if not all of them, might be treated as systems in their own right. Thus each sub-system has the potential culture, and just as these systems and sub-systems may be hierarchically arranged and reflect different organizational structures and functions, so might the emergent cultures and sub-cultures. As systems have sub-systems, cultures have sub-cultures. Indeed, Adams and Ingersoll (1989) have commented that it is often best to discuss organizational culture in terms of constituent sub-cultures. In this vein, it has been variously argued that organizational cultures and sub-cultures are nested (for example, Pidgeon, 1991) and undoubtedly overlapping, being mutually influential across and between levels in that nested structure.

Possibly the same arguments might be applied to the concept of organizational healthiness, with judgements on healthiness being applicable not only to the overall organization – as a system – but also to its component parts. For example in health care, judgements of healthiness might be made in relation to the hospital as the overall organization – or system – or in relation to its component parts – its units, wards, specialities and services. Indeed, this assumption has underpinned the work of Smewing (Smewing, 1997; Smewing and Cox, 1996) on the evaluation of an employee assistance programme across different units and services with a major UK hospital.

Acceptance of the argument that the concept of organizational healthiness can be applied, like that of organizational culture, not only to the organization as a system but also to its component parts, gives rise to the interesting possibility of a two-dimensional matrix. One dimension might represent the level of organizational analysis, from the whole organization through to its components and sub-components. Here it is not yet clear at what level of deconstruction the concept of organizational healthiness ceases to be reliable and valid. The other dimension is the level of granularity of health measurement, from the highest level reflecting the nature of employees' overall perceptions, through to the lowest level in terms of the perceived quality of the three functional (psychosocial) sub-systems.

Assessing organizational healthiness

The Organizational Health Questionnaire (OHQ) has been developed to help researchers and managers assess the health of organizations. The OHQ is the intellectual property of the Centre for Organizational Health & Development, now incorporated into the Institute of Work, Health, & Organizations, University of Nottingham (Smewing and Cox, 1995; Smewing, Cox and Kuk, 1994; Smewing, 1997; Kuk, 1993). In terms of the granularity of measurement referred to above, it offers three scales at the lowest (finest) level of measurement:

1   the quality of the organization as an environment for task completion;
2   the quality of the organization as a problem-solving environment; and
3   the quality of the organization as a development environment.

The OHQ also asks for information on employee absence, the impact of absence employee turnover and intention to leave.

The OHQ captures employees' perceptions and descriptions of the organization. These data are subjective but acceptable, given that those employees have 'expert' knowledge of their organization by reason of their experience of working within it and, necessarily, of their reflections on that experience. The concept of the 'expert' or 'reflective' employee is important. Experience gained unthinkingly – without any form of awareness or critical reflection – may not be sufficient for the present purpose. Fortunately, while the level of critical reflection must vary from one type of work, workplace or individual employee to another, it must be rare for an employee never to think about or discuss their work with another, or not to have received some form of instruction or training on how to do it. As a result, most employees' perceptions and descriptions of their work and work organizations as systems have at least some face validity as expert evidence or data. The full extent of the validity of a systems description derived from conversations with workers can be tested empirically.

As already stated, the OHQ was originally developed through a series of studies involving teachers in UK schools (Cox et al., 1988, 1989; Cox, Kuk and Leiter, 1992). These studies have been extended in two ways. First, the instrument's utility was tested in Singaporean (Ho, 1996) and Taiwanese (Lin, 1994) contexts. Second, its wider applicability in the UK was examined in relation to health-care organizations (Smewing and Cox, 1996; Smewing, 1997). In both cases, confirmatory analyses demonstrated the replicability of the original structure.

Summary

Essentially, this chapter has concerned the organizational context of work-related stress, employee health and coping. It has done so in terms of the concept of organizational health.

The chapter began with some consideration of the nature of organizations and the derivation of the sociotechnical theory of organization. This theory, combined with the individual health analogy, forms the basis for the

Nottingham concept of organizational health. The nature, definition and measurement of organizational health are addressed in the chapter, and the mechanisms which may link this concept to those of work-related stress, employee health and coping are explored. Its relationship to organizational culture is also discussed.

It is argued that the health of an organization can affect employee health through two mechanisms: first, through the design and management of its work systems and procedures and, second, through the experience of stress and the organization's impact on employee behaviour at work. Healthy organizations are those which, among other things, not only design and effectively manage healthy systems of work, but also seek to enhance the health of their employees, encouraging healthy work behaviour. It is argued that the health of employees in unhealthy organizations is expected to be poor, with employees under-performing and showing reduced commitment as those organizations attempt to implement inadequately designed and managed work systems.

From the arguments and evidence advanced in this chapter, it should be clear that the health of the organization is an important determinant of the experience of work-related stress and subsequently employee performance, health, and coping. As such it should be carefully attended to by employers and managers as a focus for work-related stress interventions.

Notes

[1] In previous publications, these sub-systems have been referred to as *psychosocial* as a convenient shorthand. They involve psychological, social and software aspects of the organization: people, groups, processes and systems.

[2] The concept of 'goodness' has been used in previous publications; however, that of 'quality' may well be more in keeping with contemporary thinking on organizations.

References

ADAMS, G. B. and INGERSOLL, V. H. (1989) 'Painting over old works: the culture of organisations in an age of technical rationality', in B. A. Turner (ed.) *Organisational Symbolism*, Berlin: Walther De Gruyter.

ARGYRIS, C. (1982) *Reasoning, Learning, and Action: Individual and Organization*, San Francisco: Jossey-Bass.

BEAZLEY, M. (1983) *Organization Theory*, London: Mitchell Beazley.

BEER, S. (1985) *Diagnosing the system for organizations*, Chichester: John Wiley and Sons.

BLAU, P. (1957) 'Formal organization: dimensions of analysis', *American Journal of Sociology* **63**: 58–69.

BROWN, S. L. and EISENHARDT, K. M. (1997) 'The art of continuous change: linking complexity theory and time-paced evolution in relentlessly shifting organizations', *Administrative Science Quarterly* **42**: 1–34.

BROWNING, L. D., BEYER, J. M. and SHELTER, J. C. (1995) 'Building cooperation in a competitive industry – sematech and the semiconductor industry', *Academy of Management Journal* **38**: 113–151.

BRYANS, P. and CRONIN, T. P. (1983) *Organization Theory*, London: Mitchell Beazley.

BUCHANAN, D. A. (1979) *The Development of Job Design Theories and Techniques*, Farnborough: Saxon House.

BUCHANAN, D. A. and HUCZYNSKI, A. A. (1985) *Organizational Behaviour*, Englewood Cliffs, New Jersey: Prentice Hall.

CHEN, S. (1997) 'A new paradigm for knowledge-based competition: building an industry through knowledge sharing', *Technology Analysis and Strategic Management* **9**: 437–452.

COX, S. and COX, T. (1996) *Safety, Systems and People*, London: Butterworth-Heinemann.

COX, T. and GRIFFITHS, A. (1995) 'The nature and measurement of work-stress: theory and practice', in J. Wilson and N. Corlett (eds) *The Evaluation of Human Work: A Practical Ergonomics Methodology*, London: Taylor and Francis.

COX, T. and HOWARTH, I. (1990) 'Organizational health, culture and helping', *Work and Stress* **4**: 107–110.

COX, T. and LEITER, M. (1992) 'The health of healthcare organizations', *Work and Stress* **6**: 219–227.

COX, T., BOOT, N., COX, S. and HARRISON, S. (1988) 'Stress in schools: an organizational perspective', *Work and Stress* **4**: 353–362.

COX, T., COX, S. and BOOT, N. (1989) 'Stress in schools: a problem solving approach', in M. Cole and S. Walker (eds) *Stress and Teaching*, Milton Keynes: Open University Publications.

COX, T., KUK, G. and LEITER, M. P. (1992) 'Burnout, health, work stress and organizational healthiness', in W. Schaufeli and C. Maslach (eds) *Professional Burnout: Recent Developments in Theory and Research*, New York: Hemisphere.

COX, T., LEATHER, P. and COX, S. (1990) 'Stress, health and organizations', *Organizational Health Review* **23**: 13–18.

DUDEKSHRIBER, L. (1997) 'Leadership qualities of occupational therapy department program directors and the organizational health of their departments', *American Journal of Occupational Therapy* **51**: 369–377.

DUNCAN, W. J. (1981) *Organizational Behaviour*, 2nd edn. Boston, MA: Houghton Mifflin.

EMERY, F. E. and TRIST, E. L. (1960) 'Sociotechnical systems', in C. W. Churchman and M. Verhulst (eds) *Management Science, Models and Techniques*, London: Pergamon Press.

EMERY, F. E. and TRIST E. L. (1981) 'Sociotechnical systems', in Open Systems Group (eds) *Systems Behaviour*, London: Harper and Row.

FLOOD, R. L. and JACKSON, M. C. (1991) *Creative Problem Solving: Total Systems Intervention*, Chichester: John Wiley and Sons Ltd.

HACKMAN, J. R. (1986) 'The psychology of self-management in organizations', in M. S. Pallak and R. Perloff (eds) *Psychology and Work: Productivity, Change and Employment*, Washington, DC: American Psychological Association.

HAPP, M. B. (1993) 'Sociotechnical systems-theory – analysis and application for nursing administration. *Journal of Nursing Administration* **23**: 47–54.

HENDRY, C. (1991) 'Corporate strategy and training', in J. Stevens and R. MacKay (eds) *Training and Competitiveness*, London: Kogan Page, 79–110.

HERBST, P. G. (1954) 'The analysis of social flow systems', *Human Relations* **7**: 327–336.

HERBST, P. G. (1962) *Autonomous Group Functioning*. London: Tavistock Publications.

HO, J. (1996) 'School organizational health and teacher stress in Singapore', unpublished PhD thesis, University of Nottingham.

HOLLAND, J. (1985) 'Making vocational choices: a theory of careers', in S. Fisher and

J. Reason (eds) *Handbook of Life Stress, Cognition and Health*, Chichester: John Wiley.

HOWARTH, I (1995) 'Complexity theory and organizational health', *International Forum for Organizational Health Newsletter,* August 9–14.

HOY, W. K. and FELDMAN, J. A. (1987) 'Organizational health: the concept and its measure', *Journal of Research and Development in Education* **20**: 30–37.

HOY, W. K. and HANNUM, J. W. (1997) 'Middle school climate: an empirical assessment of organizational health and student achievement', *Educational Administration Quarterly* **33**: 290–311.

HOY, W. K., TARTER, C. J. and BLISS, J. R. (1990) 'Organizational climate, school health and effectiveness – a comparative analysis', *Educational Adminstration Quarterly* **26**: 260–279.

HOY, W. K. and WOOLFOLK, A. E. (1997) 'Teachers sense of efficacy and the organizational health of schools', *Elementary School Journal* **93**: 355–372.

JANIS, I. and MANN, L. (1977) *Decision Making: A Psychological Analysis of Conflict, Choice and Commitment,* New York: Macmillan.

JONES, P. M. (1995) 'Designing for operations – towards a sociotechnical systems and cognitive engineering approach to concurrent engineering', *International Journal of Industrial Ergonomics* **16**: 283–292.

KUK, G. (1993) 'Organisational healthiness, help seeking and the meaning of work: person-environment transaction', unpublished PhD thesis, University of Nottingham.

LEWIN, K. (1951) *Social Theory and Social Structure,* New York: Harper.

LIN, R.-F. (1994) 'Organizational healthiness, stress and well-being', unpublished PhD thesis, University of Nottinghamm.

MARCH, J. G. and SIMON, H. A. (1958) *Organizations,* New York: John Wiley.

MILES, M. (1965) 'Planned change and organizational health: figure and ground', in F. D. Carver and T. J. Sergiovanni (eds) *Organizations and Human Behaviour: Focus on Schools,* New York: McGraw-Hill.

MILLER, E. J. and RICE, A. K. (1967) *Systems of Organization: The Control of Task and Sentient Boundaries,* London: Tavistock Publications.

MILLER, W. L., CRABTREE, B. F., McDANIEL, R. and STANGE, K. C. (1998) 'Understanding change in primary care practice using complexity theory', *Journal of Family Practice* **46**: 369–376.

MURRAY, P. J. (1998) 'Complexity theory and the fifth discipline', *Systemic Practice and Action Research* **11**: 275–293.

OPEN SYSTEMS GROUP (1981) *Systems Behaviour,* London: Harper and Row.

PASMORE, W. A. and KHALSO, G. S. (1993) 'The contributions of Trist, Eric to the social engagement of social-science', *Academy of Management Review* **18**: 546–569.

PERROW, C. (1967) *Organizational Analysis: A Sociological View,* London: Tavistock.

PETERSON, M. and WILSON, J. (1998) 'A culture-work-health model: a theoretical conceptualization', *American Journal of Health Behaviour* **22**: 378–390.

PIDGEON, N. F. (1991) 'Safety culture and risk management in organizations', *Journal of Cross-Cultural Psychology* **22**: 129–140.

PORTER, L. W., LAWLER, E. E. and HACKMAN, J. R. (1975) *Behaviour in Organizations,* New York: McGraw-Hill.

RICE, A. K. (1963) *The Enterprise and its Environment,* London: Tavistock Publications.

SCHEIN, E. H. (1985) *Organizational Culture and Leadership,* San Francisco: Jossey-Bass.

SCOTT, R. W. (1981) *Organizations: Rational, Natural and Open Systems,* Englewood Cliffs, NJ: Prentice Hall International.

SCOTT, W. G. and MITCHELL, T. R. (1972) *Organization Theory,* Homewood, IL: Irwin-Dorsey.

SIMON, H. (1981) *Sciences of the Artificial,* Cambridge, MA: MIT Press.

SMEWING, C. (1997) An evaluation of the effects of an employee assistance programme: individual well-being and organizational healthiness, unpublished PhD thesis, University of Nottingham, UK.

SMEWING, C. and Cox, T. (1995) 'The effect of employee assistance on individual well-being and organizational healthiness', paper presented at the *European Congress of Work and Occupational Psychologists*, Gyor, Hungary.

SMEWING, C. and Cox, T. (1996) 'The organizational health of health care institutions in the United Kingdom', Proceedings of the IV Seminar on Organizational Psychology of Health Care, European Network of Organizational Psychologists, Munich.

SMEWING, C., Cox, T. and KUK, G. (1994) 'Employee assistance, organizational health and staff well-being in UK hospitals', paper presented at the BPS Occupational Psychology Conference.

TRIST, E. L. and BAMFORTH, K. W. (1951) 'Some social and psychological consequences of the longwall method of coal-getting', *Human Relations* **4**: 3–38.

TRIST, E. L. and MURRAY, H. (1948) *Work Organization at the Coal Face: A Comparative Study of Mining Systems*, Doc. No. 506, London: Tavistock Institute of Human Relations.

TRIST, E. L., HIGGIN, G. W., MURRAY, H. and POLLOCK, A. B. (1963) *Organizational Choice*, London: Tavistock Publications.

WALDROP, M. M. (1989) *Complexity*, New York: Simon and Schuster.

WARING, A. E. (1989) *Systems Methods for Managers: A Practical Guide*, Oxford: Blackwell.

WHITE, M., MARIN, D., BRAZEAL, D. and FRIEDMAN, W. (1997) 'The evolution of organizations: suggestions from complexity theory about the interplay between natural selection and adaptation', *Human Relations* **50**: 1383–1401.

WOODWARD, J. (1965) *Industrial Organization: Theory and Practice*, London: Oxford University Press.

CHAPTER TEN

Organizational-level interventions designed to reduce occupational stressors

RONALD J. BURKE AND ASTRID M. RICHARDSEN

Occupational stress research has been ongoing for the past twenty years and interest in this topic continues to increase (Cooper and Cartwright, 1996; Quick, Murphy and Hurrell, 1992; Cooper and Payne, 1988; Quick, Bhagat, Dalton and Quick, 1987). Several collections (e.g. Beehr & Bhagat, 1985; Cooper & Payne, 1980; Murphy, Hurrell, Sauter and Keita, 1995; Sauter and Murphy, 1985) and journals have appeared (e.g. *Stress at Work*; *Journal of Organizational Behavior*; *Journal of Occupational Health Psychology*; *Journal of Occupational and Organizational Psychology*; *Work and Stress*), which are devoted exclusively or substantially to occupational stress research findings. In addition, research findings in this area have increasingly been used to inform national policy making (*American Psychologist*, 1990a; *American Psychologist*, 1990b). For example, Levi (1990) recommends an overall international environmental and health programme in which interventions are systems-oriented, interdisciplinary, intersectorial, health-oriented, and participative. Sauter, Murphy and Hurrell (1990), at the National Institute for Occupational Safety and Health, present a comprehensive national strategy to protect and promote the psychological health of workers, while others (Keita and Jones, 1990) make recommendations for the role of psychologists in the area of occupational stress and workplace wellness. However, the research evidence to date on the success of large-scale organizational intervention programmes to reduce job stress remains sparse, despite the importance of such programmes and encouragement for their study. Interest in the implementation and evaluation of organizational interventions can be traced back to the early 1980s, and this chapter reviews what was learnt from the research that was conducted in the decade that followed.

During the 1980s, many occupational stress researchers came to advocate a person-environment fit model for understanding work stress (Harrison, 1985). According to this view, to understand the experience of work stress, one must consider the environment, both subjective and objective, that the individual is encountering (i.e. potential sources of occupational stress and their magnitude). Examples of some widely examined sources of occupational stress include role overload, role conflict, role ambiguity, job future ambiguity, underutilization of skills, and poor work relationships (Burke, 1988).

One must also consider stable individual difference characteristics and predispositions that influence both the nature and strength of occupational stressors that are perceived, coping resources and responses that are available and utilized, and emotional and physical well-being. The experiencing of occupational stress results from a person-environment interaction or trans-action, particularly instances of person-environment misfit. There is consider-able research activity guided by this model and some evidence for its validity.

The person-environment fit model also has implications for reducing the incidence of occupational stress. It follows that there are two broad approaches for minimizing the experience of work stress. One approach involves enhancing or augmenting the strengths of individuals. If individuals at work had more competence, resources and resilience, they would experience fewer adverse consequences from occupational stress. A considerable amount has been written along these lines (Ivancevich and Matteson, 1988; Murphy, 1996; Van der Hek and Plomp, 1997), and evidence that some individual-level interventions can make a difference in *temporarily* reducing adverse responses to perceived stressors (Murphy, 1988). Newman and Beehr (1979) provide an early review of individual-level coping strategies grouped into four categories.

1   Those aimed at psychological conditions:
    ● planning ahead, managing one's life,
    ● realistic assessment of self and one's aspirations.
2   Those aimed at physical or physiological conditions:
    ● diet,
    ● exercise,
    ● sleep.
3   Those aimed at changing one's behaviour:
    ● displaying less Type A behaviour,
    ● using relaxation techniques,
    ● taking time off for fun and holiday,
    ● developing close friendships for social support.
4   Those aimed at changing one's work environment:
    ● changing to a less demanding job,
    ● changing to a less demanding organization.

But there is also evidence that such interventions do not have a lasting effect. That is, once individuals encounter demands in the work setting, the benefits of these individual-level interventions disappear (Ganster, Mayes, Sime and Tharp, 1982).

Murphy (1987) reviewed the literature on worksite stress-management training and concluded that some of this did have a positive effect, but such effects diminished with the passage of time. He also felt that such training, isolated from other components of an integrated occupational health strategy, was of limited value.

A second approach involves reducing the sources and incidence of occupational stressors in the work environment. If one reduced the number and strength of occupational stressors, individuals at work would experience less stress. In addition, this approach would help a greater number of individuals simultaneously (Brief and George, 1991). There is also evidence that organizational-level interventions are more effective than individual-level interventions (Shinn, Rosario, Morch and Chestnut, 1984).

Organizational-level stress management intervention programmes refer to efforts initiated by the management of an organization that focus on specific job stressors intended to reduce negative outcomes and consequences (Ivancevich and Matteson, 1987). These include programmes designed to remove (or reduce) occupational stressors from the work environment and programmes that improve an individual's fit with his/her job and work environment. Newman and Beehr (1979) offer a comprehensive listing of organizational-level strategies grouped into three categories:

1   Changing organizational characteristics:
    ● change organizational structure,
    ● change organizational processes (e.g. reward systems, selection, training and development systems, socialization processes; job transfer and job rotation policies, more employee-oriented supervision),
    ● develop health services.
2   Changing role characteristics:
    ● redefine roles,
    ● reduce role overload,
    ● increase participation in decision making,
    ● reduce role conflict.
3   Changing task characteristics:
    ● design jobs in the light of workers' abilities and preferences,
    ● use workers' preferences in selection and placement,
    ● provide training programmes so that workers can be more skilled,
    ● individualized treatment of workers.

Occupational stress researchers often end their articles with a description of such interventions. Unfortunately, few have been systematically described and evaluated.

Reducing stress through actions aimed at work environment stressors, while being the most straightforward stress reduction intervention, can be very costly and difficult to implement in organizations (Murphy, 1988). Not only are organizations resistant to change, but there seems to be a prevalent belief in many organizations that the work environment contributes little to employee distress and that stress is a personal problem requiring worker-oriented solutions (Neal, Singer, Schwartz and Schwartz, 1984). A survey of management and workers regarding work stress indicated employees identified physical and environmental stressors (e.g. poor ergonomic design of work), as well as psychosocial stressors (e.g. lack of control over work content and process, unrealistic task demands, lack of understanding by supervisors and management) as important sources of stress. Corporate respondents placed emphasis on personality traits, lifestyle behaviours, interpersonal relationships, and familial problems as prominent sources of stress. This may be one reason for the popularity of individual stress management/health programmes offered in work settings. The over-emphasis on individually-oriented stress management interventions may have slowed down efforts by organizations to alter work conditions that contribute to stress (Ivancevich and Matteson, 1987).

This chapter will review eleven studies of organizational-level interventions in an attempt to evaluate current knowledge of their effectiveness and to encourage researchers to design scientific studies to implement and evaluate organizational-level stress management intervention programmes.

Organizational-level interventions to reduce work stressors

There have been several organizational-level interventions designed to reduce levels of work stress experienced by managers and professionals. This trend is expected to increase in keeping with the steadily accumulating knowledge of common and important sources of work-stressors. Let us now consider examples of each of these interventions.

Reducing Role Stress Through Goal Setting

Quick (1979) conducted a field study over a fourteen-month period to examine whether goal setting would serve as a role-making technique for reducing role stress. The study was carried out in the largest division of a nationally based insurance company. Fifteen executive officers and forty-six of their immediate staff participated in the study. Three data collection points were used: six months before formal training in goal setting, five months after training, and eight months after training. The training consisted of a one-day (eight-hour) session. Three dimensions of goal setting were emphasized: task goal properties (difficulty, clarity), supervisory goal behaviours (feedback quality, feedback amount), and subordinate goal behaviour (participation). The data indicated significant improvements on these dimensions of goal setting at the five-month period and significant improvements in the eighth month after training (except on task goal properties).

Role stress was assessed through measures of two role stressors (role conflict and role ambiguity) and absenteeism. The data indicated significant declines in role conflict and role ambiguity and declines in absenteeism that almost reached statistical significance. Significant reductions in role conflict and role ambiguity were present in the eighth month following training but there was no significant difference in absenteeism rates.

Reducing Role Stress by Increasing Participation in Decision Making

Jackson (1983) proposed a causal model to describe the effect of participation in decision making on perceived influence, role conflict and role ambiguity, social support, personal and job-related communications and several individual and organizational outcome variables. Lack of participation in decision making is a primary causal determinant of role conflict and role ambiguity, mediated by perceived influence and job-related communication. To establish the causal effects of participation in decision making, a longitudinal field experiment using a Solomon four-group design with two post-tests (at three and six months) was employed.

Participants were randomly assigned to intervention and control conditions. The treatment was the introduction of frequently held staff meetings. The study was conducted in an out-patient facility associated with a university hospital. Participation was increased by requiring unit supervisors to hold scheduled staff meetings at least twice per month. All nursing supervisors attended a two-day training workshop six weeks before the introduction of the intervention. The data indicated that, after six months, participation had a significant negative effect on role conflict and role ambiguity and a positive effect on perceived influence. Role conflict and role ambiguity were, in turn, positively related to emotional strain and negatively related to job satisfaction. Emotional strain was positively related to absence frequency and intention to quit. Perceived influence was positively related to job satisfaction and intention to quit. Thus, it is likely that participation in decision making was an important causal determinant of role conflict and role ambiguity which were, in turn, determinants of employee attitudinal and behavioural outcomes. Significant differences in role conflict and role ambiguity were found between participants whose work units held few or many staff meetings.

Increased Job Autonomy

Wall and Clegg (1981) undertook a work redesign study that evaluated job characteristics and emotional distress following an intervention designed to increase worker autonomy. Participants in the research were workers in one department of an organization believed by management to have low morale, low motivation and poor productivity. Their approach to the problem used action research, socio-technical systems theory and Hackman and Oldham's job characteristics model. The nature of the jobs in the department dictated a group redesign.

The study involved three phases: problem assessment, implementation of the work redesign, and both short- (six months) and long-term (eighteen months) follow up of the intervention. The problem assessment indicated low scores on three job characteristics (autonomy, feedback and task identity). The researchers also noted low levels of internal work motivation, low job satisfaction and high levels of emotional distress. The work redesign increased group autonomy, group task identity and group feedback. Increases in group autonomy involved shifting responsibility and control from the supervisors to the work teams. Each work team was given control over the pace of work, organization of rest breaks, and allocation of work assignments and overtime. The role of the supervisor was changed to provide support with daily operations.

Significant increases in group autonomy and task identity, but not on group feedback, were present at the six-month evaluation point. The long-term follow up (eighteen months) showed, once again, increases in autonomy and task identity, but not beyond the levels present at the six-month point. Emotional distress was significantly reduced at both follow-up periods. The sound experimental design, though not without limitations, provided substantial support for the conclusions of the benefits of improved autonomy and task identity on employee emotional health.

Work Schedule Autonomy

Pierce and Newstrom (1983) considered six features of flexible work schedules (core minutes, band width, band length, schedule flexibility, schedule variability and supervisory change approval). Since these variables were themselves significantly intercorrelated, two composite scales were created which reflected schedule flexibility and discretionary time. Organizational commitment, psychological stress symptoms, absenteeism, work performance and perceived time autonomy served as outcome measures.

The participants were 188 employees from the word processing or records maintenance departments in five multiple line insurance companies, working under eight different work schedules. The flexible work schedule systems ranged from a staggered start system where employees could exercise a choice of starting times four times a year to a system where there was daily free choice of starting time. In each case these flexible schedules had been in operation for at least two years.

Correlational analysis showed that psychological stress symptoms were significantly correlated with all work-schedule features except band width, most highly with perceived time autonomy, supervisory change approval, schedule variability and discretionary time. Multiple regression analyses indicated that absenteeism, work performance and organizational commitment were the only variables related to the six work-schedule features. When perceived time autonomy was added to the model, psychological stress symptoms and job satisfaction became significant. The authors concluded that perceived time autonomy moderated the relationship between flexible work schedules and worker attitudes and response.

Improving Person-Environment Fit Using Problem-Solving Groups

Campbell (1973) designed an intervention for work teams at the Goodard Manned Space Flight Center of NASA in which group members received data on their person-environment fit in relation to various occupational stressors. Subsequently, these data were used by work groups in a series of participative meetings intended to solve problems concerning their work situation in such a way that their person-environment fit would be improved. It was hypothesized that improvements in person-environment fit would show up in improved standings on risk factors of coronary heart disease.

The study was a field experiment with experimental and control groups. To measure the effects of the intervention, questionnaires were administered to the groups just prior to the intervention, immediately after the intervention, and three months after the intervention. These questionnaires contained measures of occupational stressors as well as of the psychological and physiological strains being investigated.

The central principles upon which the intervention programme was based were participation and data feedback. Members of each experimental work group were convened in a series of ten weekly meetings to identify and solve problems that the members felt were stressful in their work setting and to devise ways of improving person-environment fit. Problem solving at the sessions was improved by feeding back the data collected before the intervention on all stressors for every member of the group. A resource person was present in each experimental group to work with the supervisor in leading the session. The results of the study showed that the intervention did not have a demonstrably beneficial effect on either occupational stresses or individual strains. But, in fact, the intervention was not carried out as anticipated. First, each group was allowed to direct the course of its own programme, i.e. select for discussion those problems it felt were most stressful in the group's own work environment. Thus, while the process of the programme in each group was similar, the content areas were not. In addition, the groups spent little time actually discussing person-environment fit with respect to participation and interpersonal relations. Second, feedback from the initial person-environment fit questionnaires was given without revealing individual sources of the data, which may have affected the discussion of person-environment fit in the groups. The effect of feedback may have been more powerful if the groups had known the sources. We conclude that the concept Campbell planned was, and still is, an interesting one in need of more appropriate examination.

Reducing Psychological Burnout by Changing Orientation Practices

The Children's Aid Society (CAS) of Metropolitan Toronto developed an interesting intervention to reduce burnout among entering frontline social workers in child welfare (Falconer and Hornick, 1983). This programme was radically different in many ways from their traditional orientation practices. It involved hiring frontline workers in batches; keeping these newly hired workers in small groups of five or six individuals for their first six months of employment, gradually increasing their caseload so that it eventually reached 60 per cent of the normal caseload by the six-month point; an enhanced supervisor's role emphasizing education (accompanied by a reduction of other supervisory duties); an improved training programme (one to two days of training every two weeks); increased social support for the group; and attempts to deal with stressors found to be associated with burnout in previous research (e.g. promoting clear, consistent and specific feedback, clarifying rules, policies and roles, allowing for autonomy and innovation).

The research employed a pre-test extended post-test design with a non-equivalent comparison group. The project, designed to run for two-and-a-half years, was cut short because of budget cutbacks resulting from the general economic recession. The resulting sample sizes rendered the quantitative data of little use. Qualitative data indicated general satisfaction with the programme and beliefs that the goals of the programme were realized. Supervisors felt that frontline workers in the programme achieved a level of skill during the six months comparable to that achieved in one year under the traditional orientation programme.

Reducing Work-Family Conflict

Renshaw conducted two experiential problem-solving seminars designed to reduce work-family conflict. In the first study (Culbert and Renshaw, 1972), members of a research unit whose responsibilities required them to travel a great deal during the summer to test products under field conditions took part. Most of them had expected heavy travel schedules when they joined the unit but the complex problems of heavy travel surprised them all. Family problems arose from disconnected family relationships, lack of social life, shifts in responsibilities during absences, and needs for emotional support. The job incumbents also found that the travel resulted in their feeling tired, lonely and guilty to the point of depression about being away from home.

A seminar with three primary objectives was designed to address the problems of travel: participation in the seminar would improve problem-solving resources of the husband and wife team, focusing these resources on the couple's problems with business travel would strengthen each person's capacity to cope with separation, and new and strengthened resources for coping would be generalized to other organizational issues and problems. The format for the seminar consisted of a two-day workshop in which couples engaged in activities to strengthen their analytical and problem-solving skills. Data was collected through interviews and questionnaires. Later the same questionnaires were submitted to a control group of travellers from other units of the company, and their wives.

The empirical results were derived from analysis of two groups of six couples. First, the data revealed that those taking part in the study felt better able to cope with travel stresses after participating in the seminar. This was reflected in respondents' self-reported behaviours in relation to twelve norms hypothesized to characterize successful problem-solving relationships (e.g. awareness, consensual decision making). In addition, the majority of seminar participants showed a decrease in the rated intensity of travel stress following the workshop.

Renshaw (1975) conducted another workshop, involving only wives, which dealt with the effects of being temporarily transferred to the United States from another country for periods of one or two years. Renshaw concluded that the key ingredient in both workshops was the amount of perceived influence individuals came to believe they had over the events in a stressful situation.

Reducing the Stress of Staff Reductions

Greenhalgh (1983) described and assessed an action research project to alter an organization's procedures for achieving work force reductions in response to declining need for its services. The organizations were state-supported hospitals, that historically followed state guidelines which advocated work-force reductions through forced lay-offs. A cost-benefit study compared two alternative workforce reduction strategies (lay-off and planned attrition) and found that the lay-off strategy was less cost-effective.

An opportunity to use an attrition programme arose when the state decided to consolidate three urban hospitals. An agreement was made to accomplish the consolidation through transfer and attrition rather than the simple use of lay-off. Thus, potentially affected employees were guaranteed that they would not be summarily laid off. The consolidation plan provided that each employee was to be given at least one opportunity for continued state employment in an equivalent position within reasonable commuting distance. Such opportunities were arranged by coordinating the hiring within the urban area by various state agencies. The timing of the consolidation plan was arranged to permit the orderly transfer of employees who did not retire or voluntarily leave state service. A mechanism to retrain some workers was also established.

Evaluation data were collected from employees transferred from the two closing hospitals (one hospital remained open). Measures of job security, productivity and turnover propensity were examined. Employees in the demonstration project were similar to control employees on the last two measures, but lower on job security. In addition, employees in the hospital that remained open were relatively neglected in the demonstration. They reported lower job security, lower productivity and greater turnover propensity.

Reducing the Stress of Mergers and Acquisitions

Blake and Mouton (1984) provide a detailed case study of the application of their interface conflict-solving model to the merger of two organizations. An American company had acquired a British company; both had previously been competitors. The intervention included top teams of both organizations and involved a series of day-long meetings. The process used perception sharing, listing identification of concerns and questions, and the development of a sound operating model by groups from both organizations. Qualitative data collected two years later suggested that the merger was a success. No senior personnel had left either organization.

Schweiger and DeNisi (1991) report the results of a longitudinal field experiment examining the effects over time of a communications programme (a realistic merger preview). Data were collected in two plants, an experimental plant in which the preview was introduced and a control plant in which the merger was managed more traditionally. They collected data at four points in time: before the impending merger was announced, following the announcement of the merger but before the implementation of the preview in the experimental plant, three days after the introduction of the merger preview, and four months later.

Both plants were engaged in light manufacturing and were part of the same large company which was being merged with another. The two plants were located in different parts of the US, but were producing the same products and had the same number of employees, management structures and systems, personnel policies, and volume of output. Data were collected from questionnaire surveys. The experimental plant had 126 employees, and the control plant had 146; a smaller number completed questionnaires. Employees in both plants received a letter from the Chief Executive Officer (CEO) informing them of the merger.

The realistic merger preview had several components. Information of several kinds was provided to employees on how the merger would affect them. Employees received information and answers to questions concerning lay-offs, transfers, promotions and demotions, and changes in pay, jobs and benefits that would occur in their work groups. This information was conveyed in various ways. These included a merger newsletter, a telephone hotline during working hours which provided updated information, and weekly meetings of the plant manager with each of the plant's departments. In addition, the plant manager met individually with all individuals affected by a particular decision that was made.

A variety of variables were examined including perceived uncertainty, global stress, job satisfaction, organizational commitment, intention to quit, work performance, absenteeism, turnover, and perceptions of the company's trustworthiness, honesty and caring.

The announcement of the merger had a negative effect in both plants (increase in global stress, perceived uncertainty and absenteeism; decrease in job satisfaction, commitment and perceptions of trustworthiness, honesty and caring). The merger preview was found to reduce the negative effects of the merger, both immediately and throughout the length of the study. Interestingly, some of these measures returned to pre-merger stage levels in the experimental plant.

Reducing Psychological Burnout through High-Stimulus Organizational Development Interventions

Golembiewski, Hilles and Daly (1987) reported an orgnizational development effort using theory-driven interventions with the thirty-one member human resource staff of a single organization. The OD approach seeks to make the workplace more democratic and humane, and in this case, the staff were directly involved in identifying and solving organizational problems that contributed to burnout. The researchers used Golembiewski's phase model to assess levels of psychological burnout and also measured various aspects of the work environment, including job involvement, peer cohesion, supervisor support, autonomy, task orientation, task clarity, control, innovation, and physical comfort. Archival data were used to assess turnover rates. Data were gathered five times over a two-year period. The group initially scored high on burnout, work pressures and turnover.

A traditional organizational development approach of high stimulus interventions involving problem sensing, data collection and feedback, formation of interest groups, and action planning was used. The level of burnout was reduced and remained lower for at least four months after the last planned intervention. This improvement diminished somewhat after a further nine months and a major reorganization. Improvements in group cohesiveness and in rate of turnover persisted, and even improved, over this time period. The findings indicated that using high-stimulus designs with burnt out individuals in an active mode of adaptation (i.e. who reported high job involvement and task orientation) was effective. Conventional organizational development interventions may not be appropriate for those in a passive and withdrawn mode of adaptation to burnout.

Methodological considerations

As a group, these studies generally indicate the success of organizational-level interventions in reducing stressors. However, there are several methodological issues that pertain to these studies. Some of these issues are relevant to stress research in general and have been outlined by a number of researchers (Burke, 1987; Ivancevich and Matteson, 1987; Murphy, 1987, 1988; Frese and Zapf, 1988).

One issue concerns the relationship between stressors at work and ill health. The research literature on work stress has no shortage of reports of significant relationships between work stressors and various stress reactions. However, the results are as a rule based on correlational evidence which does not make it possible to determine causality (Burke, 1988). As Frese and Zapf (1988) have pointed out, organizational approaches to stress management only make sense if objectively assessed stressors at work are shown to produce ill-health among workers. Other researchers (Invancevich and Matteson, 1987) have called attention to the need to conduct a proper diagnosis of the stressor-stress reaction relationship before planning and implementing an intervention. The prevalence of stress among employees as well as the contribution of organizational stressors need to be established before decisions are made about the kind of stressors that can be efficiently modified in order to reduce or minimize stress. None of the studies reviewed here included such an assessment strategy, and it is difficult to evaluate the extent to which interventions were relevant to the stress reactions measured or to which they targeted the most salient stressors in the workplace.

The studies reviewed in this chapter were mostly field studies. Very few employed an experimental-level design with random assignment of workers to groups and the inclusion of a control or comparison group. Positive results in the form of outcomes such as reduced stress symptoms, role conflict, absenteeism and turnover, or increased job satisfaction, work performance, coping skills and organizational commitment, may therefore not be attributed directly to intervention-specific effects. There may be a number of non-specific factors, intervening or moderating variables that were not measured but could affect the results. Several researchers have already addressed the need for experimental or quasi-experimental designs using appropriate control groups in order to isolate training-specific effects (Murphy, 1987; Ivancevich and Matteson, 1987).

The interventions used in the studies have targeted various aspects of the organization, from orientation practices, to work redesign to increase autonomy, to programmes designed to deal with the stress of mergers. While this diversity may indicate the power of organizationally oriented interventions to reduce individual stress reactions, it makes it difficult to generalize findings from any one study. There is not enough evidence to draw other than conservative conclusions about the efficacy of any one specific intervention strategy. A related issue concerns the duration of interventions. The length of the interventions varies from one study to the next, ranging from one-day seminars to several months of worker participation. In addition to creating a problem in comparing results between studies, this may also have implications for cost-benefit considerations. For instance if brief training sessions are shown to be as effective as more costly long-term intervention strategies, this could constitute considerable savings for the organization wanting to implement stress-reduction programmes.

Several issues concern the dependent measures used in these studies. First, evaluations of effectiveness are difficult because widely different measures were used in the studies. Some have focused on increased employee health, while others have measured improvements in organizational outcomes, such as job satisfaction, employee morale, or productivity. Many authors have pointed to persistent confusion in the stress research literature about what is being measured (Burke, 1987; Frese and Zapf, 1988). There is still disagreement among researchers concerning the definition and conceptualization of work-stress in organizations. While some progress has been made in terms of standardization of measures, there is still work to be done in order to reach consensus on issues of conceptualization. Thus, it is unclear which measures are most appropriate for evaluation of organizational-level interventions.

Second, the studies almost entirely employ self-report measures, and observed relationships may be artifacts of method similarity and content overlap in the measures. However, a positive sign is that several studies have included objective organizational outcomes such as absenteeism and turnover. Future research designs and methodologies need to incorporate multiple outcome and consequence measures in order to increase our understanding of intervention-consequence mechanisms and to identify useful models of interventions (Ivancevich and Matteson, 1987).

Other issues concern the populations studied and longitudinal design. The populations included in these studies were mostly small and were either organizationally or occupationally specific. Thus, it is not clear whether results are representative of the general population in the workforce. Perhaps it is time to design scientific research methodologies to test some of the large-scale occupational health strategies that have been proposed (e.g. Levi, 1990; Sauter et al., 1990).

While many of the studies reviewed incorporate longitudinal designs, a question often raised in the stress literature is what constitutes longitudinal research (Burke, 1987; Ivancevich and Matteson, 1987; Frese and Zapf, 1988). Little is known about the optimal time span required to permit changes in outcomes and consequences to occur. Data collection after three months, six months or even a year may not be long enough. It is important for future researchers to specify reasonable time spans for data collection in order to create an empirical basis for comprehensive theories of stress management.

Organizational-versus individual-level interventions

The importance of organizational-level interventions aimed at environmental sources of professional and managerial stress, rather than individual-level interventions, emerges from a field experiment conducted by Ganster, Mayes, Sime and Tharp (1982). They evaluated a stress management training programme in a field experiment involving ninety-nine public agency employees randomly assigned to treatment (N = 60) and control (N = 39) groups. The programme consisted of sixteen hours of training spread over eight weeks. Participants were taught progressive relaxation and cognitive restructuring techniques. Dependent variables were excretion of the hormone epinephrine and the neurotransmitter norepinephrine at work, anxiety, depression, irritation, and somatic complaints, all measured at three points in time (pre-test, post-test and four months after treatment). Treatment participants exhibited significantly lower epinephrine and depression levels than did controls at the post-test; and four-month follow-up levels did not regress to initial pre-test levels. However, the treatment effects were not found in a subsequent intervention on the original control group.

Shinn, Rosario, Morch and Chestnut (1984) conducted research and came to the same conclusion. They collected data from 141 human service workers using questionnaires assessing job stressors, coping strategies and various aspects of strain (alienation, satisfaction, symptomatology). Coping (efforts to reduce stressors and strain) was assessed at three levels: by individual workers, by workers grouping together to help one another (social support), and by their employing human service agencies. Although many more individual coping responses were mentioned than group or agency-initiated responses, only the group responses were associated with low levels of strain. Unfortunately, insufficient agency-initiated responses were identified for a meaningful analysis to be undertaken. Thus, in the work setting individual coping responses may be less useful than higher-level strategies involving groups of workers or entire units or organizations. Pearlin and Schooler (1978) also conclude that chronic, organizationally generated stressors may be resistant to reduction through individual coping efforts.

Ivancevich and Matteson (1987) made the point that organization-based stress management intervention programmes incorporating well designed evaluations have rarely been undertaken. They (along with Murphy, 1987) offered suggestions for increasing researcher interest in scientifically designing, implementing and evaluating organizational-level stress management intervention programmes. Murphy (1988) also found few well designed evaluations of interventions aimed at reducing work stressors. Those he identified were found to consistently show benefits. Yet stressor reduction represents the most direct way to reduce stress since it deals with the source. This has been termed 'primary prevention' in the stress management literature.

While the studies reviewed in this chapter generally give support for the benefits of organizational-level interventions, methodological limitations suggest that research findings should be treated cautiously. It is still not clear what organizational-level stress management interventions can do in addition to or instead of individual-level interventions (Ivancevich and Matteson, 1987). This has led some researchers (Murphy, 1987; Ivancevich, Matteson, Freedman and Phillips, 1990) to suggest that an approach which includes both individual and organizational level interventions would be more powerful or effective. For example, Ivancevich et al. (1990) developed a framework for workplace stress management interventions. This framework proposes three points of intervention: changing the stress potential in a situation by reducing the intensity and number of stressors, helping individuals modify their perception or appraisal of potentially stressful situations, and helping individuals cope more effectively with their responses to stressful situations.

Some support for this point of view also comes from the burnout literature. Leiter has proposed and researched a process model of burnout that gives a central role to the conflict between personal aspirations and organizational limitations. The stressors inherent in public service work (e.g. work overload and dull routine, interpersonal conflicts with co-workers and administrators, difficult client contacts, and lack of autonomy and decision making) contribute to burnout to the degree that the organizations lack supportive resources and are unresponsive to the concerns of their employees. What is needed to prevent burnout are ways in which to empower professionals and make human service organizations more responsive to their aspirations. Although Leiter's later writings have advocated organizational change, it seems from his model that a combination of organizational and individual interventions may be employed to prevent and alleviate burnout.

The model suggests that in order to prevent the development of burnout, it is necessary to target the stressors that directly influence emotional exhaustion. This might include adjusting the workload to make it more manageable, reallocating work tasks to create more variety in the job, and designing ways to decrease interpersonal conflict. In addition, increased sense of competence through appropriate training and skill utilization, as well as training in appropriate coping skills, may increase self-efficacy expectations and reduce stress. For individuals experiencing advanced burnout, it may be useful to include other interventions such as increasing autonomy through participative decision making, providing opportunities for consultation about difficult client contacts, training supervisors to provide support and encouragement, and enhancing collegiality through opportunities for both formal and informal staff contacts. This could be accomplished through a combination of organizational and individual interventions. Two examples of organizational level interventions are the formation of groups to solve problems around interpersonal conflicts; and action planning by staff groups to deal with issues of workload, job design, and difficult client problems. Individual staff counselling and cognitive training to enhance efficacy expectations and coping skills may also be useful in alleviating advanced burnout.

Research and practice – two cultures?

A large gap exists between work-stress researchers and practitioners – the producers and the consumers of stress research findings. The most obvious illustrations of this gap can be seen in little awareness of research findings by practitioners (managers, consultants, clinicians), little intervention activity being undertaken at the organizational level, little research being undertaken to determine the effectiveness of individual-level interventions, and only modest use of work-stress research findings for intervention and policy development. Bridging this gap would appear to be particularly important for readers of this book. The result would be better informed research *and* practice.

This gap is not unique to the area of work stress, however (Kilman, 1983). The field of organizational behaviour is examining ways of conducting research that are useful for both theory and practice (see *Administrative Science Quarterly* 27: 1982; *Administrative Science Quarterly* 28: 1983; Lawler, Mohrman, Ledford and Cummings, 1985). Both researchers and consumers of research findings would benefit from an examination of this literature. It is possible to combine research and intervention in the area of work stress (Gardell, 1982) to produce findings of value in forming policy and improving the quality of work life (Kahn, 1987).

Summary

Occupational stress research continues to grow and is increasingly being used to inform national policy on worker health and well-being. Most models of occupational stress emphasize notions of person-environment fit. Two approaches for intervention to minimize adverse consequences of occupational stress follow from such models. One emphasizes the strengthening and enhancement of individuals and their resources, the other emphasizes the reduction of common work-place based sources of occupational stress. This Chapter has reviewed eleven organizational-level interventions aimed at reducing stress at work that have been examined in field studies. These interventions were generally found to have positive effects, and given the limited success of individual-level interventions in addressing occupational sources of stress, should be encouraged.

Note

Preparation of this manuscript was supported in part by the Faculty of Administrative Studies, York University. We would like to thank Jacob Wolpin for his help.

References

ADMINISTRATIVE SCIENCE QUARTERLY (1982) Special Issue, Part I: The utilization of organizational research **27**: 588–685.

ADMINISTRATIVE SCIENCE QUARTERLY (1983) Special Issue, Part II: The utilization of organizational research **28**: 63–144.

AMERICAN PSYCHOLOGIST (1990a) Special Issue: Organizational psychology **45**: 2.

AMERICAN PSYCHOLOGIST (1990b) Psychology in the public forum **45(10)**: 1137–1166.

BEEHR, T. A. and BHAGAT, R. S. (1985) *Human Stress and Cognition in Organizations*, New York: John Wiley.

BLAKE, R. R. and MOUTON, J. S. (1984) *Solving Costly Organizational Conflicts*, San Francisco: Jossey-Bass.

BRIEF, A. P. and GEORGE, J. M. (1991) 'Psychological stress and the workplace: a brief comment on the Lazarus outlook', *Journal of Social Behavior and Personality*, forthcoming.

BURKE, R. J. (1987) 'The present and future status of stress research', in J. M. Ivancevich and D. C. Ganster (eds) *Job Stress: From Theory to Suggestion*, New York: The Hayworth Press, 249–267.

BURKE, R. J. (1988) 'Sources of managerial and professional stress in large organizations', in C. L. Cooper and R. Payne (eds) *Causes, Coping and Consequences of Stress at Work*, New York: John Wiley, 77–114.

CAMPBELL, D. (1973) 'A program to reduce coronary heart disease risk by altering job stressors', unpublished doctoral dissertation, University of Michigan, Ann Arbor.

CAPLAN, R. D., COBB, S., FRENCH, J. R. P. Jr., HARRISON, R. V. and PINNEAU, S. R. Jr. (1975) *Job Demands and Worker Health*, Washington, DC: US Government Printing Office.

COOPER, C. L. and CARTWRIGHT, S. (1996) *Mental Health and Stress in the Workplace: a Guide for Employers*, London: HMSO.

COOPER, C. L. and PAYNE, R. (1980) *Current Concerns in Occupational Stress*, New York: John Wiley.

COOPER, C. L. and PAYNE, R. (1988) *Causes, Coping and Consequences of Stress at Work*, New York: John Wiley.

CULBERT, S. A. and RENSHAW, J. R. (1972) 'Coping with the stresses of travel as an opportunity for improving the quality of work and family life', *Family Process* **11**: 321–337.

FALCONER, N. E. and HORNICK, J. P. (1983) *Attack on Burnout: The importance of early training*, Children's Aid Society of Metropolitan Toronto.

FRENCH, J. R. P. Jr. and CAPLAN, R. D. (1972) 'Occupational stress and individual strain', in A. J. Marrow (ed.) *The failure of success*, New York: Amacom.

FRESE, M. and ZAPF, D. (1988) 'Methodological issues in the study of work stress: objective vs subjective measurement of work stress and the question of longitudinal studies', in C. L. Cooper and R. Payne (eds) *Causes, Coping and Consequences of Stress at Work*, New York: John Wiley, 375–411.

GANSTER, D. C., MAYES, B. T., SIME, W. E. and THARP, G. D. (1982) 'Managing occupational stress: a field experiment', *Journal of Applied Psychology* **67**: 533–542.

GARDELL, B. (1982) 'Scandinavian research on stress in working life', *International Journal of Health Services* **9**: 31–40.

GOLEMBIEWSKI, R. T., HILLES, R. and DALY, R. (1987) 'Some effects of multiple OD interventions on burnout and worksite features', *Journal of Applied Behavioral Science* **23**: 295–313.

GREENHALGH, L. (1983) 'Managing the job insecurity crisis', *Human Resources Management* **4**: 431–434.

HARRISON, R. V. (1985) 'The person-environment fit model and the study of job stress', in T. A. Beehr and R. S. Bhagat (eds) *Human Stress and Cognition in Organizations*, New York: John Wiley, 23–56.

IVANCEVICH, J. M. and MATTESON, M. T. (1987) 'Organizational level stress management interventions: a review and recommendations', in J. M. Ivancevich and D. C. Ganster (eds) *Job Stress: From Theory to Suggestion*, New York: Howarth Press, 229–248.

IVANCEVICH, J. M. and MATTESON, M. T. (1988) 'Promoting the individual's health and well-being', in C. L. Cooper and R. Payne (eds) *Causes, Coping and Consequences of Stress at Work*, New York: John Wiley, 267–299.

IVANCEVICH, J. M., MATTESON, M. T., FREEDMAN, S. M. and PHILLIPS, J. S. (1990) 'Worksite stress management interventions', *American Psychologist* **45**: 252–261.

JACKSON, S. E. (1983) 'Participation in decision making as a strategy for reducing job-related strain', *Journal of Applied Psychology* **68**: 3–19.

KAHN, R. L. (1987) 'Work stress in the 1980's: research and practice', in J. C. Quick, R. S. Bhagat, J. E. Dalton and J. D. Quick (eds) *Work Stress: Health Care Systems in the Workplace*, New York: Praeger, 311–320.

KEITA, J. P. and JONES, J. M. (1990) 'Reducing adverse reactions to stress in the workplace: psychology's expanding role', *American Psychologist* **45**: 1137–1141.

KILMAN, R. H. (1983) *Producing Useful Knowledge for Organizations*, New York: Praeger.

LAWLER, E. E., MOHRMAN, A. M., MOHRMAN, S. A., LEDFORD, G. E. and CUMMINGS, T. G. (1985) *Doing Research that is Useful for Theory and Practice*, San Francisco: Jossey-Bass.

LEITER, M. P. (1990) 'The impact of family resources, control coping, and skill utilization on the development of burnout: a longitudinal study', *Human Relations* **43**: 1067–1083.

LEVI, L. (1990) 'Occupational Stress – spice of life or kiss of death', *American Psychologist* **45**: 1142–1145.

MURPHY, L. R. (1987) 'A review of organizational stress management research: methodological considerations', in J. M. Ivancevich and D. C. Ganster (eds) *Job Stress: From Theory to Suggestion*, New York: Howarth Press, 215–227.

MURPHY, L. R. (1988) 'Workplace interventions for stress reduction and prevention', in C. L. Cooper and R. Payne (eds) *Causes, Coping and Consequences of Stress at Work*, New York: John Wiley, 301–339.

MURPHY, L. R. (1996) 'Stress management in work settings: a critical review of the health effects', *American Journal of Health Promotion* **11**: 112–135.

MURPHY, L. R., HURRELL, J. J. Jr., SAUTER, S. L. and KEITA, G. P. (1995) *Job Stress Interventions*, Washington, DC: American Psychological Association.

NEALE, M. S., SINGER, J. A., SCHWARTZ, G. A. and SCHWARTZ, J. (1984) 'Conflicting perspectives on stress reduction in occupational settings: a systems approach to their resolutions', *Report to NIOSH on P.O. No. 82-1058*, Cincinnati, Ohio: National Institute of Occupational Safety and Health.

NEWMAN, J. D. and BEEHR, T. (1979) 'Personal and organizational strategies for handling job stress: a review of research and opinion', *Personnel Psychology* **32**: 1–43.

PAYNE, R. (1988) 'Individual differences in the study of occupational stress', in C. L. Cooper and R. Payne (eds) *Causes, Coping and Consequences of Stress at Work*, New York: John Wiley, 209–232.

PEARLIN, L. and SCHOOLER, C. (1978) 'The structure of coping', *Journal of Health and Social Behavior* **19**: 2–21.

PIERCE, J. L. and NEWSTROM, J. W. (1983) 'The design of flexible work schedules and employee responses: relationships and processes', *Journal of Occupational Behavior* **4**: 247–262.

QUICK, J. C. (1979) 'Dyadic goal setting and role stress in field study', *Academy of Management Journal* **22**: 241–252.

QUICK, J. C., BHAGAT, R. S., DALTON, J. E. and QUICK, J. D. (1987) *Work Stress: Health Care Systems in the Workplace*, New York: Praeger.

QUICK, J. C., MURPHY, L. R. and HURRELL, J. J., Jr. (1992) *Stress and Well-Being at Work: Assessments and Interventions for Occupational Mental Health*, Washington, DC: American Psychological Association.

RENSHAW, J. R. (1975) 'An exploration of the dynamics of the overlapping worlds of work and family', *Family Process* **14**: 143–165.

SAUTER, S. L., MURPHY, L. R. and HURRELL, J. J., Jr. (1990) 'Prevention of work-related psychological disorder: a national strategy proposed by the National Institute for Occupational Safety and Health', *American Psychologist* **45**: 1146–1158.

SAUTER, S. L. and MURPHY, L. R. (1995) *Organizational Risk Factors for Job Stress*, Washington, DC: American Psychological Association.

SCHWEIGER, D. M. and DENISI, A. A. (1991) 'Communication with employees following a merger: a longitudinal field experiment', *Academy of Management Journal* **34**: 110–135.

SHINN, M., ROSARIO, M., MORCH, H. and CHESTNUT, D. E. (1984) 'Coping with job stress and burnout in the human services', *Journal of Personality and Social Psychology* **46**: 864–876.

VAN DER HEK, H. and PLOMP, H. N. (1997) 'Occupational stress management programmes', *Occupational Medicine* **47**: 133–141.

WALL, T. O. and CLEGG, C. W. (1981) 'A longitudinal study of group work redesign', *Journal of Occupational Behavior* **2**: 31–49.

CHAPTER ELEVEN

Coping with the stress of new organizational challenges: the role of the employee assistance programme

JOHN R. BERRIDGE AND CARY L. COOPER

Introduction

In the advanced industrialized economies, up to the 1980s the dominant organizational model in the business and public service sectors alike was the engineering or bureaucratic model of the smooth-running machine or system. Managers saw their role as maintaining and improving the organization's mechanisms and procedures, and if necessary 'trouble-shooting' breakdowns in the system. Should such breakdowns present a persistent pattern, then managers' response was to devise a routine or technique which restored organizational functioning to a steady state, if not to the original state as far as possible. Subsequently, managers sought to re-establish control and to achieve a degree of accuracy in forecasting future organizational operation and performance (Mintzberg, 1979). In other words, they followed Harold Geneen's famous dictum 'No surprises!' which as Chief Executive Officer (CEO) he applied draconianally at International Telephone and Telegraph (ITT), the international conglomerate (Sampson, 1973). The underlying assumption of such an organizational paradigm was that the controlled and predictable organizational machine could be created, and that it would deliver high-quality performance in economic terms. Not only that, such an organization would eventually remove the pressure upon managers arising from operational crises, breakdowns and malfunctions, as a result of quasi-omniscient control and comprehensive accurate forecasting. The evidence which came from within ITT however, ran to the contrary, with high levels of interpersonal conflict among top management, traumatic board meetings and a large proportion of personally-troubled CEOs of ITT subsidiary companies.

But it was not only the CEOs who were troubled under such organizational arrangements. All across the advanced industrialized economies, men and women were suffering from the results of the dual organizational imperatives for control and performance, while having to play all their other required non-work roles in a society which also was becoming increasingly complex and demanding (Sauter *et al.*, 1989; Cooper and Lewis, 1993). They suffered equally whether they were in entry-level jobs subject to the administrative or productive doctrines of high 'Fordism' or post-Fordist managerialism (Boreham, 1992) or whether they were executives, sandwiched between the multiple strata of the organization, subject to managerial *diktats* for performance.

The organizational orthodoxy of such a complex mechanistic firm or service enterprise was dealt a series of severe shocks from the mid-1970s onward. Successive energy crises, new technologies including IT, increased international competition, changing product and labour markets, new social values – the combined impact of all these factors on organizational structures and processes is well documented. The resultant effect on organizations was to move away from the steady state (with its above-documented stressors on employees) to a new dynamic state, which brought new and different stressors. These were centred around new forms of organization (e.g. the matrix organization), or higher performance demands ('lean and mean' firms), or market responsiveness ('flexible' employment). Movements inspired by consultants or cult books took root in firms, and acquired the status of ideologies, by requiring conformity on the part of employees. Internal organizational pressures increased – for instance the 'excellence' theories (Peters and Waterman, 1982), 'thriving on chaos' (Peters, 1987) or 'reengineering' the corporation (Hammer and Champy, 1993).

Optimistically, the advocates of such movements see challenge and growth for individuals arising as a result of these changes. Yet in many instances these new, unfamiliar and relatively untested organizational forms and processes replaced the previous anomic and coercive stresses of the steady state with new stresses of insecurity, under-performance, non-adaptability and lack of competence. It was such organizational contexts that helped to spread the modern Employee Assistance Programme (EAP), which owes its rationale as much to factors internal to the employing organization, as to the pressures of wider society or those of the specific social contexts of individuals. As organizational effectiveness began to be seen as more than the optimal disposition of machinery, materials and markets, the human element of the equation became more important.

However, the extent of acceptance and timing of spread of the EAP as a tool, used in order to obtain this optimal disposition, varied considerably between (for instance) the US and Europe. Often the determinants of its diffusion were the extent of the role of the state in social and employment matters, the degree of development and statutory support enjoyed by occupational medicine, and the role and influence of personnel management departments in organizations.

In Europe, the status and specialized orientation of the personnel function was late in its development as compared with the US (e.g. for France, see Brunstein, 1995) or was dominated by economic thinking (e.g. for Spain, see Flórez Saborido et al., 1995), or by the legal or military backgrounds of personnel specialists (e.g. for Greece, see Papalexandris, 1995). However, in the US particularly, personnel specialists more often came from a systematic training in the social sciences. They could appreciate that the employee who was troubled by work-related control systems could also be assisted by new techniques and programmes using behavioural science knowledge. They no longer regarded the employee economistically as an individually self-maximizing individual, nor as a collectively group-maximizing man organized by labour unions. No longer was the employee seen by management as a person to be corrected and coerced by economic and financial discipline. These new behavioural science-based provisions had frequently proved in the past to be effective and accepted in wider society, whether in professional therapeutic contexts, in leadership training, or in lay contexts of social influence. So industry and business adopted them, and the personnel function incorporated them into its range of activities, and along with the occupational health functional specialism, developed the employee assistance programme.

But in essence, the early EAP was arguably being used to support the continuance of steady-state administrative and production systems which were increasingly being shown as inherently damaging to employees as people, and morally or societally questionable. The EAP was a tertiary-level intervention: treating symptoms, not identifying and eradicating root causes (Cooper and Cartwright, 1994). In the new organizational paradigm of performance, market-responsiveness and flexibility, the EAP is faced with new problems expressed by troubled employees, often more complex and intangible issues than those experienced in the older, more structured and more defined state. However, it can be argued that the EAP still supports the organizational orthodoxy, albeit in its new form, with its new values of performance, innovation and development – which may not accord any more closely with the professional values on which the EAP is based.

It is against such contexts that we shall attempt in this chapter to define the EAP, briefly review its historical development, assess its coverage in various countries, describe its principal forms and models, discuss its introduction, operation and evaluation, and critically appraise its present practice and future potential.

Definitions, implications and methodologies

Definitions of EAPs are problematic since many widely differing perceptions of its philosophy and methodologies are held by the various organizational stakeholders, and many different types of programme may be found. At this stage, the authors' working definition of the EAP is:

> a programmatic intervention at the workplace, usually at the level of the individual 'troubled' employee, using behavioural science knowledge and methods, for the control of certain mainly work-related problems (notably alcoholism, drug abuse and mental ill health) that adversely affect job performance, with the objective of enabling the individual to return to making her or his full organizational contribution and to attaining full functioning in personal life
>
> (Berridge, Cooper and Highley-Marchington, 1997: 16)

In certain contexts, titles other than EAP may be used, such as employee counselling programme (ECP) which is sometimes felt to project a more professional image; or employee counselling services, when employees may object to the social connotations of 'assistance' as being a financial handout (Masi, 1984), or to the normative overtones of a 'programme'.

The concept of the 'troubled employee' is at the heart of the EAP. In advanced industrialized societies, with their dominance of the organizational culture, the stigmatization which such an individual suffers is particularly acute. For many troubled employees 'the job was the last hold to ego functioning; this is important knowledge for human service workers, who have often seen the workplace as a negative rather than an ego-reinforcing environment' (Masi, 1984). Practitioners in EAPs note Freud's assertion that work is man's strongest tie to reality (Follmann, 1978), and also that work is one of the three key bases of social identity along with marriage and parenthood (Gould and Smith, 1988). The establishment of an EAP may well represent a major paradigmatic shift in an organization's view of the troubled employee. In acting thus, 'its members are making a commitment to adopting an altered view of the value of troubled employees, and accepting the notion that treatment and rehabilitation are appropriate methods of addressing mental and physical problems' (Lewis and Lewis, 1986).

Organizations which are seeking to represent themselves as socially responsible to 'quality labour' may communicate this message by incorporating EAPs as part of a long-term commitment to employees (Luthans and Waldersee, 1989). With the calibre and effectiveness of a firm's human resources being seen as a (*if* not *the*) key competitive advantage in the 'lean' organization, within such a rationale it becomes all the more important to have a mechanism like the EAP in place. This permits the rehabilitation of employees who are suffering stress, whether it be from the external intensity of market forces or the imperatives of internal efficiency, such as restructuring, delayering, performance appraisal or performance-related pay.

Indeed, some commentators have seen the growth of a wide range of services to employees within EAPs as especially significant. From their origins in alcoholism programmes in the US in the 1940s and 1950s, providing a 'functional socio-anthropological perspective in occupational programming', EAPs have become what is in essence a social movement (Archer, 1977). As such, the EAP may henceforth be interpreted as an aspect of the employer's wider social responsibility, and a contribution to wider society. The employer thus recognizes his/her duty in respect of his employees, their jobs, and their organizational commitment and his/her contribution not merely to offset the damage caused by employment, but also to assist in tackling the more general problems of society.

The methodologies of EAPs are diverse and unstandardized, needing to be tailored to contexts, in spite of tendencies towards standardization on the part of major provider firms. The categorisation of EAPs into seven 'core technologies' (Blum and Roman, 1988) is nevertheless sufficiently representative to serve well as the basis of an examination of the unique nature of the contribution of EAPs. These are:

1   the identification of troubled employees through specific, objective documented evidence of job detriment,
2   the availability of designated EAP specialists for assistance to the employee and management,
3   the use of constructive confrontation to create a recognition of the problem,
4   the acceptance by the employee that external expert resources can aid in individual problem treatment,
5   long-term collaboration between work contexts and external resources,
6   the organizational culture accepts external resources as legitimate intervention,
7   any evaluation of success in intervention is job-based in an objective manner.

Blum and Roman see various consequences of application of the core EAP technologies, these 'core functions' include:

- enhanced retention of employees,
- reduced responsibility of supervisors for counselling,
- existence of 'due process' for troubled employees,
- reduction of employers' health-care/insurance costs,
- the operation of a gatekeeping activity to health-care provisions,
- the building of employee morale and trust.

Nevertheless, as those authors readily accept, there may well be a cultural bias towards US practice and rationale in these core technologies and functions which 'cannot be generalized to any other nation or culture . . .' and which 'may require unique modifications for each and every nation'. Factors influencing national specificity will be variations in health-care provision, in definitions of personal problems, in organizations' acceptance of undesirable behaviour at work, in the salience of work and work organizations, and in legal contexts of employment (Roman and Blum, 1992).

Historical background and development of EAPs

A welfare tradition existed in most of the older industrialized countries, rooted in the social and religious patterns of that country. However, in most cases, such traditions had little carry-over to EAPs, which grew from therapeutic and professional models, as can be illustrated from the contrasting development of welfare in Britain, the US and Australia.

In the US for instance, the 'friendly visitors' and 'social secretaries' of pre-World War I disappeared in the very different social climate of the 1920s and 1930s (Popple, 1981). Employee assistance in the US is generally held to stem from the foundation of Alcoholics Anonymous (AA) at Akron, Ohio in 1935. The AA method of problem confrontation and recognition, allied with peer group pressure and support, proved effective, and led to industry's widespread adoption of the Occupational Alcoholism movement in the 1940s (Good, 1986). There was increasing official recognition (Masi, 1984) and the acceptance of alcoholism as a disease (Jellinek, 1960), and by the 1960s the American Federation of Labor (AFL) and the Congress of Industrial Organizations (CIO) which merged in 1955 into the AFL-CIO had started to negotiate in labour contracts for alcoholism treatment benefits, recognizing that 'employees' physical and emotional wellbeing and corporate interests do coincide' (AFL-CIO, 1961). In 1970 the passage of the comprehensive Alcohol Abuse and Alcohol Treatment Act (the Hughes Act) enabled programmes to be established at both federal and state level. Although narrowly intended, the legislation encouraged the creation of professionally-staffed programmes, which rapidly took a wider view of problems and which spread in both private- and public-sector employment. This growth provided immense encouragement to employee assistance provider companies, consortia and in-house departments, and contributed to the development of university training schools in EAP counselling practice. From this stage, the EAP movement could be said to be well and truly founded in the US.

In Britain, welfare provisions have a lengthy history, grouped mainly around first occupational safety and health, and second the moral and religious philosophies of employers. The duty of the state in relation to its occupational health and safety duty, stemming from a series of Factories Acts from 1833 onwards, imposed a series of obligations on employers, who otherwise were only too willing to regard safety as merely financially negotiable, before or after the event of an accident. However, from the 1920s a minority of enlightened employers went far beyond the letter of the statutes (e.g. Lever Brothers, Marks and Spencer, John Lewis Partnership), spending significant proportions of pre-tax profits on both preventative and educative health programmes. Post-1945, the state as employer in the newly-nationalized industries readily acceded to trade unions' requests for extensive traditional welfare provisions (notably in the Post Office, telephones and mining) although these were swept away by the political ideologies of commercialization and privatization in the 1980s. The advent of a free National Health Service in 1948 reduced the incentive (and advantage) of such provision by employers. However, the NHS's formal achievements in individual counselling have been minimal.

The moral and philosophical strand of welfare is strongly associated with the Protestantism and Quakerism of nineteenth-century industrial and commercial employers. The peculiarly British form of industrial welfare was a blend of practical assistance and normative prescription – all too close to a counselling role as perceived by the many influential personnel managers who had originated from a training in social work. The first progenitor association of the present national Institute of Personnel and Development (IPD) was the Institute of Welfare Officers (1913), and the personnel management occupation has been seen by one leading commentator as being 'paranoid' in trying to shake off its detested welfare image (Torrington, 1989). Such old-fashioned paternalism was evidenced especially in health and housing provisions (e.g. by chocolate makers Cadbury, Fry and Rowntree) who built 'model villages' and who were indeed concerned with employees' lifestyles to an extent which by the 1970s some found onerous. It was the engulfment of such firms into larger international business groupings (e.g. Cadbury-Schweppes, Nestlé-Rowntree) that marked the demise of such leading examples of private sector welfare, incorporated into a wider economic logic. However, there was little continuance of old-style welfare into modern EAPs, and one commentator from the former tradition refers to EAPs as 'the occupational equipment of preventative medicine' (Eggert, 1990): a statement that rejects any historical continuance in the emergence of a new distinctive set of values and practices.

Although Britain has typically been seen by US consulting firms as an open market (EAPs included), the growth of EAP provision dates only from about 1980. At that period, early EAPs tended to be in-house, often in the British-based multinational companies such as ICI, many of which had previously possessed active welfare functions and occupational health departments. Currently, specialist-professional commercial providers dominate EAP provision, but financial services companies are showing an increasing interest in entering the market, with the intention of linking EAP provision into a wider raft of employee benefits.

In Australia, while welfare departments in the Anglo-Saxon tradition had existed previously, it was a joint effort by industry, labour and federal government agencies that in 1977 created the occupational drug and alcohol programmes which by 1984 had evolved into EAPs. Each state had an identified provider organization funded through the per capita fees of subscribing employers, as well as through public finance. Any employee of a contracted-in employer may then use the EAP for the normal range of services. Large firms that are contracted out may establish their own in-house services, and consortia provision also exists (Smith and Buon, 1991). In this manner, Australia's unique EAP coverage has resulted particularly from the socio-political context, and owes an unusual amount to the role of the trade unions in protecting their members – in contrast to the trend towards individualization of the employment relationship through the EAP, as found in Britain.

A final consideration of the developmental progression of EAPs can be made through reference to the four-stage model suggested by Osawa (1980):

Stage I: narrow focus of programme to limited range of issues; staff usually qualified by experience only; emphasis on identification and referral of problem employees.

Stage II: wide range of employee problems recognized; mainly qualified EAP workers; more in-house services, including supervisor and employee education.

Stage III: comprehensive range of services; organizational interventions started, recognizing the systemic origins of troubled employees' problems.

Stage IV: EAP staff act more as organization development consultants to management, to individual employees and (if present) to labour unions – de-emphasizing distinctions of role and interest.

Current evidence suggests that, in practice, most organizations have not progressed substantially beyond Stage II, whether in the US or elsewhere. Some provider companies of EAPs are recognizing the need for a more holistic view, but are finding a reluctance by client companies to allow them to play a wider role as organizational consultants. Indeed, the US National Association of Social Workers (NASW) 1980 Code of Ethics stated that few EAP staff will be in a position to influence or reverse corporate policies, especially where they are onerous on employees, but 'they generally can and should mobilize community services, promote self-help groups, advocate for entitlements and initiate referrals to other employers who are hiring' (Kurzmann, 1988). This stance of non-involvement in corporate issues accords with many formulations of professional EAP ethics in the traditional independent sense. For others who chafe at being limited to secondary or tertiary interventions within an EAP, such reluctance will be redolent of outmoded passive occupational models.

Nevertheless, the scope of instances of EAP practice is almost infinite, taking account of the pragmatic nature of EAP programme design, which of necessity reflects the needs of a wide range of organizations and contexts. This is axiomatic if EAPs are to 'exert a long-range and significant impact on corporate policies that affect individuals, families, society and the community at large' (Flaherty, 1988). There are many issues on which intervention potentially can occur (Berridge, 1990a) and a selection of the most frequently practised of these are listed in Figure 11.1 below. In fact, one commentator from the personnel management perspective has noted that 'the benefits you

AIDS	Grievances	Relocation
Alcohol abuse	Indebtedness	Retirement
Bereavement	Induction	Sexual harassment
Career development	Job training	Smoking
Chronic illness	Lay-off	Stress (work-related)
Demotion	Legal matters	Stress (work-extrinsic)
Disability	Literacy and education	Substance abuse
Discipline	Marital problems	Suicide
Dismissal	Mental health	Verbal abuse
Divorce	Performance evaluation	Violence
Family problems	Physical fitness	Vocational guidance
Financial advice	Promotion	Weight control
Gambling	Racial harassment	Women's career breaks
Goal setting	Redundancy	Young workers' problems

Figure 11.1 Counselling issues for an EAP (Berridge and Cooper, 1994).

can offer are limited only by your own creativity in thinking up new benefits'
(Dessler, 1988).

In an organizational context of new challenges for performance, the
incidence of stress is likely to be increased over a variety of issues. Therefore
the number and complexity of issues on which counselling intervention is
required is also likely to increase – adding pertinence to the remarks quoted
above on the limits of creativity in devising new benefits.

EAP statistics: numbers covered, services provided

The statistics of EAPs are very much a subject for debate, since there is little
international agreement on the definition of an EAP and standardized data
are not reliably available. However, in most countries the pattern appears to
be following that in the US, where very rapid growth in the number of
employees covered is succeeded by a plateau effect as the characteristics of
a mature industry appear.

In the US, in 1972 there were 300 company-sponsored EAPs in place
mainly in large companies (Cohen, 1991). By 1979–81 the figure was 5,000,
covering about 50 per cent of the *Fortune* top 500 firms (Roman, 1981). By
1987–89, some 8,000–10,000 programmes were operating, involving 75 per
cent–80 per cent of the *Fortune* top 500 (Maiden and Hardcastle, 1988;
Luthans and Waldersee, 1989). A 1991 estimate (Feldman, 1991a) indicated
12,000 programmes, but by then it appeared that the plateau had been
attained among top firms, and that growth was in the medium to small firm
sector. These figures for programmes mean, according to one commentator
(Feldman, 1991b), that in 1980 some 12 per cent of US workers were covered

by an EAP, rising to about 36 per cent (around 28 million workers) by the start of the 1990s.

Outside the US, authoritative usage statistics for EAPs are difficult to obtain, because the concept and practice of employee assistance is so culturally pervaded that it manifests itself in many different forms. Large enterprises tend to be the most frequently-found adopters of EAPs. Consequently those countries characterized by small and medium sized enterprises (SMEs) tend to display a lower take-up rate of EAPs, as found for instance in Denmark or Ireland.

In countries which have a long-rooted tradition of alternative forms of employee assistance (such as Germany, where there is a tradition of corporate responsibility for employee welfare shared with trade unions, dating from the mid-nineteenth century), EAPs have not been established widely, except for TNEs wishing to boost their recruitment image. A similar pattern of low take-up is found in France (where 'safety valves' exist through the statutory 'code du travail') or previously in Britain, where the almost-free National Health Service was previously able to supply a safety net. In Australia, the EAP movement is vigorous, emerging out of state-level and publicly-funded initiatives to deal with substance abuse (Schmidenberg and Cordery, 1990) and a distinctive model exists in New Zealand, including a role for industrial chaplaincy-based counsellors (Elkin, 1992). In Canada, distinctive models have arisen, often motivated by the principles of occupational alcoholism programmes (Shain and Groeneveld, 1980).

Hence, it might be tentatively hypothesized that EAPs are more easily adapted to national cultural contexts in the Anglo-Protestant tradition of self-determination, and individualistic employment. However, the individualization model of HRM (Bacon and Storey, 1993) has gradually been taking root in other countries with different cultures. For instance, in France the new model of 'gestion des ressources humaines' (GRH) displays numerous facets of the Anglo-Saxon origins of this factor-led view of organizational operation and exploitation of resources in the pursuit of competitive advantage strategies (Porter, 1990). Accordingly, EAP approaches to individual employee motivation, satisfaction and commitment are being displayed in many European countries through proxy indicators such as stress consultancy start-ups, and inquiries to existing multinational providers in the fields of employee benefits, and to academic consultants for advice and implementation. Such interest indicates at least a curiosity, at most an urgent need to identify and meet employee dissatisfactions and stresses arising from the new internal corporate climates and changing social contexts.

Several trends can be subsumed from these informed impressions. First of these is the considerable acceptance of EAPs among large firms in advanced industrialized economies listed, Ireland and perhaps Japan excepted, where distinctive issues of industrial and social structure apply. Conversely, there is a low level of EAP penetration among smaller firms, even in the US where the EAP industry is the most mature, although no evidence exists that their

employees do not experience troubles. Second is the fact that a valid role for
EAPs is being seen even in those countries which have a long-standing tradition of state legislation or support in occupational health, or as in Germany
where systematic corporate responsibility for employee welfare is long-
established. Third, the EAP industrial sectoral emphasis has moved from
manufacturing (its traditional blue-collar origin) to public administration, to
transport and communications and to services particularly including financial
specialisms. Apart from reflecting the de-industrialization of economies, the
shift can also be interpreted as mirroring the creation for employees of new
performance-led stressors in qualitatively-dominated service employment.

Models of employee assistance programmes

Many commentators on EAPs claim to offer a model of their subject, but few
models have been developed at a critical level that would be widely and
critically acceptable in the behavioural sciences. Most would-be model
builders in effect offer typologies or taxonomies, relying on relatively simple
categorization. Several reasons may be suggested for the present shortcomings as demands upon EAP theory. First, EAPs are a 'crossroads'
discipline, drawing on inputs mainly from psychology, sociology, medicine,
anthropology and politics: typically for such a contemporary professional
area, these inputs lack coordinating theory, and arguably do not yet need it
for risk of stultification of the infant discipline. Second, the lay origins of
EAPs in the occupational alcoholism movement and its early leadership,
consisting of zealous but (for instance) untrained reformed alcoholics, contributed to a de-emphasis on theory in favour of lay social action. Third, the
ownership of EAPs is contested, lying as they do in disputed occupational
psychology territory between the (often external) professional therapists and
the internal stakeholders – the personnel managers, the occupational-health
staff members and the company financial officers. Given the widely different
values and methods of operation of these disparate professional groups, they
are unlikely to accept a common EAP model, and pragmatism and com-
promise are to be expected.

Nonetheless four models can and will be outlined and evaluated briefly.

A therapy-based model is suggested by Hellan (1986), in which four types
are identified:

Type 1   both lay assessment and referral (if needed) in-house – reflecting
         the origins of EAPs in the industrial alcoholism movement – the
         'origins' model;
Type 2   in-house professional assessment and referral – the 'classical
         EAP mode' (as it might be termed) of the self-contained,
         organizationally-integrated provision;
Type 3   open-ended internal and external assessment and treatment

model, often following self-referral – the 'community health' model;

Type 4   closed-end full service model – typified by the cost-effective EAP, using external contractors for short-term counselling (the 'contractual' model).

These models are useful in identifying the key actors in the EAP process, but the range of interpretations that may be made about their values and practice modes is so wide as to reduce each model's utility, other than in preliminary classification.

A procedural model is proposed by Masi and Friedland (1988): its four EAP types are based on the physical location of the EAP procedures for assessment and consultation. These are:

1   **In-house**
    The organization itself creates the policies; all staff are eligible as direct employees and operation of the EAP programme is employer-controlled.
2   **Out-of-house**
    The organization may determine policies, but it contracts provision of the EAP service to a single external commercial or professional provider.
3   **Consortium**
    Several companies (often smaller firms) pool resources to develop a commonly-controlled collaborative programme, usually at a location external to all, provided by a public service agency or a commercial enterprise.
4   **Affiliate**
    Similar to (3) in many respects, but coordinated through a local professional provider on an independent expert basis.

The utility of Masi and Friedland's model lies particularly in highlighting the extent of integration of the EAP into the organization, reflecting its values, procedures and other systems of social control, or the extent to which the dominant referent is external-professional.

A *cui bono* ('who gains') approach is taken by Straussner (1988), based upon a comparative study of in-house and external EAPs and their relative benefits to the principal stakeholders. This is effectively a structural-functionalist evaluative approach and is not strictly a classificatory model. Straussner initially identifies the basic distinctions between in- and out-of-house EAPs, leading to an implicit typology of employer-favouring programmes, employee-favouring programmes and finally EAP staff-favouring programmes. Her conclusion is that in the choice of EAP, the balance of socio-economic benefit is in favour of in-house provision, which possesses the most cost attractiveness to management and is intrinsically more independently coherent and cost-effective than the other types. A more extensive development of the three-fold typology could provide a political model for the analysis of EAPs, incorporating numerous evaluative inputs.

Finally, Shain and Groeneveld (1980), whose intention was to create a 'practice model', have come the nearest to meeting the sociological ideal-typical conceptualization of the EAP. Such an approach derives not so much from a desire to create the perfect model for emulation and imitation, but rather to identify the essential elements which necessarily could and should be present and the systematic relationships between them. The main elements posited are: programme specification and education; problem identification, confrontation, referral, exploration, acceptance, ownership, assistance, resolution and feedback; programme evaluation and adaptation. It is against such a model that practice can be evaluated, not judgementally, but in terms of the extent to which it contains the postulated elements and the pattern of their interaction. Seen in this light, Shain and Groeneveld (1980) have offered a useful non-prescriptive model adopting an open-system perspective, even though they may have insufficiently elaborated the complex interactions between the elements.

Such a criticism may also be made of the 'core technologies' and 'core functions' developed by Blum and Roman (1988) and mentioned earlier in this chapter. Helpful though these insights are, it should be acknowledged that their authors have not claimed that they have all the characteristics of a system or model, or that they are universally applicable (Roman and Blum, 1992). Sadly, much of the other thinking on the subject has remained however, at the level of categorization, perhaps to the detriment of professional practice, and also of understanding of EAPs on the part of managers in the organizations that consider adopting them.

Operationalizing the EAP

Selecting an EAP

The expanding network of EAPs has produced considerable variety in structural types and operational charactcristics (Flaherty, 1988), leaving the choice process relatively open. No 'best' EAP type is universally recognized and the decision parameters are not clearly specified in any professional-normative model. It is likely that most managerial decision-makers will either follow a heuristic method based on their organization's customary decisional processes, or will be influenced by the recommendations of a preferred contractor-provider company. These are growing in number and, as with most service consultancies, are also growing in the sophistication of their marketing methods and in the routinization of their delivery procedures (Berridge, Cooper and Highley-Marchington, 1997).

An aid in making choices based on a strategic and tactical contingency analysis is proposed by Fleisher and Kaplan (1988) in order to assist in determining whether the choice of EAP provision should be in- or out-of-house, company- or contractor-serviced.

They postulate three sets of parameters which need to be scrutinized, namely:

1   **The organizational context**
    This comprises the size of the organization, the workforce disposition, the organization's mission and culture, and the level of resources which can be assured for the EAP in the medium to long term.
2   **The EAP's design components**
    These include the target population (for instance, whether including dependents or part-time employees), the proposed range of services (broad or narrow brush provision, such as only alcoholism or a wider range of issues), administrative constraints (including corporate departmental affiliation and reporting mechanisms within the firm) and funding methods (whether fully supported internally, or partially reliant on co-payments or insurance support).
3   **The desired outcomes**
    These may also cover confidentiality and ensurance of privacy, the degree of accessibility of the EAP and the extent of uptake, the level of inclusion of wider social and community interests, the need for flexibility of provision available for different clients, the methods of accountability of counsellors, and the evaluation of outcomes, both qualitatively and in quantitative terms.

Fleisher and Kaplan offer few normative guidelines for the eventual choice: the company will therefore need to use a rational-economic approach, or less likely a cost-benefit analytical method. This decision aid clearly is of value in avoiding pitfalls of overlooking programmatic components, such as the nationwide distribution organization which opted for an in-house EAP, or the high-security defence agency using a community-based contracted-out affiliate-provided EAP. However, the relatively simple three-fold EAP typology used by those authors renders it less capable of discriminating application in most cases, and the inclusion of desired outcomes raises questions over the logical consistency of the authors' contingency orientation.

The Introduction of an EAP

The nature of an individual organization's chosen EAP format and the method of its introduction are unlikely to be capable of standardization, given an orientation to local particularities and problems (Berridge, Cooper and Highley-Marchington, 1997). Ensurance that full consideration has been given to all the essential elements and their interaction also represents another strong argument for working from a theoretical model, such as the practice model of Shain and Groeneveld (1980). The following seven-stage sequence can indicate a potential pattern of programme introduction, rather than represent a prescriptive path, within a conventional decisional schema:

1   *Definition of organizational philosophy*: an EAP embodies an affirmative statement about the firm's beliefs concerning its responsibilities towards its personnel: the firm's actions, both in terms of people and materials, need to accord with the statement in order to give it credibility. The EAP is not a 'band-aid', to be applied as an act of belated social conscience, or as a substitute for more fundamental action to tackle root causes or underlying corporate contradictions.

2   *Organizational need assessment*: the extent and nature of troubled employees' problems and behaviour requires expert or consultant survey assessment. This is carried out in order to determine the nature of demand for the programme, its processes and outcomes, its likely cost and the implications for other existing forms of employee benefit or administrative procedures, especially in human resource management (HRM).

3   *Formulation of policy*, based on the philosophy and the needs assessment, and ensuring that all affected groups have been consulted, particularly labour unions, insurers, legal counsel, and if needed, local agencies.

4   *Construction of procedures* for identification, referral, assistance and follow-up of troubled employees. This stage will entail consultation with managers and supervisors, whose role is central in the programme. But in addition the roles within the EAP of internal lay and professional counsellors will need negotiation and definition. Confidentiality issues should be clarified at this stage, as should rights of access, refusal to participate, and withdrawal from the programme.

5   *Resource assurance* involves the satisfactory provision of physical facilities, issues of staffing and qualifications, and financial support through budgets, cost reporting and monitoring systems.

6   *Introduction of the programme*, jointly carried out with labour unions or staff representatives if recognized, involves an educative communications campaign directed at employees, their families and local agencies, health-care providers, and perhaps the community. The training of managers, supervisors and labour union officials takes place at this stage, as also does that of support departments such as HRM, occupational health and safety, and data processing for the maintenance of records.

7   *Monitoring, control and evaluation*: in order to avoid future misunderstandings or conflicts, the criteria, methods and expectations of EAP performance need unambiguous definition, as do the roles of those to be concerned in monitoring, whether financial, professional or managerial. Health education and outreach provisions can be brought in at this stage.

This description of the stages of introduction of an EAP is typically programmatic, yet it needs to be implemented with the perceptiveness required of any major cultural change, rather than in a routinized manner. It stresses managerial and organizational aspects more than do many accounts of practice or normative statements that can be found in the literature. For instance, Masi (1984) places more emphasis on the bases of assessment and referral, on professional standards and the need for professional management of the programme, on the community referral network, and on the many facets of confidentiality, including legal issues. It is probably utopian to expect that the creation and introduction of an EAP will be without problems. It represents a major intervention in an organization's cultural and social processes, and challenges the behaviour patterns of many of the existing interest groups. However, a structured approach and a clear definition of expectations may help to mitigate future conflicts, and also have an educative effect.

Evaluation of EAPs

There has been much controversy over the appropriate method for evaluating EAPs. A few writers question whether employers, having made a philosophical commitment to an EAP, then require any economic justification of their investment (Decker et al., 1986). One report suggests that, based on a large sample of firms, few companies knew whether they were obtaining a conventional return on investment in the EAP, and further that firms were unable to record the benefits obtained from an EAP either quantitatively or qualitatively (Bower et al., 1989).

It is largely accepted that there should be economic and financial evaluation of EAPs, but little agreement exists on the suitable methods. In narrowly economic-financial terms the EAP has received close scrutiny and both fierce justification and opposition from many employers. Many large US programmes have produced apparent justification in terms of annual dollar savings, both in aggregate and in terms of annual expenditure. Their proponents are admittedly managerial in sympathy, but impressive statistics are advanced. The General Motors programme was quoted as saving $37 million per year by assisting up to 100,000 employees each year (Feldman, 1991a). A study by the Paul Revere Life Insurance Company is cited as showing a saving of between $4 and $23 in claims expenses for every dollar of premium expenditure diverted into the EAP (Intindola, 1991).

As with many social intervention activities, cost-benefit analysis (CBA) is probably the most widely advocated method of evaluation (Shain and Groeneveld, 1980; Masi, 1984; Durity, 1991; Masi and Friedland, 1988), although the latter authors recommend the variant of cost-effectiveness analysis (CEA). Another authority, while deploring the resistance of human resource professionals to measurement of their performance, suggests that utility analysis is particularly suitable, since it demands that costs and expected costs of decisions are taken into account. The utility of an EAP is therefore the degree to which it reduces the problems of the individuals participating in it, beyond that which would have occurred if it had not been in place (Cascio, 1987). One school of thought even suggests that the EAP should be seen in entrepreneurial terms, as a business in its own right, having to satisfy all normal operational criteria, including economic, financial and political aspects.

All the above methods of evaluation differ from those typically preferred by analysts from within professional counselling circles, who often advocate peer review. This typically entails a group of independent professionals who scrutinize representative case histories against professional-technical criteria using 'qualitative and quantitative methods to evaluate the quality and appropriateness of the program' (Masi, 1984). Other methods of review include assessment of the EAP's goal achievement measured through employee attitude surveys, and statistical case sampling by a multidisciplinary team of occupational health professionals (Cohen, 1991). In all such evaluation the independence of the evaluator needs to be combined with the maintenance of both confidentiality and the integrity of programme data. The reconciliation of these requirements, along with the more customary cost-effective demands of management renders evaluation of EAPs extremely problematic and open to criticism from all the stakeholders from their varying viewpoints (Berridge, Cooper and Highley-Marchington, 1997).

The roles of the key participants in an EAP

While a systemic consideration of an EAP regards all roles as necessary, this section will concentrate on four participants who are central to the EAP process. These are the EAP counsellor, the manager or supervisor, the union official or representative, and top management.

The EAP counsellor's role is complex. It can include 'referral agent, legislative analyst, researcher/evaluator, mediator, liaison, ombudsman [sic], programme development specialist, teacher/trainer, benefits administrator and consultant' (Gould and McKenzie, 1984). The risks in such a multiple role are considerable, and include overload, intrarole conflict, and lack of professional role clarity. Equally, there is a need for the EAP counsellor role to acquire professional and organizational legitimacy, distinct from that of the lay 'alcohol worker' or that of the business consultant or organizational development change agent. Some see in addition the need to 'eschew . . . the former image of social worker or do-gooder' (Shain and Groeneveld, 1980). A contribution of insight is provided by Steel (1988), with his discussion of the 'constructive broker role [which] includes the notion of balancing diverse ideologies' of organized management and labour, as well as those held by individual employees. Otherwise, 'EAPs will likely become absorbed into other groups, such as broader-focused human resource management units within . . . personnel divisions'. The absence of career paths for EAP counsellors has attracted comment, since only a limited progression would exist within EAPs in all but the largest companies (Steel, 1988).

The role of managers and supervisors in operating an EAP has been much discussed. Many commentators recognize that 'clearly, pragmatics not philanthropy motivates most managerial behavior' and that 'management frequently adopts the EAP because it is good for business' (Appelbaum and Shapiro, 1989). Equally, in the North American model of the EAP, identification of the 'problem employee' results from managerial identification of low or impaired job performance. In theory, 'supervisors, because they see their workers every day, remain in the best position to monitor employee behaviour' (Beilinson, 1991). Yet many see dangers in the supervisor's participation, suggesting that 'managers do not want to face the fact that there are problems in their workplace' (Watts, 1988). Supervisors are alleged to cover up the evidence of troubled employees. 'It's much easier to close your eyes to the problem employee, demote them, promote them, or put them on detailed work' (Watts, 1988). Most authorities agree on the need to train supervisors in terms of recognizing the 'problem employee', but there is less unanimity over whether a supervisor should confront the troubled employee and, therefore, needs those skills also. More concern, however, is expressed over such a relic of early alcoholism programmes, and whether supervisors are effectively acting as diagnosticians without the requisite training (Masi, 1984). In addition, worries are voiced that 'poorly-trained supervisors are labelling essentially well employees with their favourite illness label, and then railroading [them] into inappropriate treatment' (Luthans and Waldersee, 1989). In extreme cases it has even been suggested that the EAP may acquire some of the characteristics of an 'organizational witch-hunt' (Berridge, 1990a). Yet it also has to be recognized that 'EAP policy directives which require supervisors and managers to act counter-intuitively and against over-riding role requirements seem doomed to failure' (Schmidenburg et al., 1992). However, much as EAPs are a discontinuity in the organizational culture, they must be integrated with some form of organizational analysis to ensure that supervisors see them as credible and hence support them.

The involvement of labour unions or other bodies representing employees is important to the successful introduction of an EAP. The imposition by management of an EAP is likely to be self-defeating. The early suspicion of labour unions was based historically on a mistrust of management's motives for such programmes. They saw EAPs largely as an additional dimension of control and discipline, contributing towards the intensification of work, and the maximization of returns on human assets. Later commentators on the individualization of work, particularly within a culture of soft, tactical HRM have posited a scenario of increasing dependency on the part of the employee upon the organization (Bacon and Storey, 1993). There is the potential for this dependency to occur in the context of an EAP, resulting in the weakening of union ties and loyalties.

However, unions also had some positive motives for welcoming EAPs, such as a desire to promote members' total health (AFL-CIO, 1961), and for reasons of increased quality and confidentiality of occupational health advice to workers, 'an alternative must be found to the present company doctor system' (Glasser, 1976). In the British context of EAPs, where self-referral tends to replace managerial confrontation, the role of the union representative or shop steward is central in persuading the union member to self-enrol in the EAP ahead of any managerial suspicions of impaired performance, or use of disciplinary procedures. A more recent US study shows that labour unions have not lost their wish to extend EAP benefits for members, in spite of reservations over practice. In 1987, 63 per cent of a large sample of unions had EAPs incorporated to some extent in their collective labour contracts with employers, and EAPs were rated third in terms of desired future inclusion, after extended dental and vision care provisions (Kemp, 1989).

It is important that an EAP is perceived to be independent. This has led some labour unions to sponsor their own EAPs, in Britain using a nationwide telephone service and a commercial contractor-provider, or in the US basing the EAP around 'the union office or hiring hall, with heavy reliance on referral to community-based treatment facilities' (Smith, 1988a). The same author notes that both companies and unions potentially can form EAPs, but the 'programs developed independently in a highly charged adversarial environment common to many organizations are likely to be regarded with suspicion by whichever group has been excluded from the program development' (Smith, 1988a). Two other perceptive commentators note that 'unfortunately, the tendency may be for each side to press for the domination of the routine administration of the program – for management to press for final control, and for unions to resist any form of discipline' (Sonnenstuhl and Trice, 1986).

The role of top management is often cited, almost ritualistically, as crucial in lending support to the introduction and continuance of an EAP. Apart from the trite observation that the personal life and operational effectiveness of top executives and managers might benefit from their own participation in the EAP, these people are likely to be those having most impact on the organizational capacity of the organization to perform at a high level and to innovate. In most cases, confidentiality will prevent any publicity within the company concerning the benefit that top managers obtain from the EAP, and it may well occur that they will use advisory services outside the corporate EAP. However the payback will be highest with such persons, and their subsequent support could represent a political lever of considerable power.

Critiques of EAPs

The rapid growth of EAPs among companies bears witness to their high level of acceptance by business and industry, and a willingness on the part of many counselling practitioners to work within them and HRM managers to operate alongside them. The principles and practice of EAPs have apparently represented no unsurmountable obstacles for most of these groups, and EAPs have taken an accepted place in professional associations of occupational psychology and personnel management. Nevertheless, in spite of this widespread acceptance, a range of critical arguments may be heard on many aspects of the philosophical and therapeutic bases of EAPs, on their organizational impact, on their legal position, and on their role in the broader activity of the management of people at work.

Many commentators from an industrial social work perspective have observed that EAP counsellors are attempting to ally a mental-health driven therapy to a performance- and profit-dominated system of production of goods and services. An implicit unstated process of mutual exploitation may well have been created around EAPs, whereby counsellors pursue their professional aims, and enterprises seek to attain their economic goals in an uneasy alliance. The balance of interests may well only be maintained because of the lack of fundamental analysis of either group's function and activities on the part of the other. This opinion is supported by Hellan (1986), who notes that EAPs are a 'mysterious corporate institution, misunderstood by both management and employees, yet valued by both'. Professional bodies such as the American Association of Industrial Social Workers provide certification for EAP workers. Yet it is widely found that the dialogue and interactions between counsellors and management are 'fraught with the inherent conflict of essential elements such as treatment goals, problem causality, client progress and confidentiality' (Smith, 1988b).

Managers of the HRM function probably experience fewer conflicts and express fewer basic criticisms of EAPs than do mental-health professionals. EAPs mainly assist the personnel specialist in moving problems forward, in an alternative but routinized manner: they provide both more acceptable and more efficient methods of modifying the job behaviour of problem employees than methods based exclusively on disciplinary procedures. However, an early critic (Bartell, 1976) sees EAPs as serving the interests of the human resources function within the overall organizational power structure, representing the ultimate deterrent, requiring continuing conformity to the programme criteria, as a condition of avoiding termination (Luthans and Waldersee, 1989). For the personnel department there is the added advantage that the EAP deterrent is effectively wielded by another specialist function, often external to the organization – thereby avoiding the personnel department being perceived negatively in taking hard decisions of terminating the employment of employees with problems.

In the wider organizational context, the EAP has been criticized as a form of unwarranted organizational intervention, insufficiently taking into account the internal dynamics of the enterprise and the wider organizational context. Numerous authors have noted the tendency of EAPs to be referred to, especially by management writers, as a form of organizational programming.

The EAP then takes its place as another method of job performance control, designed with special reference to problem employees. The degree of benefit which the firm gains will relate to the extent to which this new element is integrated into the complete control structure. But, a leading practitioner-writer has commented, based on survey evidence from Britain, that 'the centre of gravity of the counselling universe is moving inexorably to the workplace' (Reddy, 1992), and this is reinforced by qualitative and quantitative findings (ICAS, 1993). Not surprisingly, several critics have commented adversely on this approach, stating that 'an EAP is not a loose piece of program material . . . it is a whole system of organizational intervention . . . not a minor adjustment: it is a change in management methods' (Shain and Groeneveld, 1980). This point is reinforced by Blum and Roman (1988), who report that the EAPs that are perceived to be the most effective are those which are fully integrated into the organization and its management control system. Writing from a counsellor training viewpoint, Carroll has attempted to identify the distinctiveness and importance of the organizational context in four aspects of counsellors' training. These are:

- identifying the organization's boundaries and roles;
- knowing the organizational ecology sufficiently well to be effective within it;
- being able to negotiate with the organization over interventions;
- using assessment procedures within the organizational setting.

He recognizes the role multiplicity that such a notion creates, but feels that the contemporary EAP counsellor can be trained to handle the 'demarcation lines, acceptable boundaries and supportive relationships' that will be entailed (Carroll, 1995).

In spite of the several aspects of incompatibility between the EAP philosophies and those of organizational development (OD), it is often argued that the logical endpoint of the EAP is that of an organizational change consultancy. This position reflects in many ways the philosophical position of the Tavistock Institute in London, in which fundamental organizational interventions are required, in order to make therapeutic work with the individual anything more than peripheral tinkering. The fourth stage of Osawa's (1980) development progression for EAP posits a similar sequence, as has already been outlined earlier in this chapter. This OD stance, based on an open-systems perspective, has been argued – by several writers from both the social work and organization theory schools of thought – to be a viable wider foundation for EAP. One study reports, however, that few organizations have in reality formally adopted an OD role for their EAPs. Managerial commentators reject this option in many cases, arguing that counsellors' training gives them little or no appreciation or skills in the economic or political reality of business (Afield, 1989). So, while it is recognized that more theoretical underpinnings for such a synthesis will be needed eventually (Smith, 1988a), it is probably far distant in most enterprises.

From a legal viewpoint, EAPs have received much critical attention. One cluster of pragmatic managerial comments (Good, 1986) condemns EAPs as encouraging malingering among employees, coddling lawbreakers and societal deviants, and exposing employers to security risks through their continued employment. Informed legal comment in the US has stressed the EAP as a positive response to the Federal Rehabilitation Act 1973 and a general trend in labour relations judgements towards help, not punishment. Yet it may be asked how extensive would be the EAP network without workers' compensation legislation, labour arbitrators and unions? One legal commentator has argued that EAPs are the 'legal expedient of providing employees with a chance, so the employer who follows EAP to the letter meets arbitration criteria when firing becomes inevitable . . . avoiding fruitless and costly reinstatement of problem employees' (Nobile, 1991). In Britain, the legal protection offered to an employer by an EAP is as yet untested. However, landmark cases in 1994 and 1995 such as Walker vs Northumberland County Council, 17 November 1994, (in which a very large sum was awarded to an employee who had suffered extreme job stress) or Johnson vs Camden and Islington Health Authority, 25 April 1995 (in which a junior doctor received a small out-of-court non-prejudicial settlement in a case massively supported by the British Medical Association and opposed by official financial and legal might) both seem to argue convincingly for EAP support as a matter of legal prudence and good HRM on the part of the employer (Earnshaw and Cooper, 1996).

To date, there have been few legal suits over EAPs in the US on issues of due process, professional malpractice or negligence. Yet warnings are heard that 'EAPs appear to have no concrete legal advantages, and an organization may actually increase its legal vulnerability by establishing one' (Luthans and Waldersee, 1989). Insurance companies have added their voice in noting the increased exposure to financial loss which an employer could face as a co-defendant in an EAP suit (Head, 1988). In the litigious climate of the US, the employer could face suits for topics as diverse as incorrect licensing of counsellors or agencies, equality of access, wrongful assessment, wrongful referral (or non-referral), refusal to reinstate, wrongful discharge, privacy and confidentiality, to name a few of the myriad issues (Nobile, 1991; Loomis, 1986; Monroe, 1988).

Conclusions

Based on the analysis developed in this paper, the future for EAPs seems to lie in their capacity to:

1   satisfy the employing organization's needs for programmatic control of the human factor of production, in parallel with the enhancement of the production imperative for reliability and quality;

2 supply a therapeutic service for the troubled employee, which offers benefits of a specific nature, at a level, form and (perhaps also) cost not obtainable through other counselling or health-care provisions in the community or wider society;

3 be capable of positive evaluation, especially in respect of legal and financial costs, or labour relations and administrative procedures;

4 provide a professionally rewarding and acceptable area for the professional practice of certain defined aspects of social work;

5 be modified in line with the changing expectations of the organizations, professionals, clients and society;

6 develop a coherent theoretical foundation and ideology to these activities, and to support them with a suitable licensing and regulatory framework.

Up to now, the company-based model of the EAP has flourished most strongly in North America. Doubts had been expressed by several commentators (Hunt, 1989; Orlans, 1986) whether similar growth and proliferation of EAPs can occur elsewhere. In spite of the different legal and social welfare provisions in (for instance) Europe or the Pacific rim countries, evidence seems to suggest that employee problems in other countries are little different from those in the US (Megranahan, 1989; Schmidenberg and Cordery, 1990). It may be, however, that the pattern of referral will represent the major difference between the various business cultures. In Europe, evidence is already gathering that supervisor referrals are likely to be far less acceptable to employees than they are in the US. Instead, self-referral seems likely to provide both the level of confidentiality sought by employees, and the extent of non-involvement with personal problems which is felt appropriate by employers.

Whatever form is adopted in practice, the potential seems to exist in these countries outside the US for EAPs to have a valued role for employers and employees alike in tackling work-related problems. But as found with both personnel management and social work which are culturally-pervaded to a marked extent, it seems likely that national, industrial and even corporate variants of EAPs will develop. Indeed they may increasingly characterize EAP provision, predominating over the standardized model, whether professionally defined or commercially promoted.

References

AFIELD, W. E. (1989) 'Running amok: employers losing control of EAP costs, management', *Business Insurance* **23**: 27–30.

AFL-CIO (American Federation of Labor – Congress of Industrial Organizations) (1961) *The Worker's Stake in Mental Health,* Publication No. 69, April.

APPELBAUM, S. H. and SHAPIRO, B. T. (1989) 'The ABC of EAPs', *Personnel* **66**: 36–46.

ARCHER, J. (1977) 'Occupational alcoholism: a review of issues and a guide to the literature', in C. J. Shramm (ed.) *Alcoholism and its Treatment in Industry*, Baltimore, MD: Johns Hopkins University Press.

BACON, N. and STOREY, J. (1993) 'Individualisation of the employment relationship and the implications for trade unions', *Employee Relations*, **15(1)**: 5–17.

BARTELL, T. (1976) 'The human relations ideology', *Human Relations* **8**: 39.

BEILINSON, J. (1991) 'Are EAPs the answer?', *Personnel* **68**: 3–4.

BERRIDGE, J. (1990a) 'The EAP – employee counselling comes of age', *Employee Counselling Today* **2**: 14–17.

BERRIDGE, J. (1990b) 'The EAP and employee counselling', *Employee Relations* **13**: 4.

BERRIDGE, J. and COOPER, C. L. (1993) 'Stress and coping in US organizations: the role of the employee assistance programme', *Work and Stress* **7(1)**: 89–102.

BERRIDGE, J. and COOPER, C. L. (1994) 'The employee assistance programme: its role in organizational coping and excellence', *Personnel Review* **23(7)**: 3–80.

BERRIDGE, J., COOPER, C. L. and HIGHLEY-MARCHINGTON, C. (1997) *Employee Assistance Programmes and Workplace Counselling*, Chichester: John Wiley and Sons.

BLUM, T. C. and ROMAN, P. M. (1988) 'Purveyor organizations and the implementation of employee assistance programs', *Journal of Applied Behavioral Science* **24**: 397–411.

BOREHAM, P. (1992) 'The myth of post-Fordist management', *Employee Relations* **14(2)**: 13–24.

BOWERS, M., DECENZO, D., WALTON, C. and GRAZER, W. (1989) 'What do employers see as the benefits of assistance programs?', *Risk Management* **36**: 46–50.

BRUNSTEIN, I. (ed) (1995) *Human Resource Management in Western Europe*, Berlin: W. de Gruyter.

CARROLL, M. (1995) 'The counsellor in organizational settings: some reflections', *Employee Counselling Today* **7(1)**: 23–29.

CASCIO, W. F. (1987) *Costing Human Resources: the Financial Impact of Behavior in Organizations*, 2nd edn. Boston, MA; PWS-Kent Publishing Company.

COHEN, P. (1991) 'Does your EAP measure up?', *Personnel* **68**: 92.

COOPER, C. L. and CARTWRIGHT, S. (1994) 'Healthy mind, healthy organisation: a proactive approach to occupational stress', *Human Relations* **47**: 455–471.

COOPER, C. L. and LEWIS, S. (1993) *The Workplace Revolution*, London: Kogan Page.

DECKER, J. T., STARRETT, R. and REDHORSE, J. (1986) 'Evaluating the cost-effectiveness of employee assistance programs', *Social Work* **31**: 391–393.

DESSLER, G. (1988) *Personnel Management*, 4th edn., Englewood Cliffs NJ: Prentice Hall.

DURITY, A. (1991) 'Has anybody seen a cost-benefit analysis?', *Personnel* **68**: 5–6.

EAP International (1992) 'EAP international survey: eight prominent EAP experts profile the role of employee programmes throughout the industrialised nations', *EAP International* **1(1)**: 15–18.

EARNSHAW, J. and COOPER, C. (1996) *Stress and Employer Liability*, London: Institute of Personnel and Development.

EGGERT, M. (1990) 'Welfare at work', in A. Cowling and C. Mailer (eds) *Managing Human Resources*, 2nd edn, London: Edward Arnold, 150–168.

ELKIN, G. (1992) 'Industrial chaplains: low profile counselling at work', *Employee Counselling Today* **4(3)**: 17–25.

FELDMAN, S. (1991a) 'Today's EAPs make the grade', *Personnel* **68**: 3–40.

FELDMAN, S. (1991b) 'Trust me: earning employee confidence', *Personnel* **68**: 7.

FLAHERTY, V. E. (1988) 'Influencing management policy', in G. M. Gould and M. L. Smith (eds) *Social Work in the Workplace*, New York: Springer 265–279.

FLEISHER, D. and KAPLAN, B. H. (1988) 'Employee assistance/counselling typologies', in G. M. Gould and M. L. Smith (eds), *Social Work in the Workplace*, New York: Springer, 31–44.

FLÓREZ SABORIDO, I., GONZÁLEZ-RENDÓN, M. and ALCAIDE CASTRO, M. (1995) 'Human resources management in Spain' in I. Brunstein (ed), op. cit.

FOLLMAN, J. F. (1978) *Helping the Troubled Employee*, New York: AMACOM.

GLASSER, M. (1976) 'Workers' health', *American Journal of Public Health*, June.

GOOD, R. K. (1986) 'Employee assistance – a critique of three corporate drug abuse policies', *Personnel Journal* **65**: 96–107.

GOULD, G. and MCKENZIE, C. (1984) 'The expanding role of industrial social work', *Social Work Papers* 18, Los Angeles, CA: Los Angeles School of Social Work.

GOULD, G. M. and SMITH, M. L. (1988) *Social Work in the Workplace*, New York: Springer.

HAMMER, M. and CHAMPY, J. (1993) *Re-engineering the Corporation*, London: Nicholas Brealey.

HARRIS, M. M. and FENNELL, M. L. (1988) 'Perception of an employee assistance program and employees' willingness to participate', *Journal of Applied Behavioral Science* **24**: 4.

HEAD, G. (1988) 'EAPs: employee assistance perils', *National Underwriter* **92**: 25–27.

HELLAN, R. T. (1986) 'An EAP update: a perspective for the '80s', *Personnel Journal* **65**: 51–54.

HUNT, D. D. (1989) 'Anglicizing an American import', *Personnel Administrator* **34**: 22–26.

INDEPENDENT COUNSELLING AND ADVISORY SERVICE (ICAS) (1993) *EAP and Counselling Provision in UK Organisations*, Milton Keynes.

INTINDOLA, B. (1991) 'EAPs still foreign to many small businesses', *National Underwriter* **95**: 21.

JELLINEK, E. M. (1960) *The Disease Concept of Alcoholism*, New Haven, CT: College and University Press.

KEMP, D. R. (1989) 'Major unions and collectively-bargained fringe benefits', *Public Personnel Management* **18**: 505–510.

KURZMANN, P. (1988) 'The ethical basis for social work in the workplace' in G. M. Gould and M. L. Smith (eds) *Social Work in the Workplace*, New York: Springer, 16–27.

LEWIS, J. A. and LEWIS, M. D. (1986) *Counseling Programs for Employers in the Workplace*, Belmont, CA: Wadsworth.

LOOMIS, L. (1986) 'Employee assistant programs: their impact on arbitration and litigation of termination cases', *Employee Relations Law Journal* **12**: 275–288.

LUTHANS, F. and WALDERSEE, R. (1989) 'What do we really know about EAPs?', *Human Resource Management* **28**: 385–401.

MAIDEN, R. P. and HARDCASTLE, D. A. (1988) 'Social work education: professionalizing EAPs', *EAP Digest* **7**: 1.

MASI, D. A. (1984) *Designing Employee Assistance Programs*, New York: AMACOM.

MASI, D. A. and FRIEDLAND, S. J. (1988) 'EAP actions and options', *Personnel Journal* **67**: 61–67.

MEGRANAHAN, M. (1989) *Counselling – Practical Guide for Employees*, London: Institute of Personnel Management.

MINTZBERG, H. (1979) *The Structuring of Organizations*, Englewood Cliffs, NJ: Prentice Hall.

MONROE, J. L. Jr. (1988) 'Employee assistance programs – the legal issues', *Employment Relations Today* **15**: 239–243.

NOBILE, R. J. (1991) 'Matters of confidentiality', *Personnel* **68**: 11–12.

ORLANS, V. (1986) 'Counselling services in organizations', *Personnel Review* **15(5)**: 19–23.

OSAWA, M. N. (1980) 'Development of social services in industry: why and how?', *Social Work* **25**: 464–470.

PAPALEXANDRIS, N. (1995) 'Human resource management', in I. Brunstein (ed.), op.cit.

PETERS, T. J. (1987) *Thriving on Chaos*, New York: Alfred Knopf.

PETERS, T. J. and WATERMAN, R. H. (1982) *In Search of Excellence*, New York: Harper and Row.

POPPLE, P. R. (1981) 'Social work practice in business and industry', *Social Science Review* 55.

PORTER, M. E. (1990) *The Competitive Advantage of Nations*, New York: Free Press.

REDDY, M. (1992) Keynote Address to BAC Conference, September 1992.

REDDY, M. (1993) EAPs and Counselling Provision in UK Organisations 1993, Milton Keynes: ICAS.

ROMAN, P. M. (1981) 'Executive caravan survey results, Labor-Management', *Alcoholism Journal,* November–December.

ROMAN, P. M. and BLUM, T. C. (1992) 'The cores of employee assistance programming: cross-national applications and limitations', *EAP International* **1(1)**: 4–8.

SAMPSON, A. (1973) *The Sovereign State: the Secret History of ITT*, London: Hodder and Stoughton.

SAUTER, S. L., HURRELL, J. and COOPER, C. L. (1989) *Job Control and Worker Health*, Chichester and New York: John Wiley and Sons.

SCHLENGER, W. E. and HAYWOOD, B. J. (1976) 'Occupational programming, problems in research and evaluation', *Alcohol Health and Research World,* Spring: 18–22.

SCHMIDENBERG, O. C. and CORDERY, J. L. (1990) 'Managing employee assistance programmes', *Employee Relations* **12**: 7–12.

SCHMIDENBERG, O. C., BLAZE-TEMPLE, D. and CORDERY, J. L. (1992) 'Supervisors and EAPs: a clash between the real and the ideal', *EAP International* **1(1)**: 9–13.

SHAIN, M. and GROENEVELD, J. (1980) *Employee assistance programs, philosophy, theory, practice*, Lexington, MA: Lexington Books.

SMITH, R. L. and BUON, T. (1991) 'EAPs down under', *EAP Digest*, March-April: 31–33.

SMITH, M. L. (1988a) 'Social work in the workplace', in G. M. Gould and M. L. Smith (eds) *Social Work in the Workplace*, New York: Springer, 3–15.

SMITH, M. L. (1988b) 'With a view to the future', in G. M. Gould and M. L. Smith (eds) *Social Work in the Workplace*, New York: Springer, 343–348.

SONNELSTUHL, W. J. and TRICE, H. M. (1986) *Strategies for Improving Employee Assistance Programs*, Key Issue, No. 30, New York State School, Industrial and Labor Relations, Cornell University, Ithaca, NY: ILR Press.

STRAUSSNER, S. L. A. (1988) 'Comparison of in-house and contracted-out employee assistance programs', *Social Work* **33**: 53–55.

STEELE, P. D. (1988) 'Employee assistance programs in context: an application of the role of the constructive broker', *Journal of Applied Behavioral Science* **24**: 365–382.

TORRINGTON, D. (1989) 'Human resource management and the personnel function', in J. Storey (ed.), *New Perspectives on Human Resource Management*, London: Routledge.

CHAPTER TWELVE

Contemporary organizational realities and professional efficacy: downsizing, reorganization, and transition

RONALD J. BURKE AND MICHAEL P. LEITER

Work stressors of potential interest to both researchers and job incumbents vary with changing circumstances over time (Burke, 1988). Four structural career crises have begun to receive increased research attention during the past few years: mergers and acquisitions; retrenchments and budget cutbacks; job future ambiguity; and occupational 'locking-in', in which individuals have almost no opportunity to modify their job or change to another. Together, these developments are changing career structures throughout post-industrialized countries. These four sources of work stressors share some common features. First, they all arise from attempts by organizations to reconfigure themselves as a way of increasing productivity with fewer resources in the face of adverse economic conditions. Second, being fairly recent areas of research, relatively little empirical work has been completed. Third, these areas have vast implications for both psychological practice and intervention at both individual and organizational levels. These features suggest that structural career crises provide an important perspective on the relationship of people with their work.

Each of these structural career crises has effects on employees' capacity to develop competence in the workplace. This chapter considers the influence of contemporary organizational realities on professional efficacy. It begins by reviewing available literature regarding the four newly emerging areas of work stressors discussed above. It then examines what is known about professional efficacy in work setting – antecedents and consequences. Links between these two bodies of research are highlighted with a discussion of psychological contracts. Approaches to individual coping and organizational interventions are then proposed.

Mergers and Acquisitions

Marks and Mirvis (1986) reported that 3,284 US companies were acquired by other companies in 1985, with the dollar value of these acquisitions estimated at $150 billion. They also provided statistics showing that between 50 and 80 per cent of all mergers were financial disappointments.

Mirvis (1985) illustrated that mergers and acquisitions heightened work stressors in a case study of negotiations during the integration of a small manufacturing firm into a multi-billion dollar conglomerate following its acquisition. Mirvis traced the strategic and tactical conflicts that surfaced between the two firms during the first year of negotiations following the sale to the parties' emotional reactions to the combination. Work stressors that were heightened during this period (and preceding the sale) included uncertainty and threat (Gill and Foulder, 1978), loss of personal and organizational identities, and conflict because of incompatibilities in company management, business systems, organizational cultures and goals for the combination itself (Sales and Mirvis, 1984; Sinetar, 1981).

Marks and Mirvis (1985a) found the 'merger syndrome' (a defensive, fear-the-worst reaction) to be a common response to the uncertainty and stress of a merger. The top levels of the acquired organization report disbelief, uncertainty, fear and stress. Lower levels in the organization circulate rumours of mass lay-offs and forced relocations, pay freezes, loss of benefits, and plant closings. This response is not limited to hostile takeovers. Crisis management is the order of the day in most mergers (Marks and Mirvis, 1985b). Senior management seal themselves off, becoming less accessible, limiting their lines of communication, and leaving their staff uninformed about what is going on. Barmash (1971) in his book titled, *Welcome to our Conglomerate – You're Fired*, referred to this as 'mushroom treatment.' 'Right after the acquisition they kept us in the dark. Then they covered us with manure. Then they cultivated us. After that they let us stew for awhile. They canned us' (*ibid.*, 46).

Merger syndrome is manifested by increased centralization and decreased communication that leaves employees in the dark about what is happening in the organization. Rumours are fuelled and run rampant through the grapevine and rumour mill. In addition, 'worst case' scenarios are developed and employees become preoccupied with the merger. The result is that employees are distracted, productivity decreases and key people leave the company. Marks and Mirvis (1985a) cite a *Wall Street Journal* survey which found that nearly 50 per cent of executives in acquired firms sought other jobs within one year and another 25 per cent planned to leave within three years.

It is apparent that mergers are stressful not only for the executives embroiled in the pre- and post-merger discussions and plans, but also for the employees involved. Executives are under stress because they have more work to do in less time, and face a great deal of uncertainty about the future. They worry about job loss due to redundancy or about simply not fitting into the newly organized structure. Furthermore, they are concerned about having to demonstrate their value and skills to a new superior (Marks and Mirvis, 1985b).

Lower-level employees are under stress for some of the same reasons. They have uncertainties about the impact of the merger on specific job factors: Will there be lay-offs or dismissals? Will I have a new boss? Will I have new duties? Do I keep the same title or will it be different? Is a transfer to a new location pending? Will the present compensation package be affected? Will the benefit plan change?

Budget Cuts

The late 1970s, early 1980s and early 1990s were characterized by economic slowdown, plant closings, lay-offs, and budget cutbacks. This mood of austerity has affected private and public sector organizations alike, and is expected to continue. Most organizations are working towards balanced budgets and fiscal reponsibility—becoming 'leaner and meaner' (Hirschhorn, 1983; Levine, 1980).

These initiatives, however, have been found to increase work stressors of job incumbents (Jick, 1985; Jick and Murray, 1982; Murray and Jick, 1985). Jick and Murray (1982) identified two dimensions: severity (size of cuts, frequency of cuts, impact on goals, programme or survival prospects, existing organizational slack, and availability of alternate funds), and time pressure (amount of forewarning, response time, duration of cuts, and clarity of information). The greater the severity and time pressure associated with budget cuts, the greater the likelihood of experienced stress in an organization.

The organizational consequences of cutbacks have also been researched. Jick (1983) identified the stressful outcomes as role confusion, job insecurity, work overload, career plateauing, poor incentives, office politics and conflict, lack of participation in decision-making, tense organizational climate, ideological disagreement, and job and personal life conflicts. Cameron, Whetten and Kim (1987) identified twelve dysfunctional organizational consequences including centralization, the absence of long-range planning, the curtailment of innovation, scapegoating, resistance to change, turnover, decreased morale, loss of slack, the emergence of special interest groups (politics), loss of credibility of top management, conflict and in-fighting, and across-the-board rather than prioritized cuts.

Consequences at the individual level have also been studied. Rosselini (1981) observed that federal budget cuts had a significant role in a recent

increase in federal employees' usage of health services. In the light of anticipated staff reductions, almost triple the number of federal employees were
treated at the Department of Health and Human Services for stress-related
symptoms such as dizziness, stomach upsets and high blood pressure.
Blundell (1978) reported that government employees in Denver whose staff
had been reduced and reorganized were found to be so fearful and concerned
about their future that productivity suffered.

Schlenker and Gutek (1987) examined the effects on professionals in a large
social service agency of being reassigned to non-professional jobs. It was
possible to study the effects of work role loss separately from the loss of one's
employment and salary since one half of the social workers were abruptly
reassigned to non-professional jobs while keeping the same salary and
benefits. The sample of 132 included sixty-six reassigned workers and sixty-
six non-reassigned workers. Data were collected about nine months following
the reassignments to allow sufficient time for workers to adjust to their demotions. Individuals who were reassigned (demoted) reported significantly less
self-esteem, significantly less job and life satisfaction, and significantly
greater intention to turnover. No differences were found on measures of
professional role involvement, professional role identification (commitment
to social work), or work-related depression.

Noer (1993) used the term *lay-off survivor sickness* to capture the malaise
among survivors in the wake of massive organizational change. Noer
conducted focus groups with survivors in an organization that had undergone
significant downsizing via lay-offs. The groups reported a perception of
reduced risk taking and lowered productivity. These groups commonly
reported feeling insecure, uncertain, fearful, depressed, tired, stressed,
frustrated, angry, distrustful, guilty. Noer contends that time does not heal
these wounds. Focus groups were held in the same organization five years later
with long-term survivors. He observed a similar set of responses from the
groups at that time.

Job Future Ambiguity

One of the most dramatic changes in organizations during the past few years
has been from traditionally secure managerial and professional jobs to
insecure ones (Hunt, 1986). Several factors have come together to produce
this phenomenon. These include an increase in acquisitions and mergers,
retrenchment and decline due to increasing competitive pressures, a loss of
managerial and professional jobs in the manufacturing sectors as jobs move
to third-world countries or the service and information sectors, increasing
competition for managerial and professional jobs from younger, better
educated men and women, and the introduction of new technology which
makes old skills, knowledge and attitudes obsolete (Gill, 1985; Handy, 1984;
Levine, 1980; Meagre, 1986).

Research has indicated that this dramatic change in expectations of previously guaranteed employment has had a marked psychological impact on those affected. These effects can be seen in managers and professionals who have lost their jobs, and those who are insecure about their jobs. Managers and professionals who lose their jobs indicate increases in depression and anxiety, along with poorer emotional health and social functioning (Fineman, 1983; Fryer and Payne, 1986; Kaufman, 1982; Latack and Dozier, 1986; Leana and Ivancevich, 1987). The consistently negative effects of unemployment are attributed to loss of psychological and economic benefits inherent in managerial and professional jobs.

Although the number of managers and professionals who have actually lost jobs may be relatively small, some writers (Greenhalgh and Rosenblatt, 1984) postulate a 'ripple effect.' This is, managers and professionals who are currently employed but see that it is increasingly harder to get and hold managerial and professional jobs will become increasingly insecure about their position. They become more concerned about maintaining their own jobs, losing promotion and career development opportunities within their organization, and relinquishing prospects of finding another job.

Research on the stressful effects of job insecurity is just beginning to emerge (Brockner, 1987; Greenhalgh, 1983a, 1983b; Hartley and Klandermans, 1987; Jick, 1985 and Roskies and Louis-Guerin, 1990). The little data that exists indicates that the effects of job insecurity appear to be similar to job loss itself. Depolo and Sarchielli (1987) compared the emotional well-being of individuals who had lost their jobs with 'survivors' in the same organization. They found no difference between the two groups. The level of emotional well-being was extremely low in both groups. Cobb and Kasl (1977) conducted a longitudinal study of job loss and reported that workers were in greater distress anticipating job loss than when they actually lost their jobs.

Several studies have examined the effects of insecurity on work commitment and job behaviour. Studies by Greenhalgh (1982) and Jick (1985) show that survivors exhibit levels of reduced work commitment and effort. Other research findings show increased resistance to change among survivors (Fox and Staw, 1979; Greenhalgh, 1982). Thus job insecurity may become a critical factor in accelerating organizational decline.

Ashford (1988) found that high levels of individual strain were associated with ambiguity about prospective work roles and job activities, and continuing employment within the organization and whether they would still have jobs. As well, she observed that individuals having a high tolerance for ambiguity and reporting greater control over their job showed significantly smaller deterioration in psychological well-being.

Nelson, Cooper and Jackson (1995) examined the effects of two events (privatization, restructuring) – which occurred during a sixteen-month period – on employee satisfaction and well-being. These two events were preceded by a 25 per cent staff reduction during the preceding six years, coupled with significant changes in jobs within the organization. The magnitude of the change was substantial. Data were collected at three points in time over a twenty-month period: prior to privatization, following privatization and following reorganization. Job satisfaction was observed to fall following privatization but then increase during the period of reorganization. In addition, individuals who viewed the work environment as uncertain and who displayed higher levels of external control, reported significantly smaller increases in job satisfaction following the reorganization. Symptoms of both mental and physical distress also increased during the period of privatization.

Occupational Locking-In

Kay (1975) identified several factors associated with increasing discontent in middle management ranks. One of these was a boxed-in feeling when individuals had almost no opportunity to move from their present jobs or when the only position for which they were qualified was the one they currently held. Quinn (1975) used the term locking-in to refer to the same phenomenon, and outlined three components of locking-in: (a) low probability of securing another job as good as or better then the present one, (b) little opportunity to modify a presently disliked employment situation by securing a change in job assignments, and (c) low likelihood that a worker who was dissatisfied with the job could take psychological refuge in the performance of other roles not linked to this job. Thurley and Word-Penney (1986) suggested that this phenomenon would be present for a considerable time.

Burke (1982) conducted a study in which self-reported locking-in of 127 administrators was related to personal and situational demographic characteristics and personality variables, occupational and life demands and satisfactions, and lifestyle and emotional and physical well-being measures. Individuals reporting greater locking-in were older, less educated, had more children, longer organizational tenure, made fewer previous geographic moves and were less interested in further promotion. Locking-in was also associated with interpersonal passivity, emotional instability, an external locus of control, and reduced Type A propensities. Work experiences and satisfactions were generally unrelated to degree of locking-in, although administrators who felt more locked-in reported skill under-utilization at work and greater life dissatisfaction. Locking-in, however, was not associated with negative emotional and physical health consequences.

Wolpin and Burke (1986) replicated this study with seventy-two senior administrators of probation, parole and after care services from a single government department. Locking-in was found to be related to both personality and social competence variables. Managers reporting greater locking-in were more passive and had lower levels of initiative and drive. In addition, more locked-in managers appeared to lack interpersonal skills and sensitivities. More locked-in managers in this sample reported less marital satisfaction, greater negative feeling states (depression and worthlessness) and less life satisfaction. Finally, more locked-in administrators exhibited poorer self-reported physical health.

Personal and professional efficacy

The link between self-efficacy and psychological burnout can be found in Hall's work on psychological success (Hall, 1976). Hall's model of psychological success proposed that work motivation and satisfaction are enhanced when a person successfully and independently achieves challenging and personally meaningful goals. Such achievement leads to feelings of psychological success which in turn encourages the individual to become more job involved, to set even more challenging goals, and to experience greater self-esteem. Hall suggested that organizational characteristics associated with psychological success include challenge, autonomy, control, feedback about results, and supervisor support. Hall's views on the characteristics associated with psychological failure – the flip side of psychological success – are particularly relevant to the structural career crises considered in this chapter. Hall suggested that psychological failure is associated with emotional withdrawal from work (lower work standards, apathy, disinterest), valuing financial rewards to the neglect of intrinsic rewards, becoming defensive and rigid, battling the organization and leaving the organization.

The personal impact of structural career crises is similar in depth and scope to the crises Cherniss (1980) identified in human-service professionals during their initial career positions. He described this collapse of the professional mystique as a precursor of burnout. 'New professionals expect their work consistently to confirm their competence. They assume that professionals exercise their skills autonomously, and that their work should regularly present them with exciting, meaningful encounters in the context of supportive, collegial relationships with other professionals' (Leiter, 1991: 550). Encounters with bureaucratic and institutional constraints on professional autonomy emphasize that considerations other than competence determine professional opportunities. Often, career viability requires that employees broaden their focus from an exclusive concern with making an impact through competent service provision to include a working knowledge of political and organizational issues. A refusal or inability to adapt career expectations may result in intolerable conflict between career aspirations and organizational limitations that individuals resolve through a profound adjustment of expectations, a career change, or burnout. In a ten-year follow-up study of human service providers, Cherniss (1989) described resolutions, ranging from an equilibrium based upon revised career expectations to career shifts entirely out of human services or into service professions seen as more influential or more extrinsically rewarding.

Structural career crises involve a similar confrontation of individual career expectations with organizational limitations. In both, people lose confidence in the social conditions on which they depend to pursue their career goals. That is, employees, despite willingness and ability to work, find that the institutional framework is inadequate or antagonistic to their career development. As with the initial career crisis, employees experience conflict between career assumptions and organizational reality. With the initial crises, the violated assumptions derive from socialization prior to and during professional education. With structural career crises, the violated expectations for ongoing opportunities to work are based upon direct experience. Structural career crises have an impact beyond employees' financial and social security; they present significant challenges to their expectations of personal efficacy through their work as well. As the socio-economic framework supporting lifelong career structures undergoes fundamental changes, these assumptions become less viable and their disappointment more common.

Psychological Contracts

Structural career crises reflect a profound shift in the psychological contract of employees with employers that Rousseau (1990) defined as, 'an individual's beliefs regarding reciprocal obligations' (*ibid.*, 390). The psychological contract is promissory and reciprocal, including expectations that each party has of the other in the employment relationship. Rousseau (1990) noted that psychological contracts for Masters in Business Administration graduates entering the private sector included assumptions regarding rewards for hard work, job security and prospects for career advancement.

A unilateral change in the psychological contract produces an experience of inequity, as the employee perceives a disruption of the balance of relative contributions and rewards (Adams, 1965). Inequity initiates a negotiation process that may or may not be explicit. The outcome is either (1) resolution of a new psychological contract with a revised equity balance, (2) termination of the employment relationship, or (3) chronic conflict. Each of these three outcomes places distinct demands on individual employees. Robinson and Rousseau (1994) reported that from the perspective of employees, unilateral violations of the psychological contracts by employers are quite common. Such violations are associated with greater turnover, and diminished satisfaction or trust. In response to a violation of a psychological contract, people are more likely to feel wronged than disappointed. Such a violation is not simply the disappointment of expectations, but the betrayal of a relationship based upon trust (Rousseau and Parks, 1993). As a result employees experience the change as a stressor with significant personal implications with which they must cope. A potential consequence of diminished trust is deterioration of organizational communication (O'Reilly and Roberts, 1976). Without effective communication processes, employees and management lack the means to establish a new psychological contract. In the resulting isolation each party is likely to develop more extreme views of the other: employees focusing on management's betrayal of the previous contract, and management decrying employees' refusal to face external realities.

The discussion above suggests three types of psychological contract violation initiated by management. The first type occurs because of misunderstandings: management may never have intended to provide the expected conditions. Cherniss' discussion of the professional mystique is an example of this type. The second type of violation occurs because management decides to renege on its obligations. In the third type of violation, management is largely responding to external forces beyond their control. Most structural career crises are of the third type, in that external conditions change to such an extent that management cannot maintain the previous psychological contract. However, some major organizational restructurings are of the second type when they are initiated proactively by management despite their foreseeable impact on psychological contracts with employees. An attribution bias towards perceiving people as originators of major events inclines employees to interpret most violations as the second type (management reneging on its obligations), exacerbating the situation further.

Rousseau (1995) contrasted transactional contracts that put career advancement and monetary reward over job security with relational contracts that emphasized stability and mutual respect. Relational contracts describe career expectations of service professionals in terms of definitions of reasonable caseloads, level of supervisory support, recognition of professional prerogatives, and opportunities to pursue individual goals within the organizational context. Relational contracts are especially important in private and public sector organizations that strive towards providing exceptionally high quality services to demanding clients. These goals require staff to develop refined skills and to devote extraordinary effort and thoroughness to providing services. Commitment to the organization's goals maintains dedicated performance in the absence of direct supervision. Cook and Rousseau (1988) argued that service organizations often offer job security in exchange for employees' loyalty and commitment to the organization's values. The cost of the relational contract to the organization is balanced by enhanced performance from committed employees.

Leiter (1995) assessed the extent to which hospital employees felt at risk from various occupational hazards: work overload, decreasing career opportunities and job insecurity. This study found work overload to be primarily the concern of employees who were experiencing exhaustion, and who felt threatened by additional physical, cognitive and emotional demands in an organization with a reduced workforce. In contrast, diminishing career opportunities and dead-end jobs were primarily a concern for people who doubted their occupational skills and the significance of their work. On the other hand, employees who perceived their work to be skill-enhancing and to make a meaningful contribution to service recipients and to the organization felt less vulnerable to the risk of diminishing career opportunities. Vulnerability to job loss, however, was less tied to the job's contribution to skills or meaning, and more a function of exhaustion, cynicism and a perception that management was unable to address external demands effectively.

In summary, initial career crises arise from a conflict between, on the one hand, role expectations developed through professional training and on the other hand constraints of worksettings. In structural career crises, the conflict reflects a change in implicit contracts between employees and employers. Employees experience increased work demands (both quantitative and qualitative), decreased opportunities to develop their careers, and decreased job security. Some of these changes arise directly from the larger socio-economic context, through reduced resources for institutions. Others arise from the attempts of organizations to address external developments, such as strategies to increase productivity in a manner that infringes upon employees' discretionary activities. Regardless of the source, the implicit employment contract changes in a manner incompatible with the employees' expectations of success.

Coping with structural career crises

Individual Coping

Coping with the stresses of structural career crises requires (1) choosing the most appropriate alternative to the conflict over the employment contract and (2) successfully addressing the problems that arise from that choice. Whereas crises that arise from socioeconomic forces exceed the control of individuals or single organizations, direct action to eliminate the source of the problem is not a viable option. However, individuals and organizations may undertake coping responses that involve active intervention, cognitive restructuring or avoidance.

Armstrong-Stassen (1994) examined how lay-off survivors coped with a workforce reduction involving permanent lay-offs. Two dimensions of coping – control-orientated and escape coping – were investigated. Leana and Feldman (1992) suggested that for lay-off survivors, problem-focused coping strategies are likely to be more effective than are symptom-focused coping strategies. Ashford (1988) reported that during an organizational transition, the expression of emotion reduced stress both before and after the transition whereas denial about the change did not.

Armstrong-Stassen considered both the antecedents of coping (i.e. what was more likely to encourage the use of either the control or the escape capacity) and the consequences or outcomes from the use of these two coping strategies. She found that survivors who perceived a greater threat to their job security were more likely to use both escape and control coping strategies. In addition, survivors using more control coping reported more favourable work outcomes (commitment, job performance, lower turnover intentions). Survivors using more escape coping strategies reported less favourable work outcomes.

Whelan (1995) noted that the use of Employee Assistance Programmes (EAPs) increases during major organizational transitions. Although EAP interventions help individuals to manage stress symptoms more effectively, they do not provide a means to have an impact on the overall management of the transition. Educational and retraining programmes in conjunction with career counselling may open alternatives to enduring the impact of transitions on a specific job. In general, more effective individual coping can increase an individual's capacity to survive a major transition in an organization.

In summary, on an individual level, cognitive coping plays an important role in maintaining self-esteem by emphasizing the separation of occupational role and personal identity. A related role of cognitive coping is removing personal blame for the loss of jobs, by viewing the elimination of positions as the function of major social forces rather than of individual failure. Individual action coping plays a role when individuals adapt the way in which they perform an ongoing job in order to address new demands, or conduct searches for alternative or supplementary employment.

Organizational Coping Initiatives

A structural career crisis such as a merger, is a process rather than an event, and processes require planning, managing and monitoring in order to occur effectively. Communication and preparation are critical in this regard.

Communication

Complete, open and early communications are a necessity for all involved in both the pre- and post-merger situations. Anxiety, uncertainty and other feelings that employees have as a result of the announcement of a merger come from the lack of reliable information about the future. Employees need to know where they stand as soon as possible. If management is straightforward and honest with employees from the beginning, it is likely that employees will, in return, give efficient levels of performance during the transition period. A good communications programme should be developed before the merger is initiated. If done properly, it will help to minimize problems in the post-merger situation.

The kinds of information that should be communicated in a merger situation will depend on the intended receiver of the message. The press and other news media will probably want information regarding the reasons for the proposed combination, along with some general financial information. Management will probably want much more detail about what the organization plans to do and what part they will play in the new combination. Employees will likely be concerned with aspects of production, relocation and job rationalization.

In a pre- and post-merger situation, regular communication channels, such as company newsletters and magazines, and formal meetings, need to be expanded to include informal face-to-face sessions with top management. Daily updates that detail the progress of the merger can be posted at coffee stations and bulletin boards. In most organizations the demands of the situation will exceed the capacity of existing communication channels.

This intense communication permits the development of a shared understanding of the challenges facing the organization, that in turn facilitates confidence in new resolutions. Organizational learning requires individuals to communicate their experience in order to build a shared perspective from which to operate in the new situation. Effective organizational learning is difficult in the best of times; it is even more challenging under financial constraints. The day-to-day communication flow within the direct interactions among management personnel and between management and the employees who report to them is rarely sufficient for the task. Formal surveys are a valuable supplement to direct communication in large organizations. They provide a means of assessing defined constructs throughout an organization. They provide management with information about the perspectives and concerns of its employees.

Preparation

Preparation can and should take on a variety of forms. Courses, seminars, workshops and psychological consultation can help prepare executives for the burdens and intensities of a merger situation. Executive teams can prepare employees for the stresses that are likely to occur and can help project a more realistic scenario for the combination.

Preparation also includes the development of joint transition teams who are staffed by personnel from both firms. These teams can work together to counteract the natural 'we versus they' feelings that arise in a merger situation. Preparation simply means thinking about and planning for the many problems that can crop up in a merger, including personality mismatches between management teams, clashes of cultures, absence of common goals, the natural temptation for one firm to smother the other, lack of communication, erosion of morale and productivity, and lack of procedures to manage the merger. The ultimate preparation for top management is thinking strategically through the human implications as well as the business and financial aspects of the combination. The eventual success of company-sponsored communications and integration programmes depends to a large degree on the attitudes of managers and employees. If the merger is hostile and traumatic, the best laid communications plans may not be effective.

Schweiger and DeNisi (1991) considered the impact of a realistic merger preview, involving a programme of realistic information, on employees of an organization that had just announced a merger. Employees from one plant received the merger preview while those in another plant received only limited information. Data were collected at four points in time: before the merger was announced, following the announcement but before the realistic merger preview programme was introduced, and twice following the realistic merger programme. The study extended for a five-month period overall. Both objective and self-report data were obtained.

The announcement of the merger was associated with significant increases in global stress, perceived uncertainty and absenteeism; decreases in job satisfaction, commitment and perceptions of the company's trustworthiness, honesty and caring, and no change in self-reported performance. There were no differences between the two plants following the announcement of the merger, but the experimental plant was significantly lower on perceived uncertainty and significantly higher on job satisfaction, commitment, and perceptions of the company's trustworthiness, honesty and caring following the realistic merger preview programme. These differences were also present three months later.

Reducing the Stress of Staff Reductions: Integrated Models

Several integrated models have been described to help organizations cope with the stress of major restructurings. Bridges (1986) described the personal experience of employees undergoing significant organizational change as following a sequence: a focus on the resentment and betrayal associated with losing the past gives way to a neutral zone that in turn develops into a forward-looking focus on new beginnings. Many aspects of the process that Bridges described are part of the negotiation of a new psychological contract throughout an organization. A neutral, less emotionally charged state is necessary for employees and management to move past blame, suspicion and misinformation to negotiate, implicitly or explicitly, a new working relationship characterized by reciprocity.

The willingness of employees to negotiate a new resolution requires confidence in the effectiveness of management's approach to addressing external forces and in the genuineness of management's efforts towards negotiating a renewed contract that is mutually beneficial. An acceptable and enduring resolution can only occur when the organization is addressing external demands effectively. That is, management must be seen to have an understanding of political realities and to be taking full advantage of opportunities for the organization that may arise. Although individual organizations are not able to control these large scale environmental developments, they may adapt to the changes with varying degree of success. Otherwise, employees cannot be confident that a new resolution is sufficiently secure to warrant their commitment. It is also important that employees perceive management to be acting fairly to permit the establishment of a new equilibrium of relative inputs and outcomes. Without trust in management's intentions, employees fear that the new resolution will be one of exploitation rather than one of reciprocity.

The importance of management's success at addressing external demands is demonstrated in a study by Greenhalgh (1982) that described and assessed an action research programme to alter an organization's procedures for achieving workforce reductions in response to declining need for its services. The organizations were state-supported hospitals. Following state guidelines, they had historically conducted workforce reductions through forced lay-offs. A cost-benefit study compared two alternative workforce reduction strategies (lay-off and planned attrition) and found that the lay-off strategy was less cost-effective.

An opportunity to use an attrition programme arose when the state decided to consolidate three urban hospitals. An agreement was made to accomplish the consolidation through transfer and attrition rather than the simple use of lay-off. Thus potentially affected employees were guaranteed that they would not be summarily laid off. The consolidation plan provided for each employee to be given at least one opportunity for continued state employment in an equivalent position within reasonable commuting distance. Such opportunities were arranged by coordinating the hiring within the urban areas by various state agencies. The timing of the consolidation plan was arranged to permit the orderly transfer of employees who did not retire or voluntarily leave state service. A mechanism to retrain some workers was also established.

Evaluation data were collected from employees transferred from the two closing hospitals (one hospital remained open). Measures of job security, productivity and propensity to quit were examined. Employees in the demonstration project were similar to control employees on the last two measures, but lower on job security. In addition, employees in the hospital that remained open, were relatively neglected in the demonstration. They reported lower job security, lower productivity and greater propensity to leave.

Blake and Mouton (1984) provide a detailed case study of the application of their interface conflict-solving model to the merger of two organizations. An American company had acquired a British company; both had previously been competitors. The intervention included top teams of both organizations and involved a series of day-long meetings. The process used perception-sharing, identification of concerns and questions and development of a sound operating model by groups from both organizations. Qualitative data collected two years later suggested that the merger was a success. No senior personnel had left either organization.

Noer (1993) considers lay-off survivor sickness – a condition that affects people who remain with an organization following downsizing. This syndrome is characterized by fear, insecurity, frustration, resentment, anger, sadness, depression, guilt, distrust, a sense of unfairness and betrayal. Bunker (1994), building on Noer's work, offers a model to aid understanding of individuals and organizational responses to downsizing and other types of organizational transitions. His R (response) factor model uses two response dimensions – learning readiness/comfort with change, and ability to learn/capacity for learning – to produce four primary responses: entrenched, overwhelmed, 'bs'-ing and learning. Individuals in the entrenched group are anxious and angry and rely on previously learnt strategies. Those in the overwhelmed group are depressed and powerless and tend to withdraw from what is happening. Those in the 'bs' group (individuals who overestimate their strengths and underestimate their weaknesses) appear confident with change and want to move forward but because they do not learn well, they move away from organizational goals. Individuals in the learning group feel challenged and eager to pursue new learning opportunities. Bunker estimates that as many as 60 per cent of employees in a downsized or changed organization are entrenched or overwhelmed. He suggests that 10 to 15 per cent fall in the 'bs'-ing group and between 15 to 25 per cent comprise the learning group.

Bunker proposes that individuals in each group require different kinds of help to become successful survivors. The overwhelmed need to be developed in place and supported by supervisors and peers. The entrenched require structured, measured learning activities, encouragement, developmental job opportunities and a risk-free environment to apply new learning. The 'bs'-ers need straight, regular feedback and a focus on development. Finally, learners benefit from demanding developmental assignments and supports and rewards for undertaking key roles in the transition process.

Noer (1993) offers a four-level process for handling lay-offs and their effects. The first level of intervention addresses the lay-off process itself. Organizations that effectively manage the lay-off process will reduce, but not eliminate, lay-off survivor sickness. The second level of intervention addresses the grieving process by providing an opportunity for catharsis in the releasing of repressed feelings and emotions. The third level of intervention helps survivors regain their sense of control, confidence, self-esteem and efficacy. The fourth level of intervention develops organizational policies, procedures and structures which will prevent future lay-off survivor sickness. This includes the use of job enrichment and employee participation, employee autonomy, non-traditional career paths, short-term job planning, and the encouragement of employee independence and empowerment.

Coping on an organizational level is built upon organizational learning, or the development of organizational practices on the basis of shared experience (Nevis, DiBella and Gould, 1995). It requires that the people in the organization adapt the way that they work with one another to address demands effectively. This change includes the renegotiation of psychological contracts as financial constraints make the previous relational contracts non-viable. Employees fear that the move away from a relational contract is towards a desperation contract rather than a transactional contract. This fear is fuelled by fact that the benefits of transactional contracts are not readily apparent: pay is not higher (often lower), and career opportunities internally and externally are rare.

An alternative is a new relational contract built upon a shared commitment to weather the storm: 'times are tough all around, but we will see this through together because of our shared commitment to the organization's mission'. In place of the established relational offer of job security, management offers its best intentions to maintain a stable work environment, to support professional development and to fulfil the organization's mission. This approach requires a clear understanding of organizational goals to avoid the work overload that can accompany attempts to maintain output during major staff reductions. This resolution offers the best prospects for negotiating economic and social agreements of the most benefit to those concerned: employees, service recipients, management, and the larger community. The establishment of a new relational contract requires a rich flow of information among employees and management.

In summary, structural career crises require coping on the level of the individual and the organization (see Figure 12.1). Education and communication are important at every level of intervention. Particularly important is the development of a problem-solving perspective that encourages action-oriented coping in the form of building a new psychological contract.

Conclusion

We believe that contemporary organizational realities continue to take their toll on employers at all levels, through what Noer (1993) refers to as a widespread malaise. A focal point of this malaise is a crisis in personal and professional efficacy. Poorly managed structural career crises result in individuals becoming less effective than they desire to be and their employing organizations becoming less vibrant and competitive than they need to be. A collaborative solution is possible and required.

Individual

Education

EAP

Information seeking

Career counselling

Retraining

Organization

Communication

Controlling speed of change

Conflict and resistance management

Human resource interventions

Building a new psychological contract

Figure 12.1. Coping interventions.

We also think that individuals and their employing organizations can do a much better job adopting and changing these realities. Our sense is that management has a particularly important role to play in averting the destructive pattern highlighted in the research studies we review. Their success in this regard has consequences for the larger culture as well as for organizational survival. Valid insight into the personal impact of structural career crises, coupled with strategic action, is vital in improving the quality of work experiences and performance of these organizations.

Enduring systematic changes in the nature of work are shaping psychological contracts of employees throughout North America. While coping with these changes challenges the skills and endurance of individuals, it places even greater pressure upon the social environment of organizations. The unilateral imposition of new employment relationships works only in the short term, and then only when management has the resources and the means to monitor closely employees' contributions. A fragmented, uncommitted group of employees will tend to pursue their personal interests, which may only partially overlap with the interests of the larger organization.

Financial constraints and changing political ideologies of government are putting the management of organizations in post-industrialized nations under great pressure. Their principal source of funding and legitimacy are demanding changes in the kinds of services offered, the cost of these services, and the way in which people are managed. Demands for increased cost-effectiveness or productivity quickly translate into changes in the psychological contract of organizations with their employees. In fact, political agendas often include reducing the professional prerogatives of public service employees as an end in itself. Management requires an accurate and rich information flow regarding the experience of their employees in order to manage a smooth transition to a new, mutually acceptable, psychological contract.

Figure 12.2 depicts the central concepts of this chapter. Broad socio-economic conditions are background variables that set the stage for the four forms of structural career crises. These, in turn, have an impact on individuals who remain with the organization (survivors) and those who depart (job losers). Research questions suggested by this model include those aimed at differentiating the impact of each of the four types of structural career crisis on individuals, both survivors and job losers. This research would lead directly to intervention studies examining the impact of a variety of coping strategies as depicted in Figure 12.1 on these relationships. Of particular interest are programmes that assist individuals to maintain their sense of professional efficacy, and which help organizations and employees to build productive, enduring and viable psychological contracts.

Downsizing in the private sector over the previous decade has shown mixed success. Attempts to increase productivity through retrenchment have often decreased cost-effectiveness or productivity. A major factor in the success of such undertakings is the involvement and commitment of employees, whose active participation is essential to the performance of human-service organizations.

Background variables
Increased competitiveness
Economic recession
Reduced government spending

Exogenous variables
Job future ambiguity
Retrenchment and budget cutbacks
Occupational locking-in
Mergers and acquisitions

Performance and Health Consequences

Survivors	*Job losers*
Guilt	Resentment
Workload expansion	Lack of structure
Changes in social networks	Changes in social networks
Increased exhaustion	Increased cynicism
Lowered accomplishment	Coping with job loss
(Psychological failure)	(Psychological failure)

Figure 12.2. Conceptual model.

Keeping employees engaged in the organization's mission during major organizational changes requires that management and employees actively develop a new psychological contract that includes a recognition of political/economic realities, but maintains a relational commitment.

Notes

Preparation of this chapter was supported in part by the Faculty of Administrative Studies, York University and by the Social Sciences & Research Council of Canada. Bruna Gaspini and Phyllis Harvie contributed to the preparation of the manuscript.
    The authors are listed alphabetically, having contributed equally to the chapter.

References

ADAMS, J. S. (1965) 'Inequity in social exchange', in L. Berkowitz (ed.) *Advances in Experimental Social Psychology*, New York: Academic Press, **2**: 267–299.
ARMSTRONG-STASSEN, M. (1994) 'Coping with transition: a study of layoff survivors', *Journal of Organizational Behavior* **15**: 597–621.

ASHFORD, S. J. (1988) 'Individual strategies for coping with stress during organizational transitions', *Journal of Applied Behavioral Sciences* **24**: 19–36.

BARMASH, T. (1971) *Welcome to our Conglomerate – You're Fired*, New York: Delacorte Press.

BLAKE, R. R. and MOUTON, J. S. (1984) *Solving Costly Organizational Conflicts*, San Francisco: Jossey-Bass.

BLUNDELL, W. E. (1978) 'As the ax falls so does productivity of grim US workers', *Wall Street Journal*.

BRIDGES, W. (1986) 'Managing organizational transitions', *Organizational Dynamics* **15**: 24–33.

BROCKNER, J. (1987) 'The effects of work layoffs on survivors: a psychological analysis', in R. P. McGlynn (ed.) *Interfaces in psychology*, vol. 5, Lybbock, TX: Texas Tech Press.

BUNKER, K. A. (1994) 'The "R factor" in downsizing', *Issues and Observations* **14**: 8–9.

BURKE, R. J. (1982) 'Occupational locking-in: some empirical findings', *Journal of Social Psychology* **118**: 177–185.

BURKE, R. J. (1988) 'Identity processes and social stress', *American Sociological Review* **56**: 836–849.

CAMERON, K. S., WHETTEN, D. A. and KIM, M. U. (1987) 'Organizational dysfunctions of decline', *Academy of Management Journal* **30**: 126–137.

CHERNISS, C. (1980) *Professional Burnout in Human Service Organizations*, New York: Praeger.

CHERNISS, C. (1989) 'Career stability in public service professionals: a longitudinal investigation based on biographical interviews', *American Journal of Community Psychology* **17**: 399–422.

COBBS, S. and KASL, S. V. (1977) *Termination: the Consequences of Job Loss*, Cincinnati: NIOSH Research Report.

COOK, R. A. and ROUSSEAU, D. M. (1988) 'Behavioral norms and expectations: a quantitative approval to the assessment of organizational culture', *Groups and Organizational Studies* **13**: 245–273.

DEPOLO, M. and SARCHIELLI, G. (1987) 'Job insecurity, psychological well-being and social representations: a case of cost sharing', in H. W. Scroiff and G. Debus (eds) *Proceedings of the West European Conference on the Psychology of Work and Organization*, Amsterdam: Elsevier Science Publishers.

FINEMAN, S. (1983) *White Collar Unemployment: Impact and Stress*, Chichester: John Wiley.

FOX, F. V. and STAW, B. M. (1979) 'The trapped administrator: effects of job insecurity and policy resistance upon commitment to a course of action', *Administrative Science Quarterly* **24**: 449–471.

FRYER, D. and PAYNE, R. (1986) 'Being unemployed: a review of the literature on the psychological experience of unemployment', in C. L. Cooper and I. Robertson (eds) *International Review of Industrial and Organizational Psychology*, New York: John Wiley.

GILL, C. (1985) *Work, Unemployment and the New Technology*, Cambridge: Policy Press.

GILL, J. and FOULDER, I. (1978) 'Managing a merger: the acquisition and its aftermath', *Personal Management* **10**: 14–17.

GREENHALGH, L. (1982) 'Maintaining organizational effectiveness during organizational retrenchment', *Journal of Applied Behavioral Science* **18**: 155–170.

GREENHALGH, L. (1983a) 'Organizational decline', *Research in the Sociology of Organizations* **2**: 231–276.

GREENHALGH, L. (1983b) 'Managing the job insecurity crisis', *Human Resources Management* **4**: 431–444.

GREENHALGH, L. and ROSENBLATT, Z. (1984) 'Job insecurity: toward conceptual clarity', *Academy of Management Review* **9**: 438–448.

HALL, D. T. (1976) *Careers in Organizations*, Pacific Pallisades, CA: Goodyear.

HANDY, C. (1984) *The Future of Work*, Oxford: Blackwell.

HARTLEY, J. and KLANDERMANS, P. G. (1987) 'Individual and collective responses to job insecurity', in H. W. Scroiff and G. Debus (eds) *Proceedings of the West European Conference on the Psychology and Work Organization*, Amsterdam: Elsevier Science Publishers.

HIRSCHHORN, L. (1983) *Cutting back*, San Francisco: Jossey-Bass.

HUNT, J. W. (1986) 'Alienation among managers – the new epidemic or the social scientists' invention?', *Personnel Review* **15**: 21–26.

JICK, T. D. (1983) 'The stressful effects of budget cuts in organizations', in L. A. Rosen (ed.) *Topics in Managerial Accounting*, New York: McGraw-Hill, 267–280.

JICK, T. D. (1985) 'As the axe falls: budget cuts and the experience of stress in organizations', in T. A. Beehr and R. S. Bhagat (eds) *Human Stress and Cognition in Organizations: An Integrated Perspective*, New York: John Wiley, 83–114.

JICK, T. D. and MURRAY, V. V. (1982) 'The management of hard times: budget cutbacks in public sector organizations', *Organization Studies* **3**: 141–170.

KAUFMAN, H. G. (1982) *Professionals in Search of Work: Coping with the Stress of Job Loss and Underemployment*, New York: John Wiley.

KAY, E. (1975) 'Middle Management', in J. O'Toole (ed.) *Work and the quality of life*, Cambridge, MA: MIT Press.

LATACK, J. C. and DOZIER, J. B. (1986) 'After the axe falls: job loss on a career transition', *Academy of Management Review* **11**: 375–392.

LEANA, C. R. and FELDMAN, D. C. (1992) *Coping with Job Loss: How Individuals, Organizations, and Communities Respond to Layoffs*, New York: Lexington Books.

LEANA, C. R. and IVANCEVICH, J. M. (1987) 'Involuntary job loss: institutional interventions and a research agenda', *Academy of Management Review* **12**: 301–312.

LEITER, M. P. (1991) 'The dream denied: professional burnout and the constraints of service organizations', *Canadian Psychology* **32**: 547–561.

LEITER, M. P. (1995) *Hamilton Civic Hospitals Survey Report*, Wolfville, NS: Acadia University, Centre for Organizational Research and Development.

LEVINE, C. H. (ed.) (1980) *Managing Fiscal Stress*, Chatham, NJ: Chatham House Publishers, Inc.

MARKS, M. L. and MIRVIS, P. H. (1985a) 'Merger syndrome: stress and uncertainty', *Mergers and Acquisitions* **20**: 50–55.

MARKS, M. L. and MIRVIS, P. H. (1985b) 'Merger syndrome: management by crisis', *Mergers and Acquisitions* **20**: 70–76.

MARKS, M. L. and MIRVIS, P. H. (1986) 'The merger syndrome', *Psychology Today* **20**: 36–42.

MEAGRE, N. (1986) 'Temporary work in Britain', *Employment Gazette* **94**: 7–14.

MIRVIS, P. H. (1985) 'Negotiations after the sale: the roots and ramifications of conflict in an acquisition', *Journal of Occupational Behavior* **6**: 65–84.

MURRAY, V. V. and JICK, T. D. (1985) 'Taking stock of organizational decline management: some issues and illustrations from an empirical study', *Journal of Management* **11**: 111–123.

NELSON, A., COOPER, C. L. and JACKSON, P. R. (1995) 'Uncertainty amidst change: the impact of privatization on employee job satisfaction and well-being', *Journal of Occupational and Organizational Psychology*, **68**: 57–71.

NEVIS, E. C., DIBELLA, A. J. and GOULD, J. M. (1995) 'Understanding organizations as learning systems', *Sloan Management Review*, Winter: 73–85.

NOER, D. (1993) *Healing the Wounds: Overcoming the Trauma of Layoffs and Revitalizing Downsized Organizations*, San Francisco: Jossey-Bass.

O'REILLY, C. and ROBERTS K. (1976) 'Relationships among components of credibility and communication behaviors in work units', *Journal of Applied Psychology* **61**: 99–102.

QUINN, R. P. (1975) *Locking-in as a moderator of the relationship between job satisfaction and mental health*, unpublished manuscript, Survey Research Center, University of Michigan, Ann Arbor.

ROBINSON, S. L. and ROUSSEAU, D. M. (1994) 'Violating the psychological contract: not the exception but the norm', *Journal of Organizational Behavior* **15**: 245–259.

ROSKIES, E. and LOUIS-GUERIN, C. (1990) 'Job insecurity in managers: antecedents and consequences', *Journal of Organizational Behavior* **114**: 617–630.

ROSSELLINI, L. (1981, 16 December) 'Federal cuts increasing workers' stress levels', *New York Times.*

ROUSSEAU, D. M. (1990) 'New hire perceptions of their own and their employer's obligations: a study of psychological contracts', *Journal of Organizational Behavior* **11**: 389–400.

ROUSSEAU, D. M. (1995) *Psychological Contracts in Organizations: Understanding Written and Unwritten Agreements*, Thousand Oaks, CA: Sage.

ROUSSEAU, D. M. and PARKS, J. M. (1993) 'The contracts of individuals and organizations', in L. L. Cummings and B. M. Staw (eds) *Research in Organizational Behavior*, vol. 15, Greenwich, CT: JAI Press, 1–43.

SALES, A. L. and MIRVIS, P. H. (1984) 'Acquisition and Collision of Cultures,' in R. Quinn and J. Kimberly (eds) *Managing Organizational Transitions*, New York: Dow Jones.

SCHLENKER, J. A. and GUTEK B. A. (1987) 'Effects of role loss on work-related attitudes', *Journal of Applied Psychology* **72**: 286–293.

SCHWEIGER, D. M. and DENISI, A. A. (1991) 'Communication with employees following a merger: a longitudinal field experiment', *Academy of Management Journal* **34**: 110–135.

SINETAR, M. (1981) 'Mergers, morale and productivity', *Personnel Journal* **60**: 863–867.

THURLEY, K. and WORD-PENNEY, C. (1986) 'Changes in the roles and functions of middle management: a literature survey of English language publications', Report to the European Foundation for the Improvement of Living and Working Conditions, London: London School of Economics.

WHELAN, J. (1995) 'The impact of occupational stress upon EAP use in a healthcare organization', in M. P. Leiter (Chair), *Burnout in the 1990s: research agenda and theory*, Invited Symposium, Canadian Society for Industrial/Organizational Psychology, Annual Convention of the Canadian Psychological Association, Charlottetown, PEI, June 1995.

WOLPIN, J. and BURKE, R. J. (1986) 'Occupational locking-in: some correlates and consequences', *International Review of Applied Psychology* **35**: 327–345.

CHAPTER THIRTEEN

Employee adjustment to an organizational change: a stress and coping perspective

DEBORAH J. TERRY AND VICTOR J. CALLAN

Many Australian organizations, like their international counterparts, are currently experiencing massive change to their structures, procedures and personnel in an effort to become more efficient and competitive. Australia's foreign debt, problems in export performance and declining international competitiveness relative to other Organization of Economic Cooperation and Development (OECD) countries are frequently cited as evidence for the need to restructure Australian industry. In addition, the establishment of a more efficient and cost-effective public service has become a major item of reform for state and federal governments. Although some organizational change has involved fine tuning, many organizations are experiencing radical transformation towards regaining strategic fit (see Dunphy and Stace, 1990).

It is well established that organizational change may create job loss, reduced status, loss of identity, interpersonal conflict at work and home, and threats to individual self-esteem and well-being (Schweiger and Ivancevich, 1985). Employees report feeling anxious and uncertain (Ashford, 1988; Kanter, 1983), especially about how the change will affect the nature of their work, career paths, co-worker relations, and reporting relationships. Despite the fact that there is evidence that organizational change may have such negative effects on employees, few researchers have directed their attention towards the identification of factors that distinguish those employees who adjust successfully to such change from those whose level of adjustment is poor. Given that the experience of change has been isolated as a central defining feature of potentially stressful events (see Holmes and Rahe, 1967), the stress and coping perspective may be an appropriate theoretical basis for this type of research. The primary aim of the present chapter is to describe a stress-coping model that outlines a set of predictors of adjustment to organizational change. Using features of Lazarus' (1990) and Folkman's (1984) cognitive-phenomenological model of stress and coping, the model proposes that an understanding of how people adjust to organizational change lies with knowledge of the event characteristics, the employees' situational appraisals, their coping responses and the extent to which they have access to personal and social resources. A description of the proposed model is presented below, followed by a discussion of the results of recent research that has been conducted to test the utility of the proposed model of adjustment to organizational change.

Event characteristics

When considering the impact of organizational change, it is necessary to take into account the characteristics of the situation. For many types of change – such as a major internal restructuring – the implications of the change process may not be constant for all employees. Thus, the extent to which employees' work demands and operations are affected by the change needs to be taken into account, a proposal that accords with the view that stress-induced distress is a function of the amount of change experienced (Holmes and Rahe, 1967). Other event characteristics that may be relevant to an understanding of the effects of organizational change include the perceived extent of employee participation in the implementation of the change. As Dunphy and Stace (1990) noted, organizational change can be implemented in either a participative (collaborative, consultative) or non-participative (directive or coercive) manner. Most models of organizational change advocate that participation is a key factor that characterizes successful change (e.g. Dunphy and Stace, 1990). In circumstances where employees perceive that they have not been consulted about the change process, levels of intolerance to the potential effects of the change are likely to be high, which will presumably impair employees' adjustment to the event. There is some support for this proposal, to the extent that employee participation in organizational decision making has been linked to favourable job attitudes and low job turnover (Spector, 1986). Related to the extent of employee participation in the change process, is the employee's perceptions of the effectiveness of leadership during the period of organizational change. As noted by Schweiger, Ivancevich and Power (1987), effective management during such periods plays an important role in reducing the levels of stress associated with the change. If leaders appear to have a clear vision of the destiny of the changed organization, it should also help employees to achieve a sense of control over the event (see Yukl, 1989). On this basis, it can be proposed that employees will adjust better to organizational change if they perceive that leadership during the change process is effective.

Adjustment to organizational change should also be influenced by the extent to which employees feel that the details of the merger process have been communicated clearly to them. Schweiger and De Nisi (1991) found that realistic communication programmes during an organizational merger helped employees cope with the uncertainty of the situation. In a similar vein, Callan and Dickson (1993) found that managers who were satisfied with the internal communication programme set up to inform staff of changes used more effective coping strategies than the dissatisfied managers. In addition to the effectiveness of the communication about the change, the source of information that employees rely on to obtain information about the situation needs to be taken into account. If employees are relying on informal sources of information during a period of organizational change, they may be vulnerable to the rumours and speculations known to be widespread during such periods (Covin, 1993; Larkin and Larkin, 1994). Because rumours are often misinformative, reliance on informal sources of information is likely to heighten the anxiety associated with the situation (DiFonzo, Bordia and Rosnow, 1994), and hence should be related negatively to adjustment.

Situational appraisals

The proposed model of adjustment to organizational change proposes further that the manner in which employees cognitively construe the event will influence their level of adjustment. As noted by Beehr and Newman (1978), to understand employees' adjustment to work stress it is necessary to consider not only the event characteristics but also their appraisals of the situation, a proposal that is consistent with the central role accorded to appraisal in current models of stress (Lazarus and Folkman, 1984; Lazarus, 1990). Situational appraisals reflect the person's subjective evaluation of the relevance of the situation to his or her level of well-being (Folkman, 1984). Central to a person's appraisal of a situation is the extent to which it is considered to be stressful, an evaluation that is referred to by Lazarus and Folkman (1984) as primary appraisal. The level of appraised stress is proposed to have a negative influence on adjustment to stress because high levels of appraised stress are likely to exceed the coping skills and resources available to the individual (Thoits, 1983). Empirical studies of the effects of stress on well-being have generally supported this proposal (Aldwin and Revenson, 1987).

According to Lazarus and Folkman (1984), people engage not only in primary appraisal, but also in secondary appraisal, a response that involves the evaluation of what can be done to manage the situation. There is some support for the view that the latter type of appraisal comprises two separate judgements: the perceived controllability of the situation and the person's efficacy expectancies (Terry, 1991a, 1994; Terry, Tonge and Callan, 1995). In relation to perceptions of event controllability, there is a substantial amount of evidence suggesting that the belief that one has control over the impact of a stressor reduces this impact on both physiological and psychological indices of well-being (see Lazarus and Folkman, 1984) and facilitates the development of effective coping strategies (Terry, 1991a, 1994). This occurs presumably because the perception that an event is amenable to personal control reduces the level of threat associated with the situation and provides a clear basis for the development of problem-oriented coping responses.

As well as appraising the event in terms of its control, people's efficacy expectancies or, in other words, their expectancies concerning the likelihood that they can perform the behaviours necessary to deal with the event (Bandura, 1977), also need to be taken into account. Employees who doubt their ability to respond to the demands of the organizational change are likely to focus attention on their feelings of incompetence, which will be accompanied by feelings of psychological distress, and a failure to deal with the situation (Bandura, 1977). In contrast, efficacious individuals are unlikely to be distressed by feelings of inadequacy and, for this reason, are expected to persist in their efforts to manage the situation. In support of these proposals, Bandura (1982) found that in threatening situations, low levels of self-efficacy had adverse effects on indices of physiological functioning, including heart rate and blood pressure. Using self-report measures of psychological well-being, Terry (1992) found evidence of a negative relationship between levels of self-efficacy and adjustment to stress in a sample of heart attack patients, whereas Terry, Tonge et al. (1995) found indirect effects (through coping) of employee perceptions of their ability to manage a stressful work encounter on measures of employee well-being and job satisfaction.

An additional dimension on which events can be appraised – and one that may be relevant to the context of organizational change – is situational uncertainty. Uncertainty reflects perceptions of ambiguity concerning the nature and likely effects of the event (Lazarus and Folkman, 1984). Uncertainty is likely to impair adjustment to stress, not only because the person has difficulty evaluating accurately the significance of the event for his or her well-being, but also because ambiguity interferes with the development of appropriate coping strategies (Lazarus and Folkman, 1984). Research has provided some evidence that the adverse effects of stress associated with organizational change are heightened in ambiguous situations (Ashford, 1988).

Coping strategies

The third major proposal of the model guiding this research is that the coping strategies that employees use in response to organizational change will influence their adjustment to the event. Coping strategies are adopted by individuals with the intention of reducing the effects of stress. A number of researchers have made the distinction between problem- and emotion-focused coping strategies (e.g. Billings and Moos, 1981; Lazarus and Folkman, 1984). Problem-focused strategies are directed towards the management of the problem, whereas emotion-focused strategies involve a failure to face the problem, dealing instead with the associated level of emotional distress. Because problem-focused strategies are directed towards the management of the problem, they are generally thought to have positive effects on adjustment. In contrast, the fact that emotion-focused strategies concentrate not on the problem, but deal with the concomitant level of distress, has meant that this type of coping is thought, in most instances, to impair adjustment to stress (Lazarus and Folkman, 1984).

Although there is support for the proposed effects of problem- and emotion-focused coping in the general stress literature (e.g. Aldwin and Revenson, 1987), the effects of coping in the occupational context are not clear. In this context, there is some evidence to suggest that a reliance on emotion-focused coping does have a negative effect on well-being (e.g. Israel, House, Schurman, Heaney, and Mero, 1989). The evidence, however, linking problem-focused strategies to well-being is weaker (see Innes and Kitto, 1989; Kirmeyer and Dougherty, 1988). On the other hand, our recent research has supported the predictions of Lazarus and Folkman (1984), with evidence of the proposed effects of both problem- and emotion-focused strategies on adjustment to work stress (Terry, Rawle and Callan, 1995; Terry, Tonge et al., 1995) and, more specifically, adjustment to organizational change (Callan, Terry and Schweitzer, 1995). Thus, on the basis of this type of evidence, problem-focused coping should have a positive relationship with measures of adjustment to organizational change. The use of emotion-focused strategies, on the other hand, should be associated with poor adjustment.

Coping resources

The fourth major prediction of the proposed model of adjustment to organizational change is that the coping resources of employees will influence their level of adjustment to this type of work stress. Coping resources are relatively stable characteristics of people's dispositions and environments, and refer to what is available to them when they develop their coping strategies (Moos and Billings, 1982). At the dispositional level, negative affect is regarded as the key enduring personality characteristic that needs to be considered in studies of adjustment to work stress. Individuals high in negative affectivity have a tendency to experience high levels of distress, a tendency that is typically assessed with measures of trait anxiety and neuroticism (Watson and Pennebaker, 1989). People high in neuroticism are particularly likely to be affected adversely by work stress (Parkes, 1990), and to rely more on emotion-focused strategies and less on problem-focused strategies than individuals who have low levels of neuroticism (Bolger, 1990; Carver, Scheier and Weintraub, 1989). One explanation for these results is that the tendency for high neurotics to focus on their emotional state will interfere with the development of goal-directed behaviour (Carver *et al.*, 1989; Watson and Pennebaker, 1989). In addition to having substantive effects in the stress process, neuroticism is an important control variable to include in studies of work stress. As noted by Watson and Pennebaker (1989), negative affectivity is likely to act as a nuisance variable in self-report (i.e. mono-method) studies of the effects of stress because negative affectivity may be tapped by measures of both stress and well-being, thus inflating the correlations between the predictors and the measures of adjustment. In the organizational context, there is some evidence that this is indeed the case. Control of the effects of negative affectivity has been shown to reduce the strength of the relationship between measures of work stress and levels of well-being (Brief, Burke, George, Robinson and Webster, 1988; Parkes, 1990).

In the stress literature, attention has also been focused on the potential role of generalized control beliefs and self-esteem. Individuals with an internal locus of control (a belief in themselves having control over their own 'destiny') are less likely to doubt the efficacy of their attempts to confront a problem than individuals with external control beliefs (beliefs that success or failure are attributable to outside forces). Thus internal dispositional control beliefs should be associated with positive adjustment to stress (Wheaton, 1983), a proposal that has received some support in the organizational context (Ashford, 1988; Kobasa, Maddi and Kahn, 1982). In addition, there is evidence that subjects with high self-esteem are less likely to experience difficulties in adjusting to both general (Folkman, Lazarus, Dunkel-Schetter, DeLongis and Gruen, 1986) and work-related stress (Ashford, 1988; Israel *et al.*, 1989) than people with low self-esteem. Such results presumably reflect the fact that people with high self-esteem have a past history of coping successfully with stress (Chan, 1977) and thus may be impervious to the ego-threatening nature of many stressors.

In addition to personal resources, people have access to resources in their social environment. Specifically, people's relationships with others are a potential source of support during times of stress. The effect of social support on adjustment to work stress has received a considerable amount of empirical attention. Research of this type has typically found that work sources of support are more important than non-work sources (e.g. Ganster, Fusilier and Mayes, 1986; Terry, Nielsen and Perchard, 1993; Terry, Rawle *et al.*, 1995), possibly because they are proximal to the source of stress and hence more able to offer relevant support than sources of support outside the organization. The two major sources of support in the work context are one's supervisor and one's work colleagues (see Caplan, Cobb, French, Harrison and Pinneau, 1975). In the context of organizational change, a supportive relationship with one's supervisor is likely to be particularly important because this person is in a key position to provide both relevant instrumental assistance and emotional support – in the form of empathy and understanding – to employees (see Schweiger *et al.*, 1987). In situations of uncertainty – such as organizational change – support from colleagues should be important, given that co-workers are an important reference point for social comparison and hence evaluation of the validity of one's response to the event (see Terry, Rawle *et al.*, 1995; Thoits, 1986).

Relationships among the proposed predictors of adjustment

When formulating the proposed predictors of adjustment to organizational change into an overall model, a number of interrelationships among the variables can be proposed. In the first instance, it can be proposed that coping strategies will mediate the effects of the event characteristics, situational appraisals and resources on adjustment – in other words, event characteristics, situational appraisals and resources should have only indirect effects on adjustment, via their effects on coping (see Figure 13.1). This model accords with the view that coping is a key mediating variable in the stress-adjustment relationship (Lazarus, 1990), and is supported by evidence that event characteristics, situational appraisals and coping resources do influence coping responses (e.g. Holahan and Moos, 1987; Parkes, 1986; Terry, 1991a, 1994; Terry, Rawle *et al.*, 1995). Second, it can be proposed that coping resources and event characteristics will have indirect effects – through situational appraisals – as well as direct effects on employees' coping responses (see Figure 13.1). According to Lazarus and Folkman's cognitive-phenomenological model of stress and coping (see Lazarus and Folkman, 1984; Lazarus, 1990), people's appraisals of a potentially stressful situation are shaped by both the characteristics of the situation and their coping resources (see Terry, 1991b, for evidence supporting this proposal). In support of the proposal that

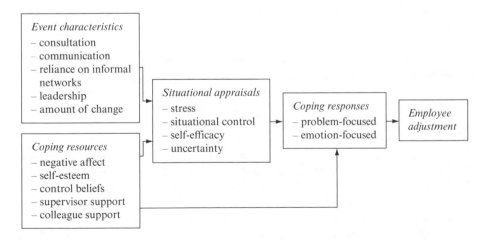

Figure 13.1 Proposed model of employee adjustment to organizational change.

both situational appraisals and coping responses are mediating variables in the stress process, Terry, Tonge *et al.* (1995) found that a model that incorporated both sets of mediating relationships provided a better fit to the data than the model that proposed that coping was the only mediating variable.

Empirical tests of proposed model of adjustment to organizational change

A Merger Between Two Airlines

As an initial test of the proposed model of adjustment to organizational change, a study was undertaken of a sample of fleet staff (pilots and engineers) involved in an airline merger. Specifically, the participants in the research had been previously employed in either one of two airline companies – a domestic (short-haul) airline and an international (long-haul) airline – that had recently merged into a new company that had retained the name of the latter airline (see Terry, Callan and Sartori, 1996). This was considered to be an appropriate context for the test of the proposed model, given that the rapid speed of the change, the large-scale nature of the change and associated level of uncertainty mean that a merger is likely to be a particularly stressful type of organizational change (Cartwright and Cooper, 1992).

On the basis of the proposed model of adjustment to organizational change, it was hypothesized: (a) that a reliance on problem-focused coping would be more adaptive than emotion-focused coping (escapism); (b) that employees would rely on problem- rather than emotion-focused coping if they perceived that they had been consulted in the change process, if they perceived that they had been kept informed about the change process, and if they judged that leadership during the period of change had been effective; (c) that employees would rely on problem- rather than emotion-focused coping when they perceived low levels of stress associated with the event, high levels of situational control and high levels of self-efficacy; and (d) that problem- rather than emotion-focused coping would be used by employees with low neuroticism and high levels of perceived social support. Participants who perceived that the change process had been implemented in a positive manner – that is, they judged that they had been consulted and kept informed during the period of change, and perceived that the leadership had been effective during this period – and who had access to adequate coping resources – low neuroticism and supportive relationships with their supervisor and work colleagues – were also expected to appraise the event as being less threatening than subjects who perceived that the change had been implemented in a negative manner and who lacked access to coping resources.

Respondents were 463 fleet staff (comprising mainly pilots) from the newly-merged airline, who ranged in age from 24 to 59 years old ($M = 39.74$, $S.D. = 8.27$). The sample consisted of 487 long-haul fleet staff (former employees of the international carrier), which represented 45 per cent of the total number of these staff members in the organization, and 175 short-haul fleet staff (55 per cent of the number of short-haul pilots in the organization). The latter employees were former employees of the domestic airline. Respondents were employed across the full range of ranks of pilots in the organization. The sample also comprised a number of flight engineers.

The proposed stress and coping model of adjustment to organizational change received some support in the study, full details of which are given in Terry, Callan and Sartori, 1996. Specifically, there was strong evidence of a link between employees' perceptions of the event characteristics and their levels of adjustment (measures of psychological distress and job satisfaction were used to assess employee adjustment). As predicted, these effects were indirect rather than direct. Participants who perceived that the merger had been implemented in a positive manner (i.e. judged that they had been consulted and kept informed during the period of change, and perceived that the leadership had been effective during this period) appraised the event as being less threatening than those participants who perceived that the merger had been implemented poorly. There was also some evidence of a link between the perception that the merger had been implemented in a positive manner and a reliance on problem-focused coping responses to manage the situation. These results support the view expressed by many commentators in the field, namely that adjustment to organizational change is better if employees perceive that they were kept adequately informed about the change process, that they were consulted about the implementation of the change and that there was effective leadership during the period of change (e.g. Callan et al., 1995; Dunphy and Stace, 1990). The results of the airline study suggest that these variables have their primary impact by influencing employees' perceptions of the levels of threat – in terms of appraised stress, control and self-efficacy – associated with the change. Thus, the results support one of the central propositions of the cognitive-phenomenological model of stress and coping, namely that the effects of event characteristics on stress responses are mediated through their effect on stress appraisals (see Lazarus and Folkman, 1984; Lazarus, 1990). Nevertheless, it does need to be acknowledged that the cross-sectional nature of the study precludes any definitive conclusions concerning the direction of the relationship between perceptions of the event characteristics and employees' appraisals of the level of threat associated with the situation.

In addition to evidence linking the perceived event characteristics to adjustment to organizational change, there was evidence that situational appraisals were relevant in this respect. In accordance with Lazarus and Folkman's (1984; Lazarus, 1990) cognitive-phenomenological model of stress and coping and current models of the determinants of coping (see for example, Terry, 1991a, 1994), these effects were, for the most part, mediated through coping responses. A reliance on problem-focused coping was associated with the perception that something could be done to manage the event

– reflected in perceptions of situational control and self-efficacy – whereas the use of escapist strategies was associated with high levels of stress. Contrary to expectations, there was also evidence of a positive relationship between appraised stress and problem-focused coping (and, as a consequence, a positive indirect effect of stress on job satisfaction), probably because the appraisal that an event has some relevance for one's well-being has a general effect of mobilizing coping responses, thus increasing one's use of all types of coping responses, irrespective of their focus (see also Terry, 1991a). In addition to its effects on coping responses, there was evidence of direct (negative) links between appraised stress and adjustment (psychological well-being and job satisfaction) and between judgements of self-efficacy to deal with the merger situation and concurrent levels of psychological well-being. These results attest to the importance of taking into account people's subjective appraisals of stress in the prediction of levels of adjustment; however, the direct relationships between stress appraisals and adjustment were probably inflated due to the effects of contemporaneous measurement.

The findings linking coping responses to adjustment were in accord with predictions – there was evidence in support of the proposal that high levels of adjustment would be associated with a reliance on problem- rather than emotion-focused coping (the effect of problem-focused coping was significant only for job satisfaction). Although adjustment to working in a newly-merged organization can be regarded as a low control situation, in that there is nothing that employees can do to change the situation, these results suggest that a failure to deal actively with the stressor is associated with poor employee adjustment (contrary to the goodness of fit model of coping effectiveness; see Conway and Terry, 1992; Forsythe and Compas, 1987). Future research should not only examine the effects of coping responses to a merger situation in a longitudinal design, but also determine whether the effects of problem- and emotion-focused coping vary across the course of the merger process.

In addition to coping responses, the present research revealed support for the importance of taking into account employees' coping resources when explaining variation in levels of adjustment to organizational change. Neuroticism had significant relationships with both types of mediating variables – that is, situational appraisals and coping responses – in the proposed model of adjustment to organizational change, as well as with levels of adjustment. Participants high in neuroticism were more likely than low neurotics to appraise the event as being threatening (high levels of appraised stress and poor efficacy expectancies), to use high levels of escapism to deal with the event, and to report low levels of psychological well-being. The pervasive nature of the effects of neuroticism underscores the importance of assessing this variable in studies of stress (see Brief et al., 1988; Watson and Pennebaker, 1989). The relationships between neuroticism and both stress appraisals and psychological well-being may be attributed to the fact that negative affect is tapped not only by measures of neuroticism, but also by measures of stress and psychological well-being (see Watson and Pennebaker,

1989), hence reflecting the contaminating role that neuroticism may play in stress research. However, the relationships between neuroticism and both efficacy expectancies and the use of escapist coping responses may reflect a more substantive influence of this variable in the stress process. A tendency to experience distress is likely to necessitate a reliance on emotion-focused coping – such as escapism – and to weaken efficacy expectancies.

Supervisor support emerged as an important component of the stress-coping model of adjustment to organizational change. Participants who considered that they had a supportive relationship with their supervisor tended to perceive that the merger had been implemented in a positive manner, to appraise the event as being low in threat (specifically, as being high in its potential for control), to be confident in their ability to deal effectively with it, to rely on active coping responses, and to exhibit high levels of adjustment to the event. The important role that supervisor support plays in the context of work stress has been reported in studies of other work stressors (see Kahn and Byosiere, 1992). It can, however, be suggested that the particularly influential role that supervisor support appeared to play in the merger study may be a reflection of the nature of the stressor under consideration. As many commentators in the area have noted, management plays a pivotal role in determining the effects of organizational change. If employees have a supportive relationship with their superior, then this is likely to mean that the channels are available for effective communication and consultation about the change, and that leadership during the period will be effective. Supervisors can also play a role in influencing employees' perceptions of the level of threat associated with the change and by assisting in the development of effective coping responses (see Terry, Rawle *et al.*, 1995; Thoits, 1986).

Contrary to expectations, there was evidence that high levels of colleague support had a negative impact on stress appraisals and coping responses. Participants high in colleague support rated the event as more stressful, perceived lower levels of self-efficacy (a relatively weak effect) and used higher levels of escapism than those who perceived lower levels of colleague support. Such results may be a reflection of the particular features of the stressor under investigation. Under conditions of situational uncertainty, a high level of colleague support – which presumably indicates that employees have close relationships with their colleagues – may actually heighten threat appraisals and the use of maladaptive coping responses because of the anxiety that is engendered in discussion about the change (see also Kaufmann and Beehr, 1986). Relatively isolated employees may have less contact with their colleagues and hence adopt a view of the situation that is not influenced by the more extreme views of others likely to emerge under conditions of uncertainty. This type of explanation is clearly speculative; however, it does accord with the suggestion that a reliance on informal channels of communication during a period of organizational change may be associated with poor levels of adjustment to the situation.

In general terms, the airline study provided some support for the usefulness of a stress and coping perspective when explaining variation in people's adjustment to large-scale organizational change. To test further the utility of the proposed model of adjustment to organizational change, a second study was undertaken. The second study was designed to test the proposed model in the context of a different type of organizational change (internal reorganization rather than a merger), and to extend the focus of the research by examining the effects of the event characteristics (amount of change and reliance on informal channels of communication) and personal resources (self-esteem and generalized control beliefs) not considered in the first study.

Public Sector Reorganization

In a second study, we examined the predictors of employees' (supervisors and managers) adjustment to organizational change in the context of the integration of two organizations in the public sector and the large-scale internal reorganization that occurred as a consequence of the integration (Terry and Callan, 1997). The integration was sudden, and directed primarily by political imperatives towards producing savings and increased efficiency. Participants were middle managers and supervisors in a large public sector organization. At the time of the research, employees were three months into the change programme. A total of 140 male middle managers ($n = 78$) and supervisors ($n = 62$) participated in the study – a 78 per cent response rate. On average, respondents were 40 years of age ($SD = 7.40$), ranging in age from 23 to 59 years.

The model tested in the public sector integration study was, in general terms, similar to the one tested in the airline model – it comprised measures of resources, event characteristics, situational appraisals, coping, and adjustment. In order to extend the focus of the airline study, some of the specific variables that were assessed did differ between the two studies. In the public sector integration study, two personal resources – self-esteem and generalized control beliefs – were assessed (neuroticism and social support were not assessed in this study). The results of the study indicated that self-esteem and generalized control beliefs were, for the most part, related to the situational appraisals or coping strategies, rather than directly to adjustment – this was with the exception of a significant (negative) relationship between self-esteem and poor social functioning. The appraisal variables – uncertainty and appraised stress – significantly predicted psychological distress. These results are consistent with the view that psychological adjustment is particularly vulnerable to appraised threat (Thoits, 1983), and provide further support for the importance of incorporating such variables into predictive models of adjustment to stress (Lazarus and Folkman, 1984; Lazarus, 1990; see also Terry, Tonge et al., 1995; Terry et al., 1996).

In order to examine more specifically the role of event characteristics, the second study incorporated a measure of the amount of change experienced, and separate measures of the perceived effectiveness of management communication about the change and reliance on informal communication channels. As expected, the amount of change experienced (event characteristic) emerged as a significant predictor of (poor) social functioning, whereas a reliance on informal sources of support was linked to both psychological distress and (poor) social functioning. Such sources of information are likely to relate negatively to adjustment because the information may be inaccurate and inconsistent, and thus stress-enhancing. There was no evidence linking the perceived effectiveness of the formal communication programme to adjustment. It is possible that even when considered to be effective, communications from higher management are seen to reflect the 'official line' and hence likely to omit relevant information. In fact, in major mergers and other forms of organizational change, employees can become quite suspicious of the real intentions of top management, thus rumours talked about among one's peers are likely to gain more credence than communications about change from top management (Napier, Simmons and Stratton, 1989; see also Covin, 1993; Larkin and Larkin, 1994).

In accord with the proposed model of adjustment to organizational change, there was evidence linking employee coping responses to both measures of adjustment. Employees who were relying on avoidant strategies reported higher levels of psychological distress and poorer social functioning than those who used only low levels of this type of coping. As in the airline merger study, there was also a weak negative relationship between the use of problem-focused coping and psychological distress. However, there was no evidence that coping strategies mediated the effects of event characteristics, situational appraisals and personal resources on adjustment. These results are inconsistent with Lazarus and Folkman's (1984; Lazarus, 1990) cognitive-phenomenological model of stress. The lack of evidence for the mediating role of coping may be attributed to the particular antecedent variables considered in the present research. As noted, in the airline study we found that the effects of efficacy expectancies were mediated through coping responses. In previous literature, similar results have been reported for social support (e.g. Manne and Zautra, 1989) and generalized control beliefs (Terry, Tonge et al., 1995). The indirect effects of the latter variable were through escapism and self-blame rather than avoidance. Although the effects of some person and situational variables on adjustment to stress appear to be mediated through coping, the results of the public sector integration study suggest that coping should not be regarded as a general mediating process in stress-adjustment models.

Despite the lack of evidence for the mediating role of coping strategies, a number of significant relationships between coping and the antecedent variables – event characteristics, situational appraisals and personal resources – were found. Consistent with previous studies of the predictors of coping strategies, avoidance (an emotion-focused strategy) was predicted not by any of the situational variables, but by a relatively enduring dispositional variable, viz. low self-esteem (see Terry, 1994). Also consistent with previous research (Terry, 1991a, 1994; Terry *et al.*, 1996) was the evidence that high levels of situational demands – that is, in terms of the amount of change experienced and the reliance on informal channels of communication – was associated with high rather than with low levels of problem-focused coping. It is possible that stress interferes with problem-focused strategies only when levels of stress are extreme, whereas lower levels of stress may serve to engender the use of problem-oriented coping. Consistent with previous research, problem-focused strategies were used more by people with internal control beliefs than by those with external control beliefs (Ashford, 1988; Parkes, 1986).

There was evidence that the effects of self-esteem on psychological distress were mediated through situational certainty – high self-esteem subjects were more likely than others to appraise situational certainty which, in turn, was associated with low distress. A mediating role for situational appraisals in the stress-adjustment relationship is consistent with Lazarus and Folkman's (1984; Lazarus, 1990) cognitive-phenomenological model of the stress process (Terry, Tonge *et al.*, 1995; Terry *et al.*, 1996). The relationship between self-esteem and situational certainty is also consistent with the evidence linking low self-esteem to the seeking of social comparative information under threat (Gibbons and Gerrard, 1989). Such findings could reflect the fact that individuals with low self-esteem experience more situational uncertainty, a condition typically regarded as the primary motive for social comparative strategies (see Festinger, 1954).

Overall, the second study provided some support for a cognitive-phenomenological model of the predictors of adjustment to organizational change. However, the study was not without its limitations. Foremost was the fact that it relied on cross-sectional data. Future research needs to examine the effects of the proposed predictors of adjustment to organizational change in a longitudinal design in an endeavour to examine the temporal effects of such variables on subsequent adjustment. In order to decontaminate the data for same-source effects, future research should use both self-report and objective measures of adjustment. The effects of other variables also need to be more fully explored, including the extent to which employees' identities are linked to their organizational membership (see Terry, forthcoming; Terry and Callan, 1998, forthcoming).

Conclusions

In summary, our findings indicate the value of a theoretical framework that simultaneously considers the effects of resources, event characteristics, situational appraisals (appraised stress and appraised uncertainty), and coping strategies (avoidance and problem-focused coping) on employee adjustment (psychological distress and poor role functioning) to organizational change. The results of the research are important in this context, given that little research has tested the utility of theory-based models of adjustment to organizational change. The findings of the research also have implications for the wider stress literature, in that they provide some additional support for models of stress that incorporate both situational appraisals and coping responses as mediating processes (see Lazarus and Folkman, 1984; Lazarus, 1990; Terry, Tonge et al., 1995).

At the applied level, the results of the research point to the need for organizations undergoing change to ensure effective leadership during the period of change and to consult with employees about the implementation of the change process. Organizations also need to ensure that managers involved in the direct supervision of employees are an adequate source of support during the period of change, and to ensure that employees are kept accurately informed about the change process. The results of the two studies, taken together, suggest that the latter goal may be difficult to achieve. During periods of large-scale change, there may be a tendency to rely on informal channels of communication. It is therefore important for organizations to ensure that this form of communication provides employees with accurate information about the change process. Given the consistent evidence linking situational appraisals and coping responses to employee adjustment to organizational change, organizational efforts that counter both the belief that the situation is threatening and the use of emotion-focused coping should serve to maintain employee well-being during the increasingly common experience of organizational change.

Note

The research reported in this chapter was supported by a research grant to the first two authors from the Australian Research Council. Thanks are due to Gloria Hynes and John Gardiner who helped with data collection and data entry.

References

ALDWIN, C. M. and REVENSON, T. A. (1987) 'Does coping help? A re-examination of the relation between coping and mental health', Journal of Personality and Social Psychology 53: 337–348.

ASHFORD, S. J. (1988) 'Individual strategies for coping with stress during organisational transitions', The Journal of Applied Behavioral Science 24: 19–36.

BANDURA, A. (1977) 'Self-efficacy: toward a unifying theory of behavioral change', Psychological Review 84: 191–215.

BANDURA, A. (1982) 'Self-efficacy mechanisms in human agency', American Psychologist 37: 122–147.

BEEHR, T. A. and NEWMAN, J. E. (1978) 'Job stress, employee health and organizational effectiveness: a facet analysis, model and literature review', Personnel Psychology 31: 665–699.

BILLINGS, A. G. and MOOS, R. H. (1981) 'The role of coping responses and social resources in attenuating the stress of life events', *Journal of Behavioral Medicine* **4**: 139–157.

BOLGER, N. (1990) 'Coping as a personality process: a prospective study', *Journal of Personality and Social Psychology* **59**: 525–537.

BRIEF, A. P., BURKE, M. J., GEORGE, J. M., ROBINSON, B. and WEBSTER, J. (1988) 'Should negative affectivity remain an unmeasured variable in the study of job stress?', *Journal of Applied Psychology* **73**: 193–199.

CALLAN, V. J., DICKSON, C. (1992) 'Managerial coping strategies during organizational change', *Asia-Pacific Journal of Human Resources* **30**: 47–59.

CALLAN, V. J., TERRY, D. J. and SCHWEITZER, R. J. (1995) 'Coping resources, coping strategies and adjustment to organizational change: direct or buffering effects', *Work and Stress* **8**: 372–383.

CAPLAN, R. D., COBB, S., FRENCH, J. R. P., VAN HARRISON, R. V. and PINNEAU, S. R., Jr. (1975) *Job Demands and Worker Health*, Washington, DC: US Department of Health, Education and Welfare.

CARTWRIGHT, S. and COOPER, C. L. (1992) *Mergers and Acquisitions: The Human Factor*, Oxford: Butterworth/Heinemann.

CARVER, C. S., SCHEIER, M. F. and WEINTRAUB, J. K. (1989) 'Assessing coping strategies: a theoretically based approach', *Journal of Personality and Social Psychology* **56**: 267–283.

CHAN, K. B. (1977) 'Individual differences in reactions to stress and their personality and situational determinants: some implications for community mental health', *Social Science and Medicine* **11**: 89–103.

CONWAY, V., TERRY, D. (1992) 'Appraised controllability as a moderator of the effectiveness of different coping strategies: a test of the goodness of fit hypothesis', *Australian Journal of Psychology* **44**: 1–7.

COVIN, T. J. (1993) 'Managing workforce reduction: a survey of employee reactions and implications for management consultants', *Organization Development Journal* **11**: 67–76.

DIFONZO, N., BORDIA, P. and ROSNOW R. L. (1994) 'Reining in rumours', *Organizational Dynamics* **23**: 47–62.

DUNPHY, D. and STACE, D. (1990) *Under New Management*, New York: McGraw-Hill.

FESTINGER, L. (1954) 'A theory of social comparison processes', *Human Relations* **7**: 117–140.

FOLKMAN, S. (1984) 'Personal control and stress and coping processes: a theoretical analysis', *Journal of Personality and Social Psychology* **46**: 839–852.

FOLKMAN, S., LAZARUS, R. S., DUNKEL-SCHETTER, C., DELONGIS, A. and GRUEN, R. J. (1986) 'Dynamics of a stressful encounter: cognitive appraisal, coping and encounter outcomes', *Journal of Personality and Social Psychology* **50**: 992–1003.

FORSYTHE, C. J. and COMPAS, B. E. (1987) 'Interaction of cognitive appraisals of stressful events and coping: testing the goodness of fit hypothesis', *Cognitive Therapy and Research* **11**: 473–485.

GANSTER, D. C., FUSILIER, M. R. and MAYES, B. T. (1986) 'Role of social support in the experience of stress at work', *Journal of Applied Psychology* **71**: 102–110.

GIBBONS, F. X. and GERRARD, M. (1989) 'Effects of upward and downward comparison on mood states', *Journal of Social and Clinical Psychology* **8**: 14–31.

HOLAHAN, C. J. and MOOS, R. H. (1987) 'Personal and contextual determinants of coping strategies', *Journal of Personality and Social Psychology* **52**: 946–955.

HOLMES, T. H. and RAHE, R. H. (1967) 'The social readjustment rating scale', *Journal of Psychosomatic Research* **11**: 213–218.

INNES, J. M. and KITTO, S. (1989) 'Neuroticism, self-consciousness, coping strategies and occupational stress in high school teachers', *Personality and Individual Differences* **10**: 303–312.

ISRAEL, B. A., HOUSE, J. S., SCHURMAN, S. J., HEANEY, C. A. and MERO, R. P. (1989) 'The relation of personal resources, participation, influence, interpersonal relationships and coping strategies to occupational stress, job strains and health: a multivariate analysis', *Work and Stress* **3**: 163–194.

KAHN, R. L. and BYOSIERE, P. (1992) 'Stress in organizations', in M. D. Dunnette and L. Hough (eds) *Handbook of Industrial and Organizational Psychology*, 2nd edn, Palo Alto, CA: Consulting Psychologists Press, 571–650.

KANTER, R. M. (1983) 'The change masters: transformations in the American corporate environment 1860–1980s', New York: Simon and Schuster.

KAUFMANN, G. M. and BEEHR, T. A. (1986) 'Interactions between job stressors and social support: some counterintuitive results', *Journal of Applied Psychology* **71**: 522–526.

KIRMEYER, S. L. and DOUGHERTY, T. W. (1988) 'Work load, tension and coping: moderating effects of supervisor support', *Personnel Psychology* **41**: 125–139.

KOBASA, S. C., MADDI, S. R. and KAHN, S. (1982) 'Hardiness and health: a prospective study', *Journal of Personality and Social Psychology* **42**: 707–712.

LARKIN, T. J. and LARKIN, S. (1994) *Communicating Change: Winning Employee Support for New Business Goals*, New York: McGraw-Hill.

LAZARUS, R. S. (1990) 'Theory based stress management', *Psychological Inquiry* **1**: 3–13.

LAZARUS, R. S. and FOLKMAN, S. (1984) *Stress, Appraisal and Coping*, New York: Springer.

MANNE, S. L. and ZAUTRA, A. J. (1989) 'Spouse criticism and support: their association with coping and psychological adjustment among women with rheumatoid arthritis', *Journal of Personality and Social Psychology* **56**: 608–617.

MOOS, R. H. and BILLINGS, A. G. (1982) 'Conceptualizing and measuring coping resources and processes', in L. Goldberger and S. Breznitz (eds) *Handbook of Stress: Theoretical and Clinical Aspects*, New York: Macmillan, 212–230.

NAPIER, N. K., SIMMONS, G. and STRATTON, K. (1989) 'Communication during a merger: the experience of two banks', *Human Resource Planning* **12**: 105–122.

PARKES, K. R. (1986) 'Coping in stressful episodes: the role of individual differences, environmental factors and situational characteristics', *Journal of Personality and Social Psychology* **51**: 1277–1292.

PARKES, K. (1990) 'Coping, negative affectivity and the work environment: additive and interactive predictors of mental health', *Journal of Applied Psychology* **75**: 399–409.

SCHWEIGER, D. M. and DE NISI, A. S. (1991) 'Communication with employees following a merger: a longitudinal field experiment', *Academy of Management Journal* **34**: 110–135.

SCHWEIGER, D. M. and IVANCEVICH, J. M. (1985) 'Human resources: the forgotten factor in mergers and acquisitions', *Personnel Administrator* **30**: 47–61.

SCHWEIGER, D. M., IVANCEVICH, J. M. and POWER, F. R. (1987) 'Executive action for managing human resources before and after acquisition', *Academy of Management Executive* **2**: 127–138.

SPECTOR, P. E. (1986) 'Perceived control by employees: a meta-analysis of studies concerning autonomy and participation at work', *Human Relations* **39**: 1005–1016.

TERRY, D. J. (1991a) 'Coping resources and situational appraisals as predictors of coping behaviour', *Personality and Individual Differences* **12**: 1031–1047.

TERRY, D. J. (1991b) 'Predictors of levels of subjective stress in a sample of new parents', *Australian Journal of Psychology* **43**: 29–37.

Terry, D.J. (1992) 'Stress, coping and coping resources as correlates of adaptation in myocardial infarction patients', *British Journal of Clinical Psychology*, **31**, 215–225.

TERRY, D. J. (1994) 'The determinants of coping: the role of stable and situational factors', *Journal of Personality and Social Psychology* **66**: 895–910.

TERRY, D. J. (forthcoming) 'Intergroup relations in the context of an organizational merger', in M. A. Griffin and P. Hart (eds) *Improving Organisational Health: Diagnosis, Intervention and Evaluation*, Melbourne, Australia: Australian Academic Press.

TERRY, D. J. and CALLAN, V. J. (1997) 'Employee adjustment to large-scale organizational change', *Australian Psychologist* **32**: 203–220.

TERRY, D. J. and CALLAN, V. J. (1998) 'Intergroup differentiation in response to an organizational merger', *Group Dynamics* **2**: 67–81.

TERRY, D. J., CALLAN, V. J. and SARTORI, G. (1996) 'Employee adjustment to an organizational merger: stress, coping and intergroup differences', *Stress Medicine* **12**: 105–122.

TERRY, D. J., CAREY, C. J. and CALLAN, V. J. (forthcoming) 'Employee adjustment to an organizational merger: an intergroup perspective', *Personality and Social Psychology Bulletin.*

TERRY, D. J., NIELSEN, M. and PERCHARD, L. (1993) 'The effects of job stress on well-being: a test of the buffering effects of social support', *Australian Journal of Psychology* **45**: 168–175.

TERRY, D. J., RAWLE, R. and CALLAN, V. J. (1995) 'The effects of social support on adjustment: the mediating role of coping', *Personal Relationships* **2**: 97–124.

TERRY, D. J. and SCOTT, W. A. (1987) 'Gender differences in correlates of marital satisfaction', *Australian Journal of Psychology* **39**: 207–221.

TERRY, D. J., TONGE, L. and CALLAN, V. J. (1995) 'Employee adjustment to stress: the role of coping resources, situational factors and coping responses', *Anxiety, Stress and Coping* **8**: 1–24.

THOITS, P. A. (1983) 'Dimensions of life events as influences upon the genesis of psychological distress and associated conditions: an evaluation and synthesis of the literature', in H. B. Kaplan (ed.) *Psychological Stress: Trends in Theory and Research*, New York: Academic Press, 33–103.

THOITS, P. A. (1986) 'Social support as coping assistance', *Journal of Consulting and Clinical Psychology* **54**: 416–423.

WATSON, D. and PENNEBAKER, J. W. (1989) 'Health complaints, stress and distress: exploring the central role of negative affectivity', *Psychological Review* **96**: 234–254.

WHEATON, B. (1983) 'Stress, personal resources and psychiatric symptoms: an investigation of interactive models', *Journal of Health and Social Behavior* **24**: 208–229.

YUKL, G. A. (1989) *Leadership in Organisations*, Englewood Cliffs, NJ: Prentice Hall.

CHAPTER FOURTEEN

Coping with work: future directions from the debates of the past

PHILIP DEWE, TOM COX AND MICHAEL LEITER

In 1982 Payne, Jick and Burke, reviewing stress research, wrote that 'the volume of organizational stress research has grown markedly in recent years as has general interest in the subject.' The 1980s were marked by intense debate over how stress should be conceptualized and investigated, although throughout this time work stress research achieved legitimacy as an area of study (Newton, 1989). By 1991 the picture appears to have been much the same. The field was still described as rapidly expanding, marked this time by vigorous disagreements about models and theories, beginning with the definition of stress itself (Kahn and Byosiere, 1991). The view that emerges from these reviews is that the attention of work-stress researchers has, somewhat understandably, been directed primarily towards identifying those conditions at work that are considered stressful, and those variables that moderate the relationship. Only now is research being directed towards the stress process – that is, those mechanisms that link the individual to the environment (see Lazarus, 1991).

The agenda for the future *must be* to agree and give empirical support to theoretical statements that emphasize the causal sequence most relevant to providing a better understanding of the work-stress process (Kahn, 1987; Kaplan, 1996). Research that obscures, or models that allow for the sequence to be merely implied, must now be regarded as necessarily limiting in scope and design (Bhagat, Quick, Quick and Dalton, 1987) and as insufficient to provide new insights and enhance the well-being of those whose working lives we are attempting to improve. The new agenda is best fulfilled by revisiting and rethinking old issues with the hope that, from such a process, future research imperatives will no longer be obscured by the debates of the past.

When reading the different reviews on work stress and coping (for example, Burke, 1987; Dewe, Cox and Ferguson, 1993; Kahn and Byosiere, 1991; Newton, 1989; Payne *et al.,* 1982) it is possible to identify a number of consistently occurring themes. This consistency is further evidence of the fact that future research can only benefit from a theoretical context that is process-oriented, providing new opportunities for research design and the measurement of constructs. One theme is the need to develop interdisciplinary research that:

- broadens the type of consequences studied (Sethi and Schuler, 1984);
- allows for more 'inclusive integrative orientations' as researchers recognize the commonality of their interests and the functional equivalence of their constructs (Kaplan, 1996: 377);
- shifts the focus away from a dominant tradition, allowing for sources of variance to be discussed in a context that provides a more complete and multidisciplinary interpretation (Bhagat *et al.,* 1987).

Another theme concerns the issue of the confounding of stress measures. Confounding is undoubtedly a difficult and complex problem. It is clear that measures must be derived in ways that, when considering different theoretical relationships, conclusions are based on qualitatively distinct measures rather than on outcomes derived from using similar measures (Burke, 1987; Kaplan, 1996). A number of views sum up this concern and reinforce the need to establish work-stress research in a theoretical context that reflects the sequence of events. These views include:

- conceptualizing stress as a process, thereby keeping the difference between independent and dependent variables distinct (Schonfeld, Rhee and Xia, 1995);
- searching for other more meaningful ways to measure variables so that researchers, by introducing more imaginative approaches, can evaluate better the utility of 'given' measures and the benefits of more sophisticated quantitative techniques (Bhagat and Beehr, 1985);
- recognizing that measurement is always best when it comes from theory and that the way in which stress is conceptualized has profound consequences for how constructs are measured and how their interrelationships are explored (Lazarus, 1990).

A third theme stems from the view that when stress is defined in transactional terms, there is the need for researchers to employ longitudinal research designs that mirror over time the unfolding of the stress process. There are a number of reasons supporting this idea that can be identified. These include the notion that since stress as a process is embedded in most of the available writing then the trend towards longitudinal studies should be encouraged if stress researchers are to establish causal relations (Burke, 1987). The second reason is that researchers could well be misled about the transactional nature of stress if the dynamics of the situation are treated as a single event and continue to be researched in a cross-sectional way (Lazarus, 1991). Third, the explanatory potential that resides in such an approach is enhanced because it provides a conceptual context for better understanding the nature of the different constructs to the stress process and how they are linked (Dewe, 1993) and because it clarifies causal relationships that are otherwise indeterminable (Kahn and Byosiere, 1991).

None of these themes is mutually exclusive and all draw attention to the need to continue to consider a number of methodological issues surrounding:

- the measurement of stressors, including the way they combine and their additive effect (Sethi and Schuler, 1984), their duration (Bhagat and Beehr, 1985), the events being measured (Breaugh and Colihan, 1994), the relevance of scale instructions (Dewe, 1991) and the role of primary appraisal (Dewe, 1993);
- the conceptualization of outcomes and the criticism that most outcome measures have been too narrowly defined (Newton, 1989), affect several quite different aspects of human functioning (Bhagat and Beehr, 1985) and fail to measure whether well-being has been significantly affected (Brief and George, 1991);
- the role of moderator variables, including their nature (Kaplan, 1996), and the context within which the moderation occurs (Kahn and Byosiere, 1991);
- the structure of coping (Trenberth, Dewe and Walkey, 1996), its measurement (Dewe, Cox and Ferguson, 1993) and the context within which coping takes place (Bhagat and Beehr, 1985);
- the nature of the population investigated, including such issues as occupationally specific versus more general samples, gender differences and work versus non-work domains (Burke, 1987).

A careful examination of all these issues suggests that the agenda for future research must be set by agreeing on a theoretical context within which work stress should be investigated. While there may be nothing particularly new in this statement, the fact that reviewers continue to suggest it illustrates that even when theoretical statements exist, the theoretical models do not guide the research agenda (Kaplan, 1996: 373). The barrier to making progress comes not from identifying what should be done but from agreeing on the context within which work-stress research should be carried out. The main gap in research stems from the need to establish a perspective that provides the framework for investigating the dynamics of work stress (Bhagat et al., 1987). Theoretical frameworks that obscure the dynamics of the stress process are necessarily limiting in scope and fail to capture the explanatory potential of the process under investigation. Considering the advances that work-stress research has made, and acknowledging the fact that work-stress research has reached a level of maturity, the goal of future researchers should be towards increasing our understanding of the sequence of events that result in the experience of stress, rather than further exploration of the causal relationship between certain stressors and a stressful outcome (Kaplan, 1996).

The preferred work-stress research model that has emerged (Tetrick and LaRocco, 1987) postulates that the perceived presence of certain working conditions (stressors) may be predictive of a variety of job-related stresses. The model also suggests that at times, certain individual differences and situational characteristics moderate the stressor-stress relationship. The research resulting from this model can be broadly divided into three types. The three types of research application are: (a) identifying, defining, classifying and measuring different work stressors, (b) exploring the relationship between the different work stressors and outcome (stress) variables, and (c) considering the moderating effect of different individual and situational variables (Dewe, 1991). Each reflects, somewhat understandably, a natural progression towards developing an understanding of the constructs involved and their interactive relationships. Yet at the same time, work-stress research has, as Newton (1989: 441) suggests, reached a hiatus as a result of the intensity of the debate over how stress should be conceptualized and how it should be investigated. What is now needed is a period of 'quiet reconstruction' where we consider not just where current models are taking research but what alternative models can provide. To do this we need to revisit three issues: definitions of stress; the need for research to be guided by theory; and measurement.

Definitions of stress

The objective here is not to revisit 'the troublesome concept of stress' (Kasl, 1983) but more to affirm that current representations of stress allow a common research pathway to emerge that captures the sequence of events and those individual processes that link the individual and the environment. The issue is no longer one of continuing to defend the utility of the term *stress* (Kaplan, 1996) or to attempt to define 'the stress of the stress process' (Lazarus, 1990) but to agree that our energy should now be redirected towards creating theoretical structures that recognize that stress is a continually changing relationship between the person and the environment (Lazarus, 1990: 4) and that no one part can be said to be stress because each is a component of a complex process and each is linked one to the other by powerful cognitive processes.

With this in mind, the direction in which we should go (see Liddle, 1994: 167) is one in which we should now be considering which of all the definitions of stress best qualifies as an organizing concept with sufficient structure to provide systematic theoretical and research inquiry that will make a significant contribution to our understanding of the stress process. If this represents the point where quiet reconstruction must take place, then we should agree that definitions of stress and resulting research that somewhat artificially

separates the different components of the stress process are no longer service-
able (Lazarus, 1991). This is particularly important when at the conceptual
level, at least, there is a growing acceptance that stress should be viewed as
relational in nature (Lazarus and Launier, 1978), involving some sort of
transaction between the individual and the environment.

This is not to deny the evolutionary stages that work-stress research has
gone through and the need first to establish some understanding of the
different constructs and their relationship. Nor is it meant to suggest that
the transactional nature of stress and its explanatory potential have gone
unrecognized in a work context. It is simply to express the point that current
work-stress research appears to continue to favour situational variables, and
fails to emphasize individual processes. It also reinforces the belief that a more
dynamic perspective in understanding work stress is well placed (Harris,
1991). The time has come when empirical and theoretical concerns should
be integrated through the context of the transaction between the individual
and the environment. This should serve as the qualifying concept for future
research and be recognized as providing the research opportunity that best
exploits the explanatory potential of the stress process.

Theoretical context

When it comes to the theoretical context within which work-stress research
should be carried out it is not surprising to find the same sorts of arguments
being put forward as those discussed above. The common theme that emerges
is one where future theoretical frameworks must allow for the specification of
those mechanisms through which the stress process develops. What will
distinguish the research context of the future will be its focus on the sequence
of events that culminate in the experience of stress, rather than the simple
relationship between certain phenomena and the stressful outcome (Kaplan,
1996: 387). What is needed is a perspective that emphasizes the dynamics of
the stress process. It is here that the gap in our knowledge is greatest. It stems,
according to Bhagat et al. (1987: 308), from the need for more systematic
research on the process of stress and coping in work settings. Again it is
important to emphasize that models of work stress have incorporated and
continue to incorporate the concept of process in their structure. By far the
most widely cited work-stress model, the person-environment fit (French,
Caplan and Van Harrison, 1982) is characterized by the idea that fit or misfit
is associated with adaptation or maladaption. Similarly Karasek's (1979) job
demands model emphasizes the adaptational consequences that flow from the
'fit' between job demands, individual resources and control, as does the
cybernetic framework proposed by Cummings and Cooper (1979).

Where these models fall short is that often the concept of fit between the
person and the environment were expressed as though the relationship were
static or linear in its function. Perhaps more importantly from the point of

view of future research is that assumptions about how the stress process may work, or how the individual and environment are linked, continue in this context to go unstated (Lazarus, 1990; 1991). The traditional use of moderator variables to explore stressor-stress relationships is a good example of this latter difficulty. When, for example, a variable is found to moderate the relationship, its effect can, more often than not, only be inferred and so the processes that link the individual and the environment continue to go unidentified and unresearched. It seems a pity that despite all the advances made in work-stress research, we still need to go outside of that body of research to consider those mechanisms that best describe the stress sequence and the individual-environment links (Kahn and Byosiere, 1991).

The unit of analysis that captures the transactional nature of stress is appraisal. It is this evaluative process that links the individual and the environment and expresses the mechanisms through which any encounter is perceived as involving harm, threat or challenge (Lazarus, 1990). The two kinds of appraisal are: primary appraisal concerning what is at stake – it is where the individual assesses the importance of and gives meaning to any encounter; and secondary appraisal where available coping options are evaluated in terms of resources and control. When the different constructs of the stress process are considered within the context of primary and secondary appraisal then a different perspective on work stress emerges from that traditionally studied. Three issues are important. One is that the fundamental condition that certain work encounters have the potential to result in stress is not being challenged. The point is that this proposition on its own is not enough to develop an understanding of the stress process (Lazarus, 1990). The second issue is that such a 'structural approach' requires only a minimum of theory (Lazarus, 1990) and with few exceptions fails to consider the meaning individuals give to events and coping behaviours. It is not that these constructs are considered unimportant but that the focus of the more traditional models somewhat unintentionally constrains the way in which such ideas can develop. Finally, research should now develop measures that encapsulate the appraisal process (Kahn and Byosiere, 1996) and which empirically discriminate these processes from established measures of stressors and stress.

Measurement

Identifying theoretically distinct constructs that taken together make up the stress process challenges researchers to develop measures of those constructs (Kaplan, 1996). It is not that stress as a transactional process has gone unresearched, or that its explanatory potential has gone untested. It is more the case that the way that work stress has been defined and modelled has diverted attention away from the appraisal process and coping. But definitions and models are also about methodology and so the challenge is not just one of thinking beyond current representations of stress (Newton, 1995) but of asking questions about where current methodologies are taking us, what alternative methodologies can provide and what it is we are measuring when we are measuring stress.

The debate is not about replacing one methodology with another, but more an attempt to reduce the 'indifferent hostility' between those who research using a quantitative methodology and those who use a qualitatitive approach. The concern that has been expressed has led to the belief that research is now methodologically rather than theoretically driven and that current research methods have assumed a level of significance such that established measurement practices may now be somewhat divorced from the phenomena they are purporting to measure, leading to the need to consider the structured reality imposed by such methods. In the end, it comes down to the issue of whose reality we are measuring when we are measuring stress, and whether the methods used reflect the shared meanings and beliefs that interpret, legitimate and justify organizational behaviour. To some, this means questioning whether measures actually capture the significance of events and their impact on well-being (Brief and George, 1991). To others it is a question of whether social and economic change have overtaken some measures to the extent that we are over-stating the importance of some events and ignoring the importance of others (Glowinkowski and Cooper, 1985). While to others it is more the need to distinguish between the perception of an encounter and the meaning that such an encounter has for an individual (Dewe, 1993), or to consider whether coping is best measured by using a qualitative approach that focuses on the stress that evokes it (Erera-Weatherley, 1996) rather than by traditional self-report methods that capture how frequently a strategy is used. The message for future researchers is that they may wish to consider a more balanced approach and begin to look for more vivid and enriched descriptions and measures of process in qualitative as well as quantitative terms, recognizing that creativity and imagination in research design may not come from a methodology that supports a conventional, standard statistical application (Bhagat and Beehr, 1985).

Any discussion of methodology and measurement cannot be divorced from the issue of application. The basic argument here is summed up by Bhagat and Beehr when they conclude that the question seems to be one of scientific rigour versus relevance of the findings; whether in fact the validity criterion has been overemphasized at the expense of a relevance criterion (1985: 419). Again the debate, not surprisingly, is similar to the one raised above though this time it is:

- acknowledging that significant differences do exist in the needs, values and orientation of researchers and practitioners;
- identifying what these differences are and the constraints they impose on the utilization of information;
- developing ways in which the gaps can be bridged (Burke, 1987; Bhagat

and Beehr, 1985). In the end the solution seems to be to look again for the 'creative' rather than the traditional solution; of 'tackling the hard stuff' (Bhagat and Beehr, 1985: 429) and of having a genuine interest in serving those whose working worlds we explore (Brief and George, 1991).

The contribution of this book

Our aim in compiling this book has been to set a context wherein research may fruitfully develop. The contributors of the various chapters have, through the concept of coping, attempted to explore some of the issues that emerge when our attention is directed towards understanding how people cope with the stresses of their working lives. The pattern of the book has followed some of the arguments raised above. It began by considering a number of theoretical and psychometric issues surrounding the construct of coping. It moved on to exploring coping in specific situations, and ended by investigating and describing ways in which different intervention strategies contribute to the well-being of workers.

The concept of coping and its role in the appraisal process offers those who research work stress an enormous opportunity to contribute to our understanding of the transactional nature of stress. Perhaps by recognizing the explanatory potential of a construct such as coping and the need to break away from those traditional ideas and methods that have for so long served us well, we will begin to recognize the fertile ground that is now being offered, and the common road forward that has for so long eluded work-stress researchers. By taking the steps necessary to explore how people cope with the stress of work, we will legitimize and reinforce the worth of coping and the transactional nature of stress and establish, at both the conceptual and empirical level, a context wherein our knowledge may advance. Our willingness to transfer this knowledge into practice will continue to reflect a genuine goal: that of improving the way work is experienced.

References

BHAGAT, R. S. and BEEHR, T. A. (1985) 'An evaluative summary and recommendations for future research', in T. A. Beehr and R. S. Bhagat (eds) *Human Stress and Cognitions in Organisations: An Integrated Perspective,* New York: John Wiley and Sons, 401–415.

BHAGAT, R. S., QUICK, J. C., QUICK, J. D., and DALTON, J. E. (1987) 'Work organisations and health care: future directions', in J. C. Quick, R. S. Bhagat, J. E. Dalton, J. D. Quick (eds) *Work Stress: Health Care Systems in the workplace,* New York: Praeger, 301–310.

BREAUGH, J. A. and COLIHAN, J. P. (1994) 'Measuring facets of job ambiguity: construct valid evidence', *Journal of Applied Psychology* **79**: 191–202.

BRIEF, A. P. and GEORGE, J. M. (1991) 'Psychological stress and the workplace: a brief comment on Lazarus' outlook', in P. L. Perrewé (ed.) *Handbook on Job Stress*, special edition, *Journal of Social Behavior and Personality* **6**: 15–20.

BURKE, R. A. (1987) 'The present and future status of stress research', in J. H. Ivancevich and D. C. Ganster (eds) *Job Stress: From Theory to Suggestion*, New York: The Haworth Press, 249–263.

CUMMINGS, T. G. and COOPER, C. L. (1979) 'A cybernetic framework for studying occupational stress', *Human Relations* **32**: 395–418.

DEWE, P. J. (1991) 'Primary appraisal, secondary appraisal and coping: their role in stressful work encounters', *Journal of Occupational Psychology* **64**: 331–351.

DEWE, P. J. (1993) 'Measuring primary appraisal: scale construction and directions for future research', *Journal of Social Behavior and Personality* **8**: 673–685.

DEWE, P. J. COX, T. and FERGUSON, E. (1993) 'Individual strategies for coping with stress at work: a review', *Work and Stress* **7**: 5–15.

ERERA-WEATHERLEY, P. I. (1996) 'Coping with stress: public welfare supervisors doing their best', *Human Relations* **49**: 157–170.

FRENCH, J. R. P., CAPLAN, R. D. and VAN HARRISON, R. (1982) *The Mechanisms of Job Stress and Strain*, New York: John Wiley and Sons.

GLOWINKOWSKI, S.P. and COOPER, C. L. (1985) 'Current issues in organisational stress research', *Bulletin of the British Psychological Society* **38**: 212–216.

HARRIS, J. R. (1991) 'The utility of the transactional approach for occupational stress research', in P. L. Perrewé (ed.) *Handbook on Job Stress*, special edition, *Journal of Social Behavior and Personality* **6**: 21–29.

KAHN, R. L. (1987) 'Work stress in the 1980s: research and practice', In J. C. Quick, R. S. Bhagat, J. E. Dalton and J. D. Quick (eds) *Work Stress: Health Care Systems in the Workplace*, New York: Praeger, 311–320.

KAHN, R. L. and BYOSIERE, S. (1991) 'Stress in organisations', in M. D. Dunnette and L. M. Hough (eds) *Handbook of Industrial and Organisational Psychology*, Chicago: Rand McNally, 571–648.

KAPLAN, H. B. (1996) 'Themes, lacunae and directions in research on psychological stress', in H. B. Kaplan (ed.) *Psychological Stress: Perspectives on Structure, Theory, Life-course and Methods*, San Diego: Academic Press, 369–403.

KARASEK, R. A. (1979) 'Job demands, job decision latitude and mental strain: implications for job redesign', *Administrative Science Quarterly* **24**: 285–308.

KASL, S. V. (1983) 'Pursuing the link between stressful life experiences and disease: a time for reappraisal', in, C. L. Cooper (ed.) *Stress Research: Issues for the Eighties*, Chichester: John Wiley and Sons, 79–102.

LAZARUS, R. S. (1990) 'Theory-based stress measurement', *Psychological Inquiry* **1**: 3–13.

LAZARUS, R. S. (1991) 'Psychological stress in the workplace', in P. L. Perrewé (ed.) *Handbook on Job Stress*, special edition, *Journal of Social Behavior and Personality* **6**: 1–14.

LAZARUS, R. S. and LAUNIER, R. (1978) 'Stress-related transactions between person and environment', in L. A. Pervin and M. Lewis (ed.) *Perspectives in International Psychology*, New York: Plenum, 287–327.

LIDDLE, H. A. (1994) 'Contextualizing resiliency', in M. C. Wang and E. W. Gorden (eds) *Educational Resilience in Inner-City America*, Hillsdale, NY: Erlbaum, 167–177.

NEWTON, T. J. (1989) 'Occupational stress and coping with stress', *Human Relations* **38**: 107–126.

NEWTON, T. J. (1995) *'Managing' Stress: Emotion and Power at Work*, London: Sage.

PAYNE, R., JICK, T. and BURKE, R. (1982) 'Whither stress research: an agenda for the 1980s', *Journal of Occupational Behavior* **3**: 131–145.

SCHONFELD, I. S., RHEE, J., and XIA, F. (1995) 'Methodological issues in occupational – stress research: research in one occupational group and wider applications', in S. L. Sauter and L. R. Murphy (eds) *Organizational Risk Factors for Job Stress*, Washington, DC: APA 323–339.

SETHI, A. S. and SCHULER, R. S. (1984) 'Organizational stress coping: research issues and future directions', in A. S. Sethi and R. S. Schuler. *Handbook of Organizational Stress Coping Strategies*, Cambridge, MA: Ballinger Publishing Co, 301–307.

TETRICK, L. E. and LaROCCO, J. M. (1987) 'Understanding, prediction, and control as moderators of the relationship between perceived stress, satisfaction, and psychological well-being', *Journal of Applied Psychology* **72**: 538–543.

TRENBERTH, L. D., DEWE, P. J. and WALKEY, F. H. (1996) 'A factor replication approach to the measurement of coping', *Stress Medicine* **12**: 71–79.

Index

Note: *Italic* page numbers refer to tables in the text.